Linear Algebra via Exterior Products

Linear Algebra via Exterior Products

Sergei Winitzki, Ph.D.

Linear Algebra via Exterior Products
Copyright (c) 2009-2010 by Sergei Winitzki, Ph.D.
ISBN 978-1-4092-9496-2, published by **lulu.com**
Version 1.2. Last change: January 4, 2010

This book is an undergraduate-level introduction to the coordinate-free approach in basic finite-dimensional linear algebra. The reader should be already exposed to the elementary array-based formalism of vector and matrix calculations. Throughout this book, extensive use is made of the exterior (anti-commutative, "wedge") product of vectors. The coordinate-free formalism and the exterior product, while somewhat more abstract, provide a deeper understanding of the classical results in linear algebra. The standard properties of determinants, the Pythagoras theorem for multidimensional volumes, the formulas of Jacobi and Liouville, the Cayley-Hamilton theorem, properties of Pfaffians, the Jordan canonical form, as well as some generalizations of these results are derived without cumbersome matrix calculations. For the benefit of students, every result is logically motivated and discussed. Exercises with some hints are provided.

Contents

Preface

In a first course of linear algebra, one learns the various uses of matrices, for instance the properties of determinants, eigenvectors and eigenvalues, and methods for solving linear equations. The required calculations are straightforward (because, conceptually, vectors and matrices are merely "arrays of numbers") if cumbersome. However, there is a more abstract and more powerful approach: Vectors are elements of abstract vector spaces, and matrices represent linear transformations of vectors. This **invariant** or **coordinate-free** approach is important in algebra and has found many applications in science.

The purpose of this book is to help the reader make a transition to the abstract coordinate-free approach, and also to give a hands-on introduction to exterior products, a powerful tool of linear algebra. I show how the coordinate-free approach together with exterior products can be used to clarify the basic results of matrix algebra, at the same time avoiding all the laborious matrix calculations.

Here is a simple theorem that illustrates the advantages of the exterior product approach. A triangle is oriented arbitrarily in three-dimensional space; the three orthogonal projections of this triangle are triangles in the three coordinate planes. Let S be the area of the initial triangle, and let A, B, C be the areas of the three projections. Then

$$S^2 = A^2 + B^2 + C^2.$$

If one uses bivectors to represent the oriented areas of the triangle and of its three projections, the statement above is equivalent to the Pythagoras theorem in the space of bivectors, and the proof requires only a few straightforward definitions and checks. A generalization of this result to volumes of k-dimensional bodies embedded in N-dimensional spaces is then obtained with no extra work. I hope that the readers will appreciate the beauty of an approach to linear algebra that allows us to obtain such results quickly and almost without calculations.

The exterior product is widely used in connection with n-forms, which are exterior products of *covectors*. In this book I do not use n-forms — instead I use vectors, n-vectors, and their exterior products. This approach allows a more straightforward geometric interpretation and also simplifies calculations and proofs.

To make the book logically self-contained, I present a proof of every basic result of linear algebra. The emphasis is not on computational techniques, although the coordinate-free approach *does* make many computations easier and more elegant.[1] The main topics covered are tensor products; exterior

[1] **Elegant** means shorter and easier to remember. Usually, **elegant** derivations are those in which

products; coordinate-free definitions of the determinant $\det \hat{A}$, the trace $\mathrm{Tr}\hat{A}$, and the characteristic polynomial $Q_{\hat{A}}(\lambda)$; basic properties of determinants; solution of linear equations, including over-determined or under-determined systems, using Kramer's rule; the Liouville formula $\det \exp \hat{A} = \exp \mathrm{Tr}\hat{A}$ as an identity of formal series; the algebraic complement (cofactor) matrix; Jacobi's formula for the variation of the determinant; variation of the characteristic polynomial and of eigenvalue; the Cayley-Hamilton theorem; analytic functions of operators; Jordan canonical form; construction of projectors onto Jordan cells; Hodge star and the computation of k-dimensional volumes through k-vectors; definition and properties of the Pfaffian $\mathrm{Pf}\hat{A}$ for antisymmetric operators $\hat{A}$. All these standard results are derived without matrix calculations; instead, the exterior product is used as a main computational tool.

This book is largely **pedagogical**, meaning that the results are long known, and the emphasis is on a clear and self-contained, logically motivated presentation aimed at students. Therefore, some exercises with hints and partial solutions are included, but not references to literature.[2] I have tried to avoid being overly pedantic while keeping the exposition mathematically rigorous.

Sections marked with a star * are not especially difficult but contain material that may be skipped at first reading. (Exercises marked with a star *are* more difficult.)

The first chapter is an introduction to the invariant approach to vector spaces. I assume that readers are familiar with elementary linear algebra in the language of row/column vectors and matrices; Appendix C contains a brief overview of that material. Good introductory books (which I did not read in detail but which have a certain overlap with the present notes) are "Finite-dimensional Vector Spaces" by P. Halmos and "Linear Algebra" by J. Hefferon (the latter is a free book).

I started thinking about the approach to linear algebra based on exterior products while still a student. I am especially grateful to Sergei Arkhipov, Leonid Positsel'sky, and Arkady Vaintrob who have stimulated my interest at that time and taught me much of what I could not otherwise learn about algebra. Thanks are also due to Prof. Howard Haber (UCSC) for constructive feedback on an earlier version of this text.

---

some powerful basic idea is exploited to obtain the result quickly.

[2]The approach to determinants via exterior products has been known since at least 1880 but does not seem especially popular in textbooks, perhaps due to the somewhat abstract nature of the tensor product. I believe that this approach to determinants and to other results in linear algebra deserves to be more widely appreciated.

0 Introduction and summary

All the notions mentioned in this section will be explained below. If you already know the definition of tensor and exterior products and are familiar with statements such as $\text{End}\, V \cong V \otimes V^*$, you may skip to Chapter 2.

0.1 Notation

The following conventions are used throughout this text.

I use the **bold emphasis** to define a new word, term, or notion, and the definition always appears near the boldface text (whether or not I write the word "Definition").

Ordered sets are denoted by round parentheses, e.g. $(1, 2, 3)$. Unordered sets are denoted using the curly parentheses, e.g. $\{a, b, c\}$.

The symbol $\equiv$ means "is now being defined as" or "equals by a previously given definition."

The symbol $\stackrel{!}{=}$ means "as we already know, equals."

A set consisting of all elements x satisfying some property $P(x)$ is denoted by $\{\, x \,|\, P(x) \text{ is true} \,\}$.

A map f from a set V to W is denoted by $f : V \to W$. An element $v \in V$ is then mapped to an element $w \in W$, which is written as $f : v \mapsto w$ or $f(v) = w$.

The sets of rational numbers, real numbers, and complex numbers are denoted respectively by $\mathbb{Q}$, $\mathbb{R}$, and $\mathbb{C}$.

Statements, Lemmas, Theorems, Examples, and Exercises are numbered only within a single subsection, so references are always to a certain statement in a certain subsection.[1] A reference to "Theorem 1.1.4" means the unnumbered theorem in Sec. 1.1.4.

Proofs, solutions, examples, and exercises are separated from the rest by the symbol ■. More precisely, this symbol means "I have finished with this; now we look at something else."

V is a finite-dimensional **vector space** over a **field** $\mathbb{K}$. Vectors from V are denoted by boldface lowercase letters, e.g. $\mathbf{v} \in V$. The **dimension** of V is $N \equiv \dim V$.

The standard N-dimensional space over real numbers (the space consisting of N-tuples of real numbers) is denoted by $\mathbb{R}^N$.

The **subspace spanned by** a given set of vectors $\{\mathbf{v}_1, ..., \mathbf{v}_n\}$ is denoted by $\text{Span}\, \{\mathbf{v}_1, ..., \mathbf{v}_n\}$.

[1] I was too lazy to implement a comprehensive system of numbering for all these items.

The vector space **dual** to V is V^*. Elements of V^* (**covectors**) are denoted by starred letters, e.g. $\mathbf{f}^* \in V^*$. A covector $\mathbf{f}^*$ acts on a vector $\mathbf{v}$ and produces a number $\mathbf{f}^*(\mathbf{v})$.

The space of linear maps (**homomorphisms**) $V \to W$ is $\text{Hom}\,(V, W)$. The space of **linear operators** (also called **endomorphisms**) of a vector space V, i.e. the space of all linear maps $V \to V$, is End V. Operators are denoted by the circumflex accent, e.g. $\hat{A}$. The **identity** operator on V is $\hat{1}_V \in \text{End}\, V$ (sometimes also denoted $\hat{1}$ for brevity).

The **direct sum** of spaces V and W is $V \oplus W$. The **tensor product** of spaces V and W is $V \otimes W$. The **exterior (anti-commutative) product** of V and V is $V \wedge V$. The exterior product of n copies of V is $\wedge^n V$. **Canonical isomorphisms** of vector spaces are denoted by the symbol $\cong$; for example, End $V \cong V \otimes V^*$.

The **scalar product** of vectors is denoted by $\langle \mathbf{u}, \mathbf{v} \rangle$. The notation $\mathbf{a} \times \mathbf{b}$ is used *only* for the traditional **vector product** (also called **cross product**) in 3-dimensional space. Otherwise, the product symbol $\times$ is used to denote the continuation a long expression that is being split between lines.

The **exterior (wedge)** product of vectors is denoted by $\mathbf{a} \wedge \mathbf{b} \in \wedge^2 V$.

Any two nonzero tensors $\mathbf{a}_1 \wedge ... \wedge \mathbf{a}_N$ and $\mathbf{b}_1 \wedge ... \wedge \mathbf{b}_N$ in an N-dimensional space are proportional to each other, say

$$\mathbf{a}_1 \wedge ... \wedge \mathbf{a}_N = \lambda \mathbf{b}_1 \wedge ... \wedge \mathbf{b}_N.$$

It is then convenient to denote λ by the "tensor ratio"

$$\lambda \equiv \frac{\mathbf{a}_1 \wedge ... \wedge \mathbf{a}_N}{\mathbf{b}_1 \wedge ... \wedge \mathbf{b}_N}.$$

The number of unordered choices of k items from n is denoted by

$$\binom{n}{k} = \frac{n!}{k!(n-k)!}.$$

The k-linear action of a linear operator $\hat{A}$ in the space $\wedge^n V$ is denoted by $\wedge^n \hat{A}^k$. (Here $0 \leq k \leq n \leq N$.) For example,

$$(\wedge^3 \hat{A}^2) \mathbf{a} \wedge \mathbf{b} \wedge \mathbf{c} \equiv \hat{A}\mathbf{a} \wedge \hat{A}\mathbf{b} \wedge \mathbf{c} + \hat{A}\mathbf{a} \wedge \mathbf{b} \wedge \hat{A}\mathbf{c} \\ + \mathbf{a} \wedge \hat{A}\mathbf{b} \wedge \hat{A}\mathbf{c}.$$

The imaginary unit ($\sqrt{-1}$) is denoted by a *roman* "i," while the base of natural logarithms is written as an *italic* "e." For example, I would write $e^{\mathrm{i}\pi} = -1$. This convention is designed to avoid conflicts with the much used index i and with labeled vectors such as $\mathbf{e}_i$.

I write an italic d in the derivatives, such as df/dx, and in integrals, such as $\int f(x)dx$, because in these cases the symbols dx do not refer to a separate well-defined object "dx" but are a part of the traditional symbolic notation used in calculus. Differential forms (or, for that matter, nonstandard calculus) *do* make "$\mathrm{d}x$" into a well-defined object; in that case I write a roman "d" in "$\mathrm{d}x$." Neither calculus nor differential forms are actually used in this book; the only exception is the occasional use of the derivative d/dx applied to polynomials in x. I will not need to make a distinction between d/dx and $\partial/\partial x$; the derivative of a function f with respect to x is denoted by $\partial_x f$.

0.2 Sample quiz problems

The following problems can be solved using techniques explained in this book. (These problems are of varying difficulty.) In these problems V is an N-dimensional vector space (with a scalar product if indicated).

Exterior multiplication: If two tensors $\omega_1, \omega_2 \in \wedge^k V$ (with $1 \leq k \leq N-1$) are such that $\omega_1 \wedge \mathbf{v} = \omega_2 \wedge \mathbf{v}$ for *all* vectors $\mathbf{v} \in V$, show that $\omega_1 = \omega_2$.

Insertions: a) It is given that $\psi \in \wedge^k V$ (with $1 \leq k \leq N-1$) and $\psi \wedge \mathbf{a} = 0$, where $\mathbf{a} \in V$ and $\mathbf{a} \neq 0$. Further, a covector $\mathbf{f}^* \in V^*$ is given such that $\mathbf{f}^*(\mathbf{a}) \neq 0$. Show that

$$\psi = \frac{1}{\mathbf{f}^*(\mathbf{a})} \mathbf{a} \wedge (\iota_{\mathbf{f}^*} \psi).$$

b) It is given that $\psi \wedge \mathbf{a} = 0$ and $\psi \wedge \mathbf{b} = 0$, where $\psi \in \wedge^k V$ (with $2 \leq k \leq N-1$) and $\mathbf{a}, \mathbf{b} \in V$ such that $\mathbf{a} \wedge \mathbf{b} \neq 0$. Show that there exists $\chi \in \wedge^{k-2} V$ such that $\psi = \mathbf{a} \wedge \mathbf{b} \wedge \chi$.

c) It is given that $\psi \wedge \mathbf{a} \wedge \mathbf{b} = 0$, where $\psi \in \wedge^k V$ (with $2 \leq k \leq N-2$) and $\mathbf{a}, \mathbf{b} \in V$ such that $\mathbf{a} \wedge \mathbf{b} \neq 0$. Is it always true that $\psi = \mathbf{a} \wedge \mathbf{b} \wedge \chi$ for some $\chi \in \wedge^{k-2} V$?

Determinants: a) Suppose $\hat{A}$ is a linear operator defined by $\hat{A} = \sum_{i=1}^{N} \mathbf{a}_i \otimes \mathbf{b}_i^*$, where $\mathbf{a}_i \in V$ are given vectors and $\mathbf{b}_i \in V^*$ are given covectors; $N = \dim V$. Show that

$$\det \hat{A} = \frac{\mathbf{a}_1 \wedge ... \wedge \mathbf{a}_N}{\mathbf{e}_1 \wedge ... \wedge \mathbf{e}_N} \frac{\mathbf{b}_1^* \wedge ... \wedge \mathbf{b}_N^*}{\mathbf{e}_1^* \wedge ... \wedge \mathbf{e}_N^*},$$

where $\{\mathbf{e}_j\}$ is an arbitrary basis and $\{\mathbf{e}_j^*\}$ is the corresponding dual basis. Show that the expression above is independent of the choice of the basis $\{\mathbf{e}_j\}$.

b) Suppose that a scalar product is given in V, and an operator $\hat{A}$ is defined by

$$\hat{A}\mathbf{x} \equiv \sum_{i=1}^{N} \mathbf{a}_i \langle \mathbf{b}_i, \mathbf{x} \rangle .$$

Further, suppose that $\{\mathbf{e}_j\}$ is an orthonormal basis in V. Show that

$$\det \hat{A} = \frac{\mathbf{a}_1 \wedge ... \wedge \mathbf{a}_N}{\mathbf{e}_1 \wedge ... \wedge \mathbf{e}_N} \frac{\mathbf{b}_1 \wedge ... \wedge \mathbf{b}_N}{\mathbf{e}_1 \wedge ... \wedge \mathbf{e}_N},$$

and that this expression is independent of the choice of the orthonormal basis $\{\mathbf{e}_j\}$ and of the orientation of the basis.

Hyperplanes: a) Let us suppose that the "price" of the vector $\mathbf{x} \in V$ is given by the formula

$$\text{Cost}\,(\mathbf{x}) \equiv C(\mathbf{x}, \mathbf{x}),$$

where $C(\mathbf{a}, \mathbf{b})$ is a known, positive-definite bilinear form. Determine the "cheapest" vector $\mathbf{x}$ belonging to the affine hyperplane $\mathbf{a}^*(\mathbf{x}) = \alpha$, where $\mathbf{a}^* \in V^*$ is a nonzero covector and α is a number.

b) We are now working in a vector space with a scalar product, and the "price" of a vector $\mathbf{x}$ is $\langle \mathbf{x}, \mathbf{x} \rangle$. Two affine hyperplanes are given by equations $\langle \mathbf{a}, \mathbf{x} \rangle = \alpha$ and $\langle \mathbf{b}, \mathbf{x} \rangle = \beta$, where $\mathbf{a}$ and $\mathbf{b}$ are given vectors, α and β are numbers, and $\mathbf{x} \in V$. (It is assured that $\mathbf{a}$ and $\mathbf{b}$ are nonzero and not parallel to each other.) Determine the "cheapest" vector $\mathbf{x}$ belonging to the intersection of the two hyperplanes.

Too few equations: A linear operator $\hat{A}$ is defined by $\hat{A} = \sum_{i=1}^{k} \mathbf{a}_i \otimes \mathbf{b}_i^*$, where $\mathbf{a}_i \in V$ are given vectors and $\mathbf{b}_i^* \in V^*$ are given covectors, and $k < N = \dim V$. Show that the vector equation $\hat{A}\mathbf{x} = \mathbf{c}$ has no solutions if $\mathbf{a}_1 \wedge ... \wedge \mathbf{a}_k \wedge \mathbf{c} \neq 0$. In case $\mathbf{a}_1 \wedge ... \wedge \mathbf{a}_k \wedge \mathbf{c} = 0$, show that solutions $\mathbf{x}$ surely exist when $\mathbf{b}_1^* \wedge ... \wedge \mathbf{b}_k^* \neq 0$ but may not exist otherwise.

Operator functions: It is known that the operator $\hat{A}$ satisfies the operator equation $\hat{A}^2 = -\hat{1}$. Simplify the operator-valued functions $\frac{1+\hat{A}}{3-\hat{A}}$, $\cos(\lambda\hat{A})$, and $\sqrt{\hat{A}+2}$ to linear formulas involving $\hat{A}$. (Here λ is a number, while the numbers 1, 2, 3 stand for multiples of the identity operator.) Compare the results with the complex numbers $\frac{1+\mathrm{i}}{3-\mathrm{i}}$, $\cos(\lambda\mathrm{i})$, $\sqrt{\mathrm{i}+2}$ and generalize the conclusion to a theorem about computing analytic functions $f(\hat{A})$.

Inverse operator: It is known that $\hat{A}\hat{B} = \lambda\hat{1}_V$, where $\lambda \neq 0$ is a number. Prove that also $\hat{B}\hat{A} = \lambda\hat{1}_V$. (Both $\hat{A}$ and $\hat{B}$ are linear operators in a finite-dimensional space V.)

Trace and determinant: Consider the space of polynomials in the variables x and y, where we admit only polynomials of the form $a_0 + a_1 x + a_2 y + a_3 xy$ (with $a_j \in \mathbb{R}$). An operator $\hat{A}$ is defined by

$$\hat{A} \equiv x\frac{\partial}{\partial x} - \frac{\partial}{\partial y}.$$

Show that $\hat{A}$ is a linear operator in this space. Compute the trace and the determinant of $\hat{A}$. If $\hat{A}$ is invertible, compute $\hat{A}^{-1}(x+y)$.

Cayley-Hamilton theorem: Express $\det\hat{A}$ through $\mathrm{Tr}\hat{A}$ and $\mathrm{Tr}(\hat{A}^2)$ for an arbitrary operator $\hat{A}$ in a *two*-dimensional space.

Algebraic complement: Let $\hat{A}$ be a linear operator and $\tilde{\hat{A}}$ its algebraic complement.

a) Show that

$$\mathrm{Tr}\tilde{\hat{A}} = \wedge^N \hat{A}^{N-1}.$$

Here $\wedge^N \hat{A}^{N-1}$ is the coefficient at $(-\lambda)$ in the characteristic polynomial of $\hat{A}$ (that is, minus the coefficient preceding the determinant).

b) For t-independent operators $\hat{A}$ and $\hat{B}$, show that

$$\frac{\partial}{\partial t}\det(\hat{A}+t\hat{B})=\mathrm{Tr}(\tilde{\hat{A}}\hat{B}).$$

Liouville formula: Suppose $\hat{X}(t)$ is a defined as solution of the differential equation

$$\partial_t\hat{X}(t)=\hat{A}(t)\hat{X}(t)-\hat{X}(t)\hat{A}(t),$$

where $\hat{A}(t)$ is a given operator. (Operators that are functions of t can be understood as operator-valued formal power series.)

a) Show that the determinant of $\hat{X}(t)$ is independent of t.

b) Show that all the coefficients of the characteristic polynomial of $\hat{X}(t)$ are independent of t.

Hodge star: Suppose $\{\mathbf{v}_1,...,\mathbf{v}_N\}$ is a basis in V, not necessarily orthonormal, while $\{\mathbf{e}_j\}$ is a positively oriented orthonormal basis. Show that

$$*(\mathbf{v}_1\wedge...\wedge\mathbf{v}_N)=\frac{\mathbf{v}_1\wedge...\wedge\mathbf{v}_N}{\mathbf{e}_1\wedge...\wedge\mathbf{e}_N}.$$

Volume in space: Consider the space of polynomials of degree at most 4 in the variable x. The scalar product of two polynomials $p_1(x)$ and $p_2(x)$ is defined by

$$\langle p_1,p_2\rangle\equiv\frac{1}{2}\int_{-1}^{1}p_1(x)p_2(x)dx.$$

Determine the three-dimensional volume of the tetrahedron with vertices at the "points" 0, $1+x$, x^2+x^3, x^4 in this five-dimensional space.

0.3 A list of results

Here is a list of some results explained in this book. If you already know all these results and their derivations, you may not need to read any further.

Vector spaces may be defined over an abstract number field, without specifying the number of dimensions or a basis.

The set $\{a+b\sqrt{41}\,|\,a,b\in\mathbb{Q}\}$ is a number field.

Any vector can be represented as a linear combination of basis vectors. All bases have equally many vectors.

The set of all linear maps from one vector space to another is denoted $\mathrm{Hom}(V,W)$ and is a vector space.

The zero vector is not an eigenvector (by definition).

An operator having in some basis the matrix representation $\begin{pmatrix}0&1\\0&0\end{pmatrix}$ cannot be diagonalized.

The dual vector space V^* has the same dimension as V (for finite-dimensional spaces).

Given a nonzero covector $\mathbf{f}^*\in V^*$, the set of vectors $\mathbf{v}\in V$ such that $\mathbf{f}^*(\mathbf{v})=0$ is a subspace of codimension 1 (a hyperplane).

The tensor product of $\mathbb{R}^m$ and $\mathbb{R}^n$ has dimension mn.

Any linear map $\hat{A} : V \to W$ can be represented by a tensor of the form $\sum_{i=1}^k \mathbf{v}_i^* \otimes \mathbf{w}_i \in V^* \otimes W$. The rank of $\hat{A}$ is equal to the smallest number of simple tensor product terms $\mathbf{v}_i^* \otimes \mathbf{w}_i$ required for this representation.

The identity map $\hat{1}_V : V \to V$ is represented as the tensor $\sum_{i=1}^N \mathbf{e}_i^* \otimes \mathbf{e}_i \in V^* \otimes V$, where $\{\mathbf{e}_i\}$ is any basis and $\{\mathbf{e}_i^*\}$ its dual basis. This tensor does not depend on the choice of the basis $\{\mathbf{e}_i\}$.

A set of vectors $\{\mathbf{v}_1, ..., \mathbf{v}_k\}$ is linearly independent if and only if $\mathbf{v}_1 \wedge ... \wedge \mathbf{v}_k \neq 0$. If $\mathbf{v}_1 \wedge ... \wedge \mathbf{v}_k \neq 0$ but $\mathbf{v}_1 \wedge ... \wedge \mathbf{v}_k \wedge \mathbf{x} = 0$ then the vector $\mathbf{x}$ belongs to the subspace Span $\{\mathbf{v}_1, ..., \mathbf{v}_k\}$.

The dimension of the space $\wedge^k V$ is $\binom{N}{k}$, where $N \equiv \dim V$.

Insertion $\iota_{\mathbf{a}^*}\omega$ of a covector $\mathbf{a}^* \in V^*$ into an antisymmetric tensor $\omega \in \wedge^k V$ has the property

$$\mathbf{v} \wedge (\iota_{\mathbf{a}^*}\omega) + \iota_{\mathbf{a}^*}(\mathbf{v} \wedge \omega) = \mathbf{a}^*(\mathbf{v})\omega.$$

Given a basis $\{\mathbf{e}_i\}$, the dual basis $\{\mathbf{e}_i^*\}$ may be computed as

$$\mathbf{e}_i^*(\mathbf{x}) = \frac{\mathbf{e}_1 \wedge ... \wedge \mathbf{x} \wedge ... \wedge \mathbf{e}_N}{\mathbf{e}_1 \wedge ... \wedge \mathbf{e}_N},$$

where $\mathbf{x}$ replaces $\mathbf{e}_i$ in the numerator.

The subspace spanned by a set of vectors $\{\mathbf{v}_1, ..., \mathbf{v}_k\}$, not necessarily linearly independent, can be characterized by a certain antisymmetric tensor ω, which is the exterior product of the largest number of $\mathbf{v}_i$'s such that $\omega \neq 0$. The tensor ω, computed in this way, is unique up to a constant factor.

The n-vector (antisymmetric tensor) $\mathbf{v}_1 \wedge ... \wedge \mathbf{v}_n$ represents geometrically the oriented n-dimensional volume of the parallelepiped spanned by the vectors $\mathbf{v}_i$.

The determinant of a linear operator $\hat{A}$ is the coefficient that multiplies the oriented volume of any parallelepiped transformed by $\hat{A}$. In our notation, the operator $\wedge^N \hat{A}^N$ acts in $\wedge^N V$ as multiplication by $\det \hat{A}$.

If each of the given vectors $\{\mathbf{v}_1, ..., \mathbf{v}_N\}$ is expressed through a basis $\{\mathbf{e}_i\}$ as $\mathbf{v}_j = \sum_{i=1}^N v_{ij}\mathbf{e}_i$, the determinant of the matrix v_{ij} is found as

$$\det(v_{ij}) = \det(v_{ji}) = \frac{\mathbf{v}_1 \wedge ... \wedge \mathbf{v}_N}{\mathbf{e}_1 \wedge ... \wedge \mathbf{e}_N}.$$

A linear operator $\hat{A} : V \to V$ and its canonically defined transpose $\hat{A}^T : V^* \to V^*$ have the same characteristic polynomials.

If $\det \hat{A} \neq 0$ then the inverse operator $\hat{A}^{-1}$ exists, and a linear equation $\hat{A}\mathbf{x} = \mathbf{b}$ has the unique solution $\mathbf{x} = \hat{A}^{-1}\mathbf{b}$. Otherwise, solutions exist if $\mathbf{b}$ belongs to the image of $\hat{A}$. Explicit solutions may be constructed using Kramer's rule: If a vector $\mathbf{b}$ belongs to the subspace spanned by vectors $\{\mathbf{v}_1, ..., \mathbf{v}_n\}$ then $\mathbf{b} = \sum_{i=1}^n b_i \mathbf{v}_i$, where the coefficients $\mathbf{b}_i$ may be found (assuming $\mathbf{v}_1 \wedge ... \wedge \mathbf{v}_n \neq 0$) as

$$b_i = \frac{\mathbf{v}_1 \wedge ... \wedge \mathbf{x} \wedge ... \wedge \mathbf{v}_n}{\mathbf{v}_1 \wedge ... \wedge \mathbf{v}_n}$$

(here $\mathbf{x}$ replaces $\mathbf{v}_i$ in the exterior product in the numerator).

Eigenvalues of a linear operator are roots of its characteristic polynomial. For each root λ_i, there exists at least one eigenvector corresponding to the eigenvalue λ_i.

If $\{\mathbf{v}_1, ..., \mathbf{v}_k\}$ are eigenvectors corresponding to *all different* eigenvalues $\lambda_1, ..., \lambda_k$ of some operator, then the set $\{\mathbf{v}_1, ..., \mathbf{v}_k\}$ is linearly independent.

The dimension of the eigenspace corresponding to λ_i is not larger than the algebraic multiplicity of the root λ_i in the characteristic polynomial.

(Below in this section we always denote by N the dimension of the space V.)

The trace of an operator $\hat{A}$ can be expressed as $\wedge^N \hat{A}^1$.

We have $\mathrm{Tr}(\hat{A}\hat{B}) = \mathrm{Tr}(\hat{B}\hat{A})$. This holds even if $\hat{A}, \hat{B}$ are maps between different spaces, i.e. $\hat{A} : V \to W$ and $\hat{B} : W \to V$.

If an operator $\hat{A}$ is nilpotent, its characteristic polynomial is $(-\lambda)^N$, i.e. the same as the characteristic polynomial of a zero operator.

The j-th coefficient of the characteristic polynomial of $\hat{A}$ is $(-1)^j (\wedge^N \hat{A}^j)$.

Each coefficient of the characteristic polynomial of $\hat{A}$ can be expressed as a polynomial function of N traces of the form $\mathrm{Tr}(\hat{A}^k)$, $k = 1, ..., N$.

The space $\wedge^{N-1}V$ is N-dimensional like V itself, and there is a canonical isomorphism between $\mathrm{End}(\wedge^{N-1}V)$ and $\mathrm{End}(V)$. This isomorphism, called **exterior transposition**, is denoted by $(...)^{\wedge T}$. The exterior transpose of an operator $\hat{X} \in \mathrm{End}\, V$ is defined by

$$(\hat{X}^{\wedge T}\omega) \wedge \mathbf{v} \equiv \omega \wedge \hat{X}\mathbf{v}, \quad \forall \omega \in \wedge^{N-1}V, \ \mathbf{v} \in V.$$

Similarly, one defines the exterior transposition map between $\mathrm{End}(\wedge^{N-k}V)$ and $\mathrm{End}(\wedge^k V)$ for all $k = 1, ..., N$.

The algebraic complement operator (normally defined as a matrix consisting of minors) is canonically defined through exterior transposition as $\tilde{\hat{A}} \equiv (\wedge^{N-1}\hat{A}^{N-1})^{\wedge T}$. It can be expressed as a polynomial in $\hat{A}$ and satisfies the identity $\tilde{\hat{A}}\hat{A} = (\det \hat{A})\hat{1}_V$. Also, all other operators

$$\hat{A}_{(k)} \equiv \left(\wedge^{N-1}\hat{A}^{N-k}\right)^{\wedge T}, \quad k = 1, ..., N$$

can be expressed as polynomials in $\hat{A}$ with known coefficients.

The characteristic polynomial of $\hat{A}$ gives the zero operator if applied to the operator $\hat{A}$ (the Cayley-Hamilton theorem). A similar theorem holds for each of the operators $\wedge^k \hat{A}^1$, $2 \le k \le N-1$ (with different polynomials).

A formal power series $f(t)$ can be applied to the operator $t\hat{A}$; the result is an operator-valued formal series $f(t\hat{A})$ that has the usual properties, e.g.

$$\partial_t f(t\hat{A}) = \hat{A} f'(t\hat{A}).$$

If $\hat{A}$ is diagonalized with eigenvalues $\{\lambda_i\}$ in the eigenbasis $\{\mathbf{e}_i\}$, then a formal power series $f(t\hat{A})$ is diagonalized in the same basis with eigenvalues $f(t\lambda_i)$.

If an operator $\hat{A}$ satisfies a polynomial equation such as $p(\hat{A}) = 0$, where $p(x)$ is a known polynomial of degree k (not necessarily, but possibly, the characteristic polynomial of $\hat{A}$) then any formal power series $f(t\hat{A})$ is reduced to a polynomial in $t\hat{A}$ of degree not larger than $k - 1$. This polynomial can be computed as the interpolating polynomial for the function $f(tx)$ at points $x = x_i$ where x_i are the (all different) roots of $p(x)$. Suitable modifications are available when *not all* roots are different. So one can compute any analytic function $f(\hat{A})$ of the operator $\hat{A}$ as long as one knows a polynomial equation satisfied by $\hat{A}$.

A square root of an operator $\hat{A}$ (i.e. a linear operator $\hat{B}$ such that $\hat{B}\hat{B} = \hat{A}$) is not unique and does not always exist. In two and three dimensions, one can either obtain all square roots explicitly as polynomials in $\hat{A}$, or determine that some square roots are not expressible as polynomials in $\hat{A}$ or that square roots of $\hat{A}$ do not exist at all.

If an operator $\hat{A}$ depends on a parameter t, one can express the derivative of the determinant of $\hat{A}$ through the algebraic complement $\tilde{\hat{A}}$ (Jacobi's formula),

$$\partial_t \det \hat{A}(t) = \mathrm{Tr}(\tilde{\hat{A}}\partial_t \hat{A}).$$

Derivatives of other coefficients $q_k \equiv \wedge^N \hat{A}^{N-k}$ of the characteristic polynomial are given by similar formulas,

$$\partial_t q_k = \mathrm{Tr}\left[(\wedge^{N-1}\hat{A}^{N-k-1})^{\wedge T}\partial_t \hat{A}\right].$$

The Liouville formula holds: $\det \exp \hat{A} = \exp \mathrm{Tr}\hat{A}$.

Any operator (not necessarily diagonalizable) can be reduced to a Jordan canonical form in a Jordan basis. The Jordan basis consists of eigenvectors and root vectors for each eigenvalue.

Given an operator $\hat{A}$ whose characteristic polynomial is known (hence all roots λ_i and their algebraic multiplicities m_i are known), one can construct explicitly a projector $\hat{P}_{\lambda_i}$ onto a Jordan cell for any chosen eigenvalue λ_i. The projector is found as a polynomial in $\hat{A}$ with known coefficients.

(Below in this section we assume that a scalar product is fixed in V.)

A nondegenerate scalar product provides a one-to-one correspondence between vectors and covectors. Then the canonically transposed operator $\hat{A}^T : V^* \to V^*$ can be mapped into an operator in V, denoted also by $\hat{A}^T$. (This operator is represented by the transposed matrix only in an *orthonormal* basis.) We have $(\hat{A}\hat{B})^T = \hat{B}^T\hat{A}^T$ and $\det(\hat{A}^T) = \det \hat{A}$.

Orthogonal transformations have determinants equal to ± 1. Mirror reflections are orthogonal transformations and have determinant equal to -1.

Given an orthonormal basis $\{\mathbf{e}_i\}$, one can define the **unit volume tensor** $\omega = \mathbf{e}_1 \wedge ... \wedge \mathbf{e}_N$. The tensor ω is then independent of the choice of $\{\mathbf{e}_i\}$ up to a factor ± 1 due to the orientation of the basis (i.e. the ordering of the vectors of the basis), as long as the scalar product is kept fixed.

Given a fixed scalar product $\langle \cdot, \cdot \rangle$ and a fixed orientation of space, the Hodge star operation is uniquely defined as a linear map (isomorphism) $\wedge^k V \to$

$\wedge^{N-k}V$ for each $k = 0, ..., N$. For instance,

$$*\mathbf{e}_1 = \mathbf{e}_2 \wedge \mathbf{e}_3 \wedge ... \wedge \mathbf{e}_N; \quad *(\mathbf{e}_1 \wedge \mathbf{e}_2) = \mathbf{e}_3 \wedge ... \wedge \mathbf{e}_N,$$

if $\{\mathbf{e}_i\}$ is *any* positively oriented, orthonormal basis.

The Hodge star map satisfies

$$\langle \mathbf{a}, \mathbf{b} \rangle = *(\mathbf{a} \wedge *\mathbf{b}) = *(\mathbf{b} \wedge *\mathbf{a}), \quad \mathbf{a}, \mathbf{b} \in V.$$

In a three-dimensional space, the usual vector product and triple product can be expressed through the Hodge star as

$$\mathbf{a} \times \mathbf{b} = *(\mathbf{a} \wedge \mathbf{b}), \ \mathbf{a} \cdot (\mathbf{b} \times \mathbf{c}) = *(\mathbf{a} \wedge \mathbf{b} \wedge \mathbf{c}).$$

The volume of an N-dimensional parallelepiped spanned by $\{\mathbf{v}_1, ..., \mathbf{v}_N\}$ is equal to $\sqrt{\det(G_{ij})}$, where $G_{ij} \equiv \langle \mathbf{v}_i, \mathbf{v}_j \rangle$ is the matrix of the pairwise scalar products.

Given a scalar product in V, a scalar product is canonically defined also in the spaces $\wedge^k V$ for all $k = 2, ..., N$. This scalar product can be defined by

$$\langle \omega_1, \omega_2 \rangle = *(\omega_1 \wedge *\omega_2) = *(\omega_2 \wedge *\omega_1) = \langle \omega_2, \omega_1 \rangle,$$

where $\omega_{1,2} \in \wedge^k V$. Alternatively, this scalar product is defined by choosing an orthonormal basis $\{\mathbf{e}_j\}$ and postulating that $\mathbf{e}_{i_1} \wedge ... \wedge \mathbf{e}_{i_k}$ is normalized and orthogonal to any other such tensor with different indices $\{i_j | j = 1, ..., k\}$. The k-dimensional volume of a parallelepiped spanned by vectors $\{\mathbf{v}_1, ..., \mathbf{v}_k\}$ is found as $\sqrt{\langle \psi, \psi \rangle}$ with $\psi \equiv \mathbf{v}_1 \wedge ... \wedge \mathbf{v}_k \in \wedge^k V$.

The insertion $\iota_{\mathbf{v}}\psi$ of a vector $\mathbf{v}$ into a k-vector $\psi \in \wedge^k V$ (or the "interior product") can be expressed as

$$\iota_{\mathbf{v}}\psi = *(\mathbf{v} \wedge *\psi).$$

If $\omega \equiv \mathbf{e}_1 \wedge ... \wedge \mathbf{e}_N$ is the unit volume tensor, we have $\iota_{\mathbf{v}}\omega = *\mathbf{v}$.

Symmetric, antisymmetric, Hermitian, and anti-Hermitian operators are always diagonalizable (if we allow complex eigenvalues and eigenvectors). Eigenvectors of these operators can be chosen orthogonal to each other.

Antisymmetric operators are representable as elements of $\wedge^2 V$ of the form $\sum_{i=1}^{n} \mathbf{a}_i \wedge \mathbf{b}_i$, where one needs no more than $N/2$ terms, and the vectors $\mathbf{a}_i$, $\mathbf{b}_i$ can be chosen mutually orthogonal to each other. (For this, we do not need complex vectors.)

The **Pfaffian** of an antisymmetric operator $\hat{A}$ in even-dimensional space is the number $\text{Pf}\,\hat{A}$ defined as

$$\frac{1}{(N/2)!}\underbrace{A \wedge ... \wedge A}_{N/2} = (\text{Pf}\,\hat{A})\mathbf{e}_1 \wedge ... \wedge \mathbf{e}_N,$$

where $\{\mathbf{e}_i\}$ is an orthonormal basis. Some basic properties of the Pfaffian are

$$(\text{Pf}\,\hat{A})^2 = \det \hat{A},$$
$$\text{Pf}\,(\hat{B}\hat{A}\hat{B}^T) = (\det \hat{B})(\text{Pf}\,\hat{A}),$$

where $\hat{A}$ is an antisymmetric operator ($\hat{A}^T = -\hat{A}$) and $\hat{B}$ is an arbitrary operator.

1 Linear algebra without coordinates

1.1 Vector spaces

Abstract vector spaces are developed as a generalization of the familiar vectors in Euclidean space.

1.1.1 Three-dimensional Euclidean geometry

Let us begin with something you already know. Three-dimensional vectors are specified by triples of coordinates, $\mathbf{r} \equiv (x, y, z)$. The operations of **vector sum** and **vector product** of such vectors are defined by

$$(x_1, y_1, z_1) + (x_2, y_2, z_2) \equiv (x_1 + x_2, y_1 + y_2, z_1 + z_2); \tag{1.1}$$

$$\begin{aligned}(x_1, y_1, z_1) \times (x_2, y_2, z_2) \equiv (& y_1 z_2 - z_1 y_2,\, z_1 x_2 - x_1 z_2, \\ & x_1 y_2 - y_1 x_2).\end{aligned} \tag{1.2}$$

(I assume that these definitions are familiar to you.) Vectors can be **rescaled** by multiplying them with real numbers,

$$c\mathbf{r} = c\,(x, y, z) \equiv (cx, cy, cz). \tag{1.3}$$

A rescaled vector is parallel to the original vector and points either in the same or in the opposite direction. In addition, a **scalar product** of two vectors is defined,

$$(x_1, y_1, z_1) \cdot (x_2, y_2, z_2) \equiv x_1 x_2 + y_1 y_2 + z_1 z_2. \tag{1.4}$$

These operations encapsulate all of Euclidean geometry in a purely algebraic language. For example, the **length** of a vector $\mathbf{r}$ is

$$|\mathbf{r}| \equiv \sqrt{\mathbf{r} \cdot \mathbf{r}} = \sqrt{x^2 + y^2 + z^2}, \tag{1.5}$$

the **angle** α between vectors $\mathbf{r}_1$ and $\mathbf{r}_2$ is found from the relation (the cosine theorem)

$$|\mathbf{r}_1|\,|\mathbf{r}_2| \cos \alpha = \mathbf{r}_1 \cdot \mathbf{r}_2,$$

while the **area** of a triangle spanned by vectors $\mathbf{r}_1$ and $\mathbf{r}_2$ is

$$S = \frac{1}{2}\,|\mathbf{r}_1 \times \mathbf{r}_2|\,.$$

Using these definitions, one can reformulate every geometric statement (such as, "a triangle having two equal sides has also two equal angles") in terms of relations between vectors, which are ultimately reducible to algebraic equations involving a set of numbers. The replacement of geometric constructions by algebraic relations is useful because it allows us to free ourselves from the confines of our three-dimensional intuition; we are then able to solve problems in higher-dimensional spaces. The price is a greater complication of the algebraic equations and inequalities that need to be solved. To make these equations more transparent and easier to handle, the theory of linear algebra is developed. The first step is to realize what features of vectors are essential and what are just accidental facts of our familiar three-dimensional Euclidean space.

1.1.2 From three-dimensional vectors to abstract vectors

Abstract vector spaces retain the essential properties of the familiar Euclidean geometry but generalize it in two ways: First, the dimension of space is not 3 but an arbitrary integer number (or even infinity); second, the coordinates are "abstract numbers" (see below) instead of real numbers. Let us first pass to higher-dimensional vectors.

Generalizing the notion of a three-dimensional vector to a higher (still finite) dimension is straightforward: instead of triples (x, y, z) one considers sets of n coordinates $(x_1, ..., x_n)$. The definitions of the vector sum (1.1), scaling (1.3) and scalar product (1.4) are straightforwardly generalized to n-tuples of coordinates. In this way we can describe n-dimensional Euclidean geometry. All theorems of linear algebra are proved in the same way regardless of the number of components in vectors, so the generalization to n-dimensional spaces is a natural thing to do.

Question: The scalar product can be generalized to n-dimensional spaces,

$$(x_1, ..., x_n) \cdot (y_1, ..., y_n) \equiv x_1 y_1 + ... + x_n y_n,$$

but what about the vector product? The formula (1.2) seems to be complicated, and it is hard to guess what should be written, say, in four dimensions.

Answer: It turns out that the vector product (1.2) *cannot* be generalized to arbitrary n-dimensional spaces.[1] At this point we will not require the vector spaces to have either a vector or a scalar product; instead we will concentrate on the basic algebraic properties of vectors. Later we will see that there is an algebraic construction (the exterior product) that replaces the vector product in higher dimensions.

Abstract numbers

The motivation to replace the real coordinates x, y, z by complex coordinates, rational coordinates, or by some other, more abstract numbers comes from many branches of physics and mathematics. In any case, the statements of linear algebra almost never rely on the fact that coordinates of vectors are real

[1] A vector product exists only in some cases, e.g. $n = 3$ and $n = 7$. This is a theorem of higher algebra which we will not prove here.

numbers. Only *certain properties* of real numbers are actually used, namely that one can add or multiply or divide numbers. So one can easily replace real numbers by complex numbers or by some other kind of numbers as long as one can add, multiply and divide them as usual. (The use of the square root as in Eq. (1.5) can be avoided if one considers only *squared* lengths of vectors.)

Instead of specifying each time that one works with real numbers or with complex numbers, one says that one is working with some "abstract numbers" that have all the needed properties of numbers. The required properties of such "abstract numbers" are summarized by the axioms of a number field.

Definition: A **number field** (also called simply a **field**) is a set $\mathbb{K}$ which is an abelian group with respect to addition and multiplication, such that the distributive law holds. More precisely: There exist elements 0 and 1, and the operations $+$, $-$, $*$, and $/$ are defined such that $a+b=b+a$, $a*b=b*a$, $0+a=a, 1*a=a, 0*a=0$, and for every $a\in\mathbb{K}$ the numbers $-a$ and $1/a$ (for $a\neq 0$) exist such that $a+(-a)=0, a*(1/a)=1$, and also $a*(b+c)=a*b+a*c$. The operations $-$ and $/$ are defined by $a-b\equiv a+(-b)$ and $a/b=a*(1/b)$.

In a more visual language: A field is a set of elements on which the operations $+$, $-$, $*$, and $/$ are defined, the elements 0 and 1 exist, and the familiar arithmetic properties such as $a+b=b+a$, $a+0=0$, $a-a=0$, $a*1=1$, $a/b*b=a$ (for $b\neq 0$), etc. are satisfied. Elements of a field can be visualized as "abstract numbers" because they can be added, subtracted, multiplied, and divided, with the usual arithmetic rules. (For instance, division by zero is still undefined, even with abstract numbers!) I will call elements of a number field simply **numbers** when (in my view) it does not cause confusion.

Examples of number fields

Real numbers $\mathbb{R}$ are a field, as are rational numbers $\mathbb{Q}$ and complex numbers $\mathbb{C}$, with all arithmetic operations defined as usual. Integer numbers $\mathbb{Z}$ with the usual arithmetic are *not* a field because e.g. the division of 1 by a nonzero number 2 cannot be an integer.

Another interesting example is the set of numbers of the form $a+b\sqrt{3}$, where $a,b\in\mathbb{Q}$ are *rational* numbers. It is easy to see that sums, products, and ratios of such numbers are again numbers from the same set, for example

$$
\begin{aligned}
&(a_1+b_1\sqrt{3})(a_2+b_2\sqrt{3})\\
&=(a_1a_2+3b_1b_2)+(a_1b_2+a_2b_1)\,\sqrt{3}.
\end{aligned}
$$

Let's check the division property:

$$
\frac{1}{a+b\sqrt{3}}=\frac{a-b\sqrt{3}}{a-b\sqrt{3}}\frac{1}{a+b\sqrt{3}}=\frac{a-b\sqrt{3}}{a^2-3b^2}.
$$

Note that $\sqrt{3}$ is irrational, so the denominator a^2-3b^2 is never zero as long as a and b are rational and at least one of a,b is nonzero. Therefore, we can divide numbers of the form $a+b\sqrt{3}$ and again get numbers of the same kind.

It follows that the set $\{a + b\sqrt{3} \,|\, a, b \in \mathbb{Q}\}$ is indeed a number field. This field is usually denoted by $\mathbb{Q}[\sqrt{3}]$ and called an extension of rational numbers by $\sqrt{3}$. Fields of this form are useful in algebraic number theory.

A field might even consist of a *finite* set of numbers (in which case it is called a **finite field**). For example, the set of three numbers $\{0, 1, 2\}$ can be made a field if we define the arithmetic operations as

$$1 + 2 \equiv 0,\; 2 + 2 \equiv 1,\; 2 * 2 \equiv 1,\; 1/2 \equiv 2,$$

with all other operations as in usual arithmetic. This is the field of integers modulo 3 and is denoted by $\mathbb{F}_3$. Fields of this form are useful, for instance, in cryptography.

Any field must contain elements that play the role of the numbers 0 and 1; we denote these elements simply by 0 and 1. Therefore the smallest possible field is the set $\{0, 1\}$ with the usual relations $0 + 1 = 1, 1 \cdot 1 = 1$ etc. This field is denoted by $\mathbb{F}_2$.

Most of the time we will not need to specify the number field; it is all right to imagine that we always use $\mathbb{R}$ or $\mathbb{C}$ as the field. (See Appendix A for a brief introduction to complex numbers.)

Exercise: Which of the following sets are number fields:

a) $\{x + iy\sqrt{2} \,|\, x, y \in \mathbb{Q}\}$, where i is the imaginary unit.

b) $\{x + y\sqrt{2} \,|\, x, y \in \mathbb{Z}\}$.

Abstract vector spaces

After a generalization of the three-dimensional vector geometry to n-dimensional spaces and real numbers $\mathbb{R}$ to abstract number fields, we arrive at the following definition of a vector space.

Definition V1: An n-dimensional vector space over a field $\mathbb{K}$ is the set of all n-tuples $(x_1, ..., x_n)$, where $x_i \in \mathbb{K}$; the numbers x_i are called **components** of the vector (in older books they were called **coordinates**). The operations of vector sum and the scaling of vectors by numbers are given by the formulas

$$(x_1, ..., x_n) + (y_1, ..., y_n) \equiv (x_1 + y_1, ..., x_n + y_n)\,,\; x_i, y_i \in \mathbb{K};$$
$$\lambda\,(x_1, ..., x_n) \equiv (\lambda x_1, ..., \lambda x_n)\,,\; \lambda \in \mathbb{K}.$$

This vector space is denoted by $\mathbb{K}^n$.

Most problems in physics involve vector spaces over the field of real numbers $\mathbb{K} = \mathbb{R}$ or complex numbers $\mathbb{K} = \mathbb{C}$. However, most results of basic linear algebra hold for arbitrary number fields, and for now we will consider vector spaces over an arbitrary number field $\mathbb{K}$.

Definition V1 is adequate for applications involving *finite*-dimensional vector spaces. However, it turns out that further abstraction is necessary when one considers infinite-dimensional spaces. Namely, one needs to do away with coordinates and define the vector space by the basic requirements on the vector sum and scaling operations.

We will adopt the following "coordinate-free" definition of a vector space.

Definition V2: A set V is a **vector space over a number field** $\mathbb{K}$ if the following conditions are met:

1. V is an abelian group; the **sum** of two vectors is denoted by the "+" sign, the zero element is the vector $\mathbf{0}$. So for any $\mathbf{u}, \mathbf{v} \in V$ the vector $\mathbf{u} + \mathbf{v} \in V$ exists, $\mathbf{u} + \mathbf{v} = \mathbf{v} + \mathbf{u}$, and in particular $\mathbf{v} + \mathbf{0} = \mathbf{v}$ for any $\mathbf{v} \in V$.

2. An operation of **multiplication by numbers** is defined, such that for each $\lambda \in \mathbb{K}$, $\mathbf{v} \in V$ the vector $\lambda\mathbf{v} \in V$ is determined.

3. The following properties hold, for all vectors $\mathbf{u}, \mathbf{v} \in V$ and all numbers $\lambda, \mu \in \mathbb{K}$:

$$(\lambda + \mu)\,\mathbf{v} = \lambda\mathbf{v} + \mu\mathbf{v}, \quad \lambda\,(\mathbf{v} + \mathbf{u}) = \lambda\mathbf{v} + \lambda\mathbf{u},$$
$$1\mathbf{v} = \mathbf{v}, \quad 0\mathbf{v} = \mathbf{0}.$$

   These properties guarantee that the multiplication by numbers is compatible with the vector sum, so that usual rules of arithmetic and algebra are applicable.

Below I will not be so pedantic as to write the boldface $\mathbf{0}$ for the zero vector $\mathbf{0} \in V$; denoting the zero vector simply by 0 never creates confusion in practice.

Elements of a vector space are called **vectors**; in contrast, numbers from the field $\mathbb{K}$ are called **scalars**. For clarity, since this is an introductory text, I will print all vectors in boldface font so that $\mathbf{v}$, $\mathbf{a}$, $\mathbf{x}$ are vectors but v, a, x are scalars (i.e. numbers). Sometimes, for additional clarity, one uses Greek letters such as α, λ, μ to denote scalars and Latin letters to denote vectors. For example, one writes expressions of the form $\lambda_1\mathbf{v}_1 + \lambda_2\mathbf{v}_2 + ... + \lambda_n\mathbf{v}_n$; these are called **linear combinations** of vectors $\mathbf{v}_1, \mathbf{v}_2, ..., \mathbf{v}_n$.

The definition V2 is standard in abstract algebra. As we will see below, the coordinate-free language is well suited to proving theorems about general properties of vectors.

Question: I do not understand how to work with abstract vectors in abstract vector spaces. According to the vector space axioms (definition V2), I should be able to add vectors together and multiply them by scalars. It is clear how to add the n-tuples $(v_1, ..., v_n)$, but how can I compute anything with an abstract vector $\mathbf{v}$ that does not seem to have any components?

Answer: Definition V2 is "abstract" in the sense that it does not explain *how* to add particular kinds of vectors, instead it merely lists the set of properties *any* vector space must satisfy. To define a *particular* vector space, we of course need to specify a particular set of vectors and a rule for adding its elements in an explicit fashion (see examples below in Sec. 1.1.3). Definition V2 is used in the following way: Suppose someone claims that a certain set X of particular mathematical objects is a vector space over some number field, then we only need to check that the sum of vectors and the multiplication of vector by a number are well-defined and conform to the properties listed in

Definition V2. If every property holds, then the set X is a vector space, and all the theorems of linear algebra will automatically hold for the elements of the set X. Viewed from this perspective, Definition V1 specifies a *particular* vector space—the space of rows of numbers $(v_1, ..., v_n)$. In some cases the vector space at hand is exactly that of Definition V1, and then it is convenient to work with components v_j when performing calculations with specific vectors. However, components are not needed for proving general theorems. In this book, when I say that "a vector $\mathbf{v} \in V$ is given," I imagine that enough concrete information about $\mathbf{v}$ will be available when it is actually needed.

1.1.3 Examples of vector spaces

Example 0. The familiar example is the three-dimensional Euclidean space. This space is denoted by $\mathbb{R}^3$ and is the set of all triples (x_1, x_2, x_3), where x_i are real numbers. This is a vector space over $\mathbb{R}$.

Example 1. The set of complex numbers $\mathbb{C}$ is a vector space over the field of real numbers $\mathbb{R}$. Indeed, complex numbers can be added and multiplied by real numbers.

Example 2. Consider the set of all three-dimensional vectors $\mathbf{v} \in \mathbb{R}^3$ which are orthogonal to a given vector $\mathbf{a} \neq 0$; here we use the standard scalar product (1.4); vectors $\mathbf{a}$ and $\mathbf{b}$ are called **orthogonal to each other** if $\mathbf{a} \cdot \mathbf{b} = 0$. This set is closed under vector sum and scalar multiplication because if $\mathbf{u} \cdot \mathbf{a} = 0$ and $\mathbf{v} \cdot \mathbf{a} = 0$, then for any $\lambda \in \mathbb{R}$ we have $(\mathbf{u} + \lambda \mathbf{v}) \cdot \mathbf{a} = 0$. Thus we obtain a vector space (a certain subset of $\mathbb{R}^3$) which is defined not in terms of components but through geometric relations between vectors of another (previously defined) space.

Example 3. Consider the set of all real-valued continuous functions $f(x)$ defined for $x \in [0, 1]$ and such that $f(0) = 0$ and $f(1) = 0$. This set is a vector space over $\mathbb{R}$. Indeed, the definition of a vector space is satisfied if we define the sum of two functions as $f(x) + f(y)$ and the multiplication by scalars, $\lambda f(x)$, in the natural way. It is easy to see that the axioms of the vector space are satisfied: If $h(x) = f(x) + \lambda g(x)$, where $f(x)$ and $g(x)$ are vectors from this space, then the function $h(x)$ is continuous on $[0, 1]$ and satisfies $h(0) = h(1) = 0$, i.e. the function $h(x)$ is also an element of the same space.

Example 4. To represent the fact that there are λ_1 gallons of water and λ_2 gallons of oil, we may write the expression $\lambda_1 \mathbf{X} + \lambda_2 \mathbf{Y}$, where $\mathbf{X}$ and $\mathbf{Y}$ are formal symbols and $\lambda_{1,2}$ are numbers. The set of all such expressions is a vector space. This space is called the space of **formal linear combinations** of the symbols $\mathbf{X}$ and $\mathbf{Y}$. The operations of sum and scalar multiplication are defined in the natural way, so that we can perform calculations such as

$$\frac{1}{2}(2\mathbf{X} + 3\mathbf{Y}) - \frac{1}{2}(2\mathbf{X} - 3\mathbf{Y}) = 3\mathbf{Y}.$$

For the purpose of manipulating such expressions, it is unimportant that $\mathbf{X}$ and $\mathbf{Y}$ stand for water and oil. We may simply work with formal expressions such as $2\mathbf{X}+3\mathbf{Y}$, where $\mathbf{X}$ and $\mathbf{Y}$ and "+" are symbols that do not mean anything by themselves except that they can appear in such linear combinations and have familiar properties of algebraic objects (the operation "+" is commutative and associative, etc.). Such formal constructions are often encountered in mathematics.

Question: It seems that such "formal" constructions are absurd and/or useless. I know how to add numbers or vectors, but how can I add $\mathbf{X}+\mathbf{Y}$ if $\mathbf{X}$ and $\mathbf{Y}$ are, as you say, "meaningless symbols"?

Answer: Usually when we write "$a+b$" we imply that the operation "+" is already defined, so $a+b$ is another number if a and b are numbers. However, in the case of formal expressions described in Example 4, the "+" sign is actually going to acquire a *new* definition. So $\mathbf{X}+\mathbf{Y}$ is not equal to a new symbol $\mathbf{Z}$, instead $\mathbf{X}+\mathbf{Y}$ is just *an expression* that we can manipulate. Consider the analogy with complex numbers: the number $1+2\mathrm{i}$ is an expression that we manipulate, and the imaginary unit, i, is a symbol that is never "equal to something else." According to its definition, the expression $\mathbf{X}+\mathbf{Y}$ cannot be simplified to anything else, just like $1+2\mathrm{i}$ cannot be simplified. The symbols $\mathbf{X}, \mathbf{Y}, \mathrm{i}$ are *not* meaningless: their meaning comes *from the rules of computations* with these symbols.

Maybe it helps to change notation. Let us begin by writing a pair (a,b) instead of $a\mathbf{X}+b\mathbf{Y}$. We can define the sum of such pairs in the natural way, e.g.

$$(2,3)+(-2,1)=(0,4).$$

It is clear that these pairs build a vector space. Now, to remind ourselves that the numbers of the pair stand for, say, quantities of water and oil, we write $(2\mathbf{X},3\mathbf{Y})$ instead of $(2,3)$. The symbols $\mathbf{X}$ and $\mathbf{Y}$ are merely part of the notation. Now it is natural to change the notation further and to write simply $2\mathbf{X}$ instead of $(2\mathbf{X},0\mathbf{Y})$ and $a\mathbf{X}+b\mathbf{Y}$ instead of $(a\mathbf{X},b\mathbf{Y})$. It is clear that we do not introduce anything new when we write $a\mathbf{X}+b\mathbf{Y}$ instead of $(a\mathbf{X},b\mathbf{Y})$: We merely change the notation so that computations appear easier. Similarly, complex numbers can be understood as pairs of real numbers, such as $(3,2)$, for which $3+2\mathrm{i}$ is merely a more convenient notation that helps remember the rules of computation. ■

Example 5. The set of all polynomials of degree at most n in the variable x with complex coefficients is a vector space over $\mathbb{C}$. Such polynomials are expressions of the form $p(x)=p_0+p_1x+...+p_nx^n$, where x is a **formal variable** (i.e. no value is assigned to x), n is an integer, and p_i are complex numbers.

Example 6. Consider now the set of all polynomials in the variables x, y, and z, with complex coefficients, and such that the combined degree in x, in y, and in z is at most 2. For instance, the polynomial $1+2\mathrm{i}x-yz-\sqrt{3}x^2$ is an element of that vector space (while x^2y is not because its combined degree is 3). It is clear that the degree will never increase above 2 when any two such polynomials are added together, so these polynomials indeed form a vector space over the field $\mathbb{C}$.

Exercise. Which of the following are vector spaces over $\mathbb{R}$?

1. The set of all complex numbers z whose real part is equal to 0. The complex numbers are added and multiplied by real constants as usual.

2. The set of all complex numbers z whose imaginary part is equal to 3. The complex numbers are added and multiplied by real constants as usual.

3. The set of pairs of the form (apples, \$3.1415926), where the first element is always the word "apples" and the second element is a price in dollars (the price may be an arbitrary real number, not necessarily positive or with an integer number of cents). Addition and multiplication by real constants is defined as follows:

$$\begin{aligned}(\text{apples}, \$x) + (\text{apples}, \$y) &\equiv (\text{apples}, \$(x+y)) \\ \lambda \cdot (\text{apples}, \$x) &\equiv (\text{apples}, \$(\lambda \cdot x))\end{aligned}$$

4. The set of pairs of the form either (apples, $\$x$) or (chocolate, $\$y$), where x and y are real numbers. The pairs are added as follows,

$$\begin{aligned}(\text{apples}, \$x) + (\text{apples}, \$y) &\equiv (\text{apples}, \$(x+y)) \\ (\text{chocolate}, \$x) + (\text{chocolate}, \$y) &\equiv (\text{chocolate}, \$(x+y)) \\ (\text{chocolate}, \$x) + (\text{apples}, \$y) &\equiv (\text{chocolate}, \$(x+y))\end{aligned}$$

(that is, chocolate "takes precedence" over apples). The multiplication by a number is defined as in the previous question.

5. The set of "bracketed complex numbers," denoted $[z]$, where z is a complex number such that $|z| = 1$. For example: $[\mathrm{i}]$, $[\frac{1}{2} - \frac{1}{2}\mathrm{i}\sqrt{3}]$, $[-1]$. Addition and multiplication by real constants λ are defined as follows,

$$[z_1] + [z_2] = [z_1 z_2], \quad \lambda \cdot [z] = \left[z e^{\mathrm{i}\lambda} \right].$$

6. The set of infinite arrays $(a_1, a_2, ...)$ of arbitrary real numbers. Addition and multiplication are defined term-by-term.

7. The set of polynomials in the variable x with real coefficients and of arbitrary (but finite) degree. Addition and multiplication is defined as usual in algebra.

Question: All these abstract definitions notwithstanding, would it be all right if I always keep in the back of my mind that a vector $\mathbf{v}$ is a row of components $(v_1, ..., v_n)$?

Answer: It will be perfectly all right *as long as* you work with *finite*-dimensional vector spaces. (This intuition often fails when working with infinite-dimensional spaces!) Even if all we need is finite-dimensional vectors, there is another argument in favor of the coordinate-free thinking. Suppose I persist in visualizing vectors as rows $(v_1, ..., v_n)$; let us see what happens. First, I introduce the vector notation and write $\mathbf{u}+\mathbf{v}$ instead of $(u_1+v_1, ..., u_n+v_n)$; this is just for convenience and to save time. Then I check the axioms of the vector space (see the definition V2 above); row vectors of course obey these axioms. Suppose I somehow manage to produce all proofs and calculations using only the vector notation and the axioms of the abstract vector space, and suppose I never use the coordinates v_j explicitly, even though I keep them in the back of my mind. Then all my results will be valid not only for collections of components $(v_1, ..., v_n)$ but also for *any* mathematical objects that obey the axioms of the abstract vector space. In fact I would then realize that I have been working with abstract vectors *all along* while carrying the image of a row vector $(v_1, ..., v_n)$ in the back of my mind.

1.1.4 Dimensionality and bases

Unlike the definition V1, the definition V2 does not include any information about the dimensionality of the vector space. So, on the one hand, this definition treats finite- and infinite-dimensional spaces on the same footing; the definition V2 lets us establish that a certain set is a vector space without knowing its dimensionality in advance. On the other hand, once a particular vector space is given, we may need some additional work to figure out the number of dimensions in it. The key notion used for that purpose is "linear independence."

We say, for example, the vector $\mathbf{w} \equiv 2\mathbf{u} - 3\mathbf{v}$ is "linearly dependent" on $\mathbf{u}$ and $\mathbf{v}$. A vector $\mathbf{x}$ is linearly independent of vectors $\mathbf{u}$ and $\mathbf{v}$ if $\mathbf{x}$ *cannot* be expressed as a linear combination $\lambda_1\mathbf{u} + \lambda_2\mathbf{v}$.

A set of vectors is **linearly dependent** if one of the vectors is a linear combination of others. This property can be formulated more elegantly:

Definition: The set of vectors $\{\mathbf{v}_1, ..., \mathbf{v}_n\}$ is a **linearly dependent set** if there exist numbers $\lambda_1, ..., \lambda_n \in \mathbb{K}$, not all equal to zero, such that

$$\lambda_1\mathbf{v}_1 + ... + \lambda_n\mathbf{v}_n = 0. \qquad (1.6)$$

If no such numbers exist, i.e. if Eq. (1.6) holds only with all $\lambda_i = 0$, the vectors $\{\mathbf{v}_i\}$ constitute a **linearly independent set**.

Interpretation: As a first example, consider the set $\{\mathbf{v}\}$ consisting of a single nonzero vector $\mathbf{v} \neq 0$. The set $\{\mathbf{v}\}$ is a linearly independent set because $\lambda\mathbf{v} = 0$ only if $\lambda = 0$. Now consider the set $\{\mathbf{u}, \mathbf{v}, \mathbf{w}\}$, where $\mathbf{u} = 2\mathbf{v}$ and $\mathbf{w}$ is any vector. This set is linearly dependent because there exists a nontrivial linear combination (i.e. a linear combination with *some* nonzero coefficients) which is equal to zero,

$$\mathbf{u} - 2\mathbf{v} = 1\mathbf{u} + (-2)\,\mathbf{v} + 0\mathbf{w} = 0.$$

More generally: If a set $\{\mathbf{v}_1, ..., \mathbf{v}_n\}$ is linearly dependent, then there exists at least one vector equal to a linear combination of other vectors. Indeed, by definition there must be at least one nonzero number among the numbers λ_i involved in Eq. (1.6); suppose $\lambda_1 \neq 0$, then we can divide Eq. (1.6) by λ_1 and express $\mathbf{v}_1$ through other vectors,

$$\mathbf{v}_1 = -\frac{1}{\lambda_1}\left(\lambda_2\mathbf{v}_2 + ... + \lambda_n\mathbf{v}_n\right).$$

In other words, the existence of numbers λ_i, not all equal to zero, is indeed the formal statement of the idea that at least some vector in the set $\{\mathbf{v}_i\}$ is a linear combination of other vectors. By writing a linear combination $\sum_i \lambda_i \mathbf{v}_i = 0$ and by saying that "not all λ_i are zero" we avoid specifying *which* vector is equal to a linear combination of others.

Remark: Often instead of saying "a linearly independent *set* of vectors" one says "a set of linearly independent *vectors*." This is intended to mean the same thing but might be confusing because, taken literally, the phrase "a set of independent vectors" means a set in which each vector is "independent" by itself. Keep in mind that linear independence is a property of a *set of vectors*; this property depends on the relationships between all the vectors in the set and is not a property of each vector taken separately. It would be more consistent to say e.g. "a set of *mutually* independent vectors." In this text, I will pedantically stick to the phrase "linearly independent set."

Example 1: Consider the vectors $\mathbf{a} = (0, 1)$, $\mathbf{b} = (1, 1)$ in $\mathbb{R}^2$. Is the set $\{\mathbf{a}, \mathbf{b}\}$ linearly independent? Suppose there exists a linear combination $\alpha\mathbf{a} + \beta\mathbf{b} = 0$ with at least one of $\alpha, \beta \neq 0$. Then we would have

$$\alpha\mathbf{a} + \beta\mathbf{b} = (0, \alpha) + (\beta, \beta) = (\beta, \alpha + \beta) \overset{!}{=} 0.$$

This is possible only if $\beta = 0$ and $\alpha = 0$. Therefore, $\{\mathbf{a}, \mathbf{b}\}$ is linearly independent.

Exercise 1: a) A set $\{\mathbf{v}_1, ..., \mathbf{v}_n\}$ is linearly independent. Prove that any subset, say $\{\mathbf{v}_1, ..., \mathbf{v}_k\}$, where $k < n$, is also a linearly independent set.

b) Decide whether the given sets $\{\mathbf{a}, \mathbf{b}\}$ or $\{\mathbf{a}, \mathbf{b}, \mathbf{c}\}$ are linearly independent sets of vectors from $\mathbb{R}^2$ or other spaces as indicated. For linearly dependent sets, find a linear combination showing this.

1. $\mathbf{a} = (2, \sqrt{2})$, $\mathbf{b} = (\frac{1}{\sqrt{2}}, \frac{1}{2})$ in $\mathbb{R}^2$

2. $\mathbf{a} = (-2, 3)$, $\mathbf{b} = (6, -9)$ in $\mathbb{R}^2$

3. $\mathbf{a} = (1 + 2\mathrm{i}, 10, 20)$, $\mathbf{b} = (1 - 2\mathrm{i}, 10, 20)$ in $\mathbb{C}^3$

4. $\mathbf{a} = (0, 10\mathrm{i}, 20\mathrm{i}, 30\mathrm{i})$, $\mathbf{b} = (0, 20\mathrm{i}, 40\mathrm{i}, 60\mathrm{i})$, $\mathbf{c} = (0, 30\mathrm{i}, 60\mathrm{i}, 90\mathrm{i})$ in $\mathbb{C}^4$

5. $\mathbf{a} = (3, 1, 2)$, $\mathbf{b} = (1, 0, 1)$, $\mathbf{c} = (0, -1, 2)$ in $\mathbb{R}^3$

The **number of dimensions** (or simply the **dimension**) of a vector space is the maximum possible number of vectors in a linearly independent set. The formal definition is the following.

Definition: A vector space is n-**dimensional** if linearly independent sets of n vectors can be found in it, but no linearly independent sets of $n+1$ vectors. The dimension of a vector space V is then denoted by $\dim V \equiv n$. A vector space is **infinite-dimensional** if linearly independent sets having *arbitrarily many* vectors can be found in it.

By this definition, in an n-dimensional vector space there exists *at least one* linearly independent set of n vectors $\{\mathbf{e}_1, ..., \mathbf{e}_n\}$. Linearly independent sets containing exactly $n = \dim V$ vectors have useful properties, to which we now turn.

Definition: A **basis** in the space V is a linearly independent set of vectors $\{\mathbf{e}_1, ..., \mathbf{e}_n\}$ such that for any vector $\mathbf{v} \in V$ there exist numbers $v_k \in \mathbb{K}$ such that $\mathbf{v} = \sum_{k=1}^{n} v_k \mathbf{e}_k$. (In other words, every other vector $\mathbf{v}$ is a linear combination of basis vectors.) The numbers v_k are called the **components** (or **coordinates**) of the vector $\mathbf{v}$ *with respect to the basis* $\{\mathbf{e}_i\}$.

Example 2: In the three-dimensional Euclidean space $\mathbb{R}^3$, the set of three triples $(1,0,0)$, $(0,1,0)$, and $(0,0,1)$ is a basis because every vector $\mathbf{x} = (x, y, z)$ can be expressed as

$$\mathbf{x} = (x, y, z) = x\,(1,0,0) + y\,(0,1,0) + z\,(0,0,1)\,.$$

This basis is called the **standard basis**. Analogously one defines the standard basis in $\mathbb{R}^n$. ■

The following statement is standard, and I write out its full proof here as an example of an argument based on the abstract definition of vectors.

Theorem: (1) If a set $\{\mathbf{e}_1, ..., \mathbf{e}_n\}$ is linearly independent and $n = \dim V$, then the set $\{\mathbf{e}_1, ..., \mathbf{e}_n\}$ is a basis in V. **(2)** For a given vector $\mathbf{v} \in V$ and a given basis $\{\mathbf{e}_1, ..., \mathbf{e}_n\}$, the coefficients v_k involved in the decomposition $\mathbf{v} = \sum_{k=1}^{n} v_k \mathbf{e}_k$ are uniquely determined.

Proof: **(1)** By definition of dimension, the set $\{\mathbf{v}, \mathbf{e}_1, ..., \mathbf{e}_n\}$ must be linearly *dependent*. By definition of linear dependence, there exist numbers λ_0, ..., λ_n, not all equal to zero, such that

$$\lambda_0 \mathbf{v} + \lambda_1 \mathbf{e}_1 + ... + \lambda_n \mathbf{e}_n = 0. \tag{1.7}$$

Now if we had $\lambda_0 = 0$, it would mean that not all numbers in the smaller set $\{\lambda_1, ..., \lambda_n\}$ are zero; however, in that case Eq. (1.7) would contradict the linear independence of the set $\{\mathbf{e}_1, ..., \mathbf{e}_n\}$. Therefore $\lambda_0 \neq 0$ and Eq. (1.7) shows that the vector $\mathbf{v}$ can be expressed through the basis, $\mathbf{v} = \sum_{k=1}^{n} v_k \mathbf{e}_k$ with the coefficients $v_k \equiv -\lambda_k / \lambda_0$.

(2) To show that the set of coefficients $\{v_k\}$ is unique, we assume that there are two such sets, $\{v_k\}$ and $\{v'_k\}$. Then

$$0 = \mathbf{v} - \mathbf{v} = \sum_{k=1}^{n} v_k \mathbf{e}_k - \sum_{k=1}^{n} v'_k \mathbf{e}_k = \sum_{k=1}^{n} (v_k - v'_k)\, \mathbf{e}_k.$$

Since the set $\{\mathbf{e}_1, ..., \mathbf{e}_n\}$ is linearly independent, all coefficients in this linear combination must vanish, so $v_k = v'_k$ for all k. ■

If we fix a basis $\{\mathbf{e}_i\}$ in a finite-dimensional vector space V then all vectors $\mathbf{v} \in V$ are uniquely represented by n-tuples $\{v_1, ..., v_n\}$ of their components. Thus we recover the original picture of a vector space as a set of n-tuples of numbers. (Below we will prove that *every* basis in an n-dimensional space has the same number of vectors, namely n.) Now, if we choose another basis $\{\mathbf{e}'_i\}$, the same vector $\mathbf{v}$ will have different components v'_k:

$$\mathbf{v} = \sum_{k=1}^{n} v_k \mathbf{e}_k = \sum_{k=1}^{n} v'_k \mathbf{e}'_k.$$

Remark: One sometimes reads that "the components are transformed" or that "vectors are sets of numbers that transform under a change of basis." I do not use this language because it suggests that the components v_k, which are numbers such as $\frac{1}{3}$ or $\sqrt{2}$, are somehow not simply numbers but "know how to transform." I prefer to say that the components v_k of a vector $\mathbf{v}$ in a particular basis $\{\mathbf{e}_k\}$ express the relationship of $\mathbf{v}$ to that basis and are therefore functions of the vector $\mathbf{v}$ and of *all* basis vectors $\mathbf{e}_j$. ■

For many purposes it is better to think about a vector $\mathbf{v}$ not as a set of its components $\{v_1, ..., v_n\}$ in some basis, but as a geometric object; a "directed magnitude" is a useful heuristic idea. Geometric objects exist in the vector space independently of a choice of basis. In linear algebra, one is typically interested in problems involving relations between vectors, for example $\mathbf{u} = a\mathbf{v} + b\mathbf{w}$, where $a, b \in \mathbb{K}$ are numbers. No choice of basis is necessary to describe such relations between vectors; I will call such relations **coordinate-free** or **geometric**. As I will demonstrate later in this text, many statements of linear algebra are more transparent and easier to prove in the coordinate-free language. Of course, in many practical applications one absolutely needs to perform specific calculations with components in an appropriately chosen basis, and facility with such calculations is important. But I find it helpful to keep a coordinate-free (geometric) picture in the back of my mind even when I am doing calculations in coordinates.

Question: I am not sure how to determine the number of dimensions in a vector space. According to the definition, I should figure out whether there exist certain linearly independent sets of vectors. But surely it is impossible to go over all sets of n vectors checking the linear independence of each set?

Answer: Of course it is impossible when there are infinitely many vectors. This is simply not the way to go. We can determine the dimensionality of a given vector space by *proving* that the space has a basis consisting of a certain number of vectors. A particular vector space must be specified in concrete terms (see Sec. 1.1.3 for examples), and in each case we should manage to find a general proof that covers all sets of n vectors at once.

Exercise 2: For each vector space in the examples in Sec. 1.1.3, find the dimension or show that the dimension is infinite.

Solution for Example 1: The set $\mathbb{C}$ of complex numbers is a two-dimensional vector space over $\mathbb{R}$ because every complex number $a + ib$ can be represented as a linear combination of *two* basis vectors (1 and i) with real coefficients a, b. The set $\{1, \mathrm{i}\}$ is linearly independent because $a + \mathrm{i}b = 0$ only when both $a = b = 0$.

Solution for Example 2: The space V is defined as the set of triples (x, y, z) such that $ax + by + cz = 0$, where at least one of a, b, c is nonzero. Suppose, without loss of generality, that $a \neq 0$; then we can express

$$x = -\frac{b}{a}y - \frac{c}{a}z.$$

Now the two parameters y and z are arbitrary while x is determined. Hence it appears plausible that the space V is *two*-dimensional. Let us prove this formally. Choose as the possible basis vectors $\mathbf{e}_1 = (-\frac{b}{a}, 1, 0)$ and $\mathbf{e}_2 = (-\frac{c}{a}, 0, 1)$. These vectors belong to V, and the set $\{\mathbf{e}_1, \mathbf{e}_2\}$ is linearly independent (straightforward checks). It remains to show that every vector $\mathbf{x} \in V$ is expressed as a linear combination of $\mathbf{e}_1$ and $\mathbf{e}_2$. Indeed, any such $\mathbf{x}$ must have components x, y, z that satisfy $x = -\frac{b}{a}y - \frac{c}{a}z$. Hence, $\mathbf{x} = y\mathbf{e}_1 + z\mathbf{e}_2$.

Exercise 3: Describe a vector space that has dimension zero.

Solution: If there are *no* linearly independent sets in a space V, it means that all sets consisting of just one vector $\{\mathbf{v}\}$ are already linearly *dependent*. More formally, $\forall \mathbf{v} \in V : \exists \lambda \neq 0$ such that $\lambda \mathbf{v} = 0$. Thus $\mathbf{v} = 0$, that is, all vectors $\mathbf{v} \in V$ are equal to the zero vector. Therefore a zero-dimensional space is a space that consists of only one vector: the zero vector.

Exercise 4*: Usually a vector space admits infinitely many choices of a basis. However, above I cautiously wrote that a vector space "has at least one basis." Is there an example of a vector space that has *only one* basis?

Hints: The answer is positive. Try to build a new basis from an existing one and see where that might fail. This has to do with finite number fields (try $\mathbb{F}_2$), and the only available example is rather dull.

1.1.5 All bases have equally many vectors

We have seen that any linearly independent set of n vectors in an n-dimensional space is a basis. The following statement shows that a basis cannot have *fewer* than n vectors. The proof is somewhat long and can be skipped unless you would like to gain more facility with coordinate-free manipulations.

Theorem: In a finite-dimensional vector space, all bases have equally many vectors.

Proof: Suppose that $\{\mathbf{e}_1, ..., \mathbf{e}_m\}$ and $\{\mathbf{f}_1, ..., \mathbf{f}_n\}$ are two bases in a vector space V and $m \neq n$. I will show that this assumption leads to contradiction, and then it will follow that any two bases must have equally many vectors.

Assume that $m > n$. The idea of the proof is to take the larger set $\{\mathbf{e}_1, ..., \mathbf{e}_m\}$ and to replace one of its vectors, say $\mathbf{e}_s$, by $\mathbf{f}_1$, so that the resulting set of m vectors

$$\{\mathbf{e}_1, ..., \mathbf{e}_{s-1}, \mathbf{f}_1, \mathbf{e}_{s+1}, ..., \mathbf{e}_m\} \tag{1.8}$$

is still linearly independent. I will prove shortly that such a replacement is possible, assuming only that the initial set is linearly independent. Then I will continue to replace other vectors $\mathbf{e}_k$ by $\mathbf{f}_2$, $\mathbf{f}_3$, etc., always keeping the resulting set linearly independent. Finally, I will arrive to the linearly independent set

$$\{\mathbf{f}_1, ..., \mathbf{f}_n, \mathbf{e}_{k_1}, \mathbf{e}_{k_2}, ..., \mathbf{e}_{k_{m-n}}\},$$

which contains all $\mathbf{f}_j$ as well as $(m-n)$ vectors $\mathbf{e}_{k_1}$, $\mathbf{e}_{k_2}$, ..., $\mathbf{e}_{k_{m-n}}$ left over from the original set; there must be at least one such vector left over because (by assumption) there are more vectors in the basis $\{\mathbf{e}_j\}$ than in the basis $\{\mathbf{f}_j\}$, in other words, because $m - n \geq 1$. Since the set $\{\mathbf{f}_j\}$ is a basis, the vector $\mathbf{e}_{k_1}$ is a linear combination of $\{\mathbf{f}_1, ..., \mathbf{f}_n\}$, so the set $\{\mathbf{f}_1, ..., \mathbf{f}_n, \mathbf{e}_{k_1}, ...\}$ cannot be linearly independent. This contradiction proves the theorem.

It remains to show that it is possible to find the index s such that the set (1.8) is linearly independent. The required statement is the following: If $\{\mathbf{e}_j \,|\, 1 \leq j \leq m\}$ and $\{\mathbf{f}_j \,|\, 1 \leq j \leq n\}$ are two bases in the space V, and if the set $S \equiv \{\mathbf{e}_1, ..., \mathbf{e}_k, \mathbf{f}_1, ..., \mathbf{f}_l\}$ (where $l < n$) is linearly independent then there exists an index s such that $\mathbf{e}_s$ in S can be replaced by $\mathbf{f}_{l+1}$ and the new set

$$T \equiv \{\mathbf{e}_1, ..., \mathbf{e}_{s-1}, \mathbf{f}_{l+1}, \mathbf{e}_{s+1}, ..., \mathbf{e}_k, \mathbf{f}_1, ..., \mathbf{f}_l\} \tag{1.9}$$

is still linearly independent. To find a suitable index s, we try to decompose $\mathbf{f}_{l+1}$ into a linear combination of vectors from S. In other words, we ask whether the set

$$S' \equiv S \cup \{\mathbf{f}_{l+1}\} = \{\mathbf{e}_1, ..., \mathbf{e}_k, \mathbf{f}_1, ..., \mathbf{f}_{l+1}\}$$

is linearly independent. There are two possibilities: First, if S' is linearly independent, we can remove any $\mathbf{e}_s$, say $\mathbf{e}_1$, from it, and the resulting set

$$T = \{\mathbf{e}_2, ..., \mathbf{e}_k, \mathbf{f}_1, ..., \mathbf{f}_{l+1}\}$$

will be again linearly independent. This set T is obtained from S by replacing $\mathbf{e}_1$ with $\mathbf{f}_{l+1}$, so now there is nothing left to prove. Now consider the second possibility: S' is linearly dependent. In that case, $\mathbf{f}_{l+1}$ can be decomposed as

$$\mathbf{f}_{l+1} = \sum_{j=1}^{k} \lambda_j \mathbf{e}_j + \sum_{j=1}^{l} \mu_j \mathbf{f}_j, \tag{1.10}$$

where λ_j, μ_j are some constants, not all equal to zero. Suppose all λ_j are zero; then $\mathbf{f}_{l+1}$ would be a linear combination of other $\mathbf{f}_j$; but this cannot happen for a basis $\{\mathbf{f}_j\}$. Therefore not all λ_j, $1 \leq j \leq k$ are zero; for example, $\lambda_s \neq 0$. This gives us the index s. Now we can replace $\mathbf{e}_s$ in the set S by $\mathbf{f}_{l+1}$; it remains to prove that the resulting set T defined by Eq. (1.9) is linearly independent.

This last proof is again by contradiction: if T is linearly *dependent*, there exists a vanishing linear combination of the form

$$\sum_{j=1}^{s-1} \rho_j \mathbf{e}_j + \sigma_{l+1} \mathbf{f}_{l+1} + \sum_{j=s+1}^{k} \rho_j \mathbf{e}_j + \sum_{j=1}^{l} \sigma_j \mathbf{f}_j = 0, \tag{1.11}$$

where ρ_j, σ_j are not all zero. In particular, $\sigma_{l+1} \neq 0$ because otherwise the initial set S would be linearly dependent,

$$\sum_{j=1}^{s-1} \rho_j \mathbf{e}_j + \sum_{j=s+1}^{k} \rho_j \mathbf{e}_j + \sum_{j=1}^{l} \sigma_j \mathbf{f}_j = 0.$$

If we now substitute Eq. (1.10) into Eq. (1.11), we will obtain a vanishing linear combination that contains only vectors from the initial set S in which the coefficient at the vector $\mathbf{e}_s$ is $\sigma_{l+1}\lambda_s \neq 0$. This contradicts the linear independence of the set S. Therefore the set T is linearly independent. ■

Exercise 1: Completing a basis. If a set $\{\mathbf{v}_1, ..., \mathbf{v}_k\}$, $\mathbf{v}_j \in V$ is linearly independent and $k < n \equiv \dim V$, the theorem says that the set $\{\mathbf{v}_j\}$ is *not* a basis in V. Prove that there exist $(n-k)$ additional vectors $\mathbf{v}_{k+1}, ..., \mathbf{v}_n \in V$ such that the set $\{\mathbf{v}_1, ..., \mathbf{v}_n\}$ is a basis in V.

Outline of proof: If $\{\mathbf{v}_j\}$ is not yet a basis, it means that there exists at least one vector $\mathbf{v} \in V$ which cannot be represented by a linear combination of $\{\mathbf{v}_j\}$. Add it to the set $\{\mathbf{v}_j\}$; prove that the resulting set is still linearly independent. Repeat these steps until a basis is built; by the above Theorem, the basis will contain exactly n vectors.

Exercise 2: Eliminating unnecessary vectors. Suppose that a set of vectors $\{\mathbf{e}_1, ..., \mathbf{e}_s\}$ **spans the space** V, i.e. every vector $\mathbf{v} \in V$ can be represented by a linear combination of $\{\mathbf{v}_j\}$; and suppose that $s > n \equiv \dim V$. By definition of dimension, the set $\{\mathbf{e}_j\}$ must be linearly dependent, so it is not a basis in V. Prove that one can remove certain vectors from this set so that the remaining vectors are a basis in V.

Hint: The set has too many vectors. Consider a nontrivial linear combination of vectors $\{\mathbf{e}_1, ..., \mathbf{e}_s\}$ that is equal to zero. Show that one can remove some vector $\mathbf{e}_k$ from the set $\{\mathbf{e}_1, ..., \mathbf{e}_s\}$ such that the remaining set still spans V. The procedure can be repeated until a basis in V remains.

Exercise 3: Finding a basis. Consider the vector space of polynomials of degree at most 2 in the variable x, with real coefficients. Determine whether the following four sets of vectors are linearly independent, and which of them can serve as a basis in that space. The sets are $\{1+x, 1-x\}$; $\{1, 1+x, 1-x\}$; $\{1, 1+x-x^2\}$; $\{1, 1+x, 1+x+x^2\}$.

Exercise 4: Not a basis. Suppose that a set $\{\mathbf{v}_1, ..., \mathbf{v}_n\}$ in an n-dimensional space V is not a basis; show that this set must be linearly dependent.

1.2 Linear maps in vector spaces

An important role in linear algebra is played by matrices, which usually represent linear transformations of vectors. Namely, with the definition **V1** of vectors as n-tuples v_i, one defines matrices as square tables of numbers, A_{ij}, that describe transformations of vectors according to the formula

$$u_i \equiv \sum_{j=1}^{n} A_{ij} v_j. \tag{1.12}$$

This transformation takes a vector $\mathbf{v}$ into a new vector $\mathbf{u} = \hat{A}\mathbf{v}$ in the same vector space. For example, in two dimensions one writes the transformation of column vectors as

$$\begin{bmatrix} u_1 \\ u_2 \end{bmatrix} = \begin{pmatrix} A_{11} & A_{12} \\ A_{21} & A_{22} \end{pmatrix} \begin{bmatrix} v_1 \\ v_2 \end{bmatrix} \equiv \begin{bmatrix} A_{11}v_1 + A_{12}v_2 \\ A_{21}v_1 + A_{22}v_2 \end{bmatrix}.$$

The **composition** of two transformations A_{ij} and B_{ij} is a transformation described by the matrix

$$C_{ij} = \sum_{k=1}^{n} A_{ik}B_{kj}. \tag{1.13}$$

This is the law of matrix multiplication. (I assume that all this is familiar to you.)

More generally, a map from an m-dimensional space V to an n-dimensional space W is described by a rectangular $m \times n$ matrix that transforms m-tuples into n-tuples in an analogous way. Most of the time we will be working with transformations within one vector space (described by square matrices).

This picture of matrix transformations is straightforward but relies on the coordinate representation of vectors and so has two drawbacks: (i) The calculations with matrix components are often unnecessarily cumbersome. (ii) Definitions and calculations cannot be easily generalized to infinite-dimensional spaces. Nevertheless, many of the results have nothing to do with components and *do* apply to infinite-dimensional spaces. We need a different approach to characterizing linear transformations of vectors.

The way out is to concentrate on the **linearity** of the transformations, i.e. on the properties

$$\hat{A}(\lambda\mathbf{v}) = \lambda\hat{A}(\mathbf{v}),$$
$$\hat{A}(\mathbf{v}_1 + \mathbf{v}_2) = \hat{A}(\mathbf{v}_1) + \hat{A}(\mathbf{v}_2),$$

which are easy to check directly. In fact it turns out that the multiplication law and the matrix representation of transformations can be *derived* from the above requirements of linearity. Below we will see how this is done.

1.2.1 Abstract definition of linear maps

First, we define an abstract **linear map** as follows.

Definition: A map $\hat{A} : V \to W$ between two vector spaces V, W is **linear** if for any $\lambda \in \mathbb{K}$ and $\mathbf{u}, \mathbf{v} \in V$,

$$\hat{A}(\mathbf{u} + \lambda\mathbf{v}) = \hat{A}\mathbf{u} + \lambda\hat{A}\mathbf{v}. \tag{1.14}$$

(Note, pedantically, that the "+" in the left side of Eq. (1.14) is the vector sum in the space V, while in the right side it is the vector sum in the space W.)

Linear maps are also called **homomorphisms** of vector spaces. Linear maps acting from a space V to the same space are called **linear operators** or **endomorphisms** of the space V.

At first sight it might appear that the abstract definition of a linear transformation offers much less information than the definition in terms of matrices. This is true: the abstract definition does not *specify* any particular linear map, it only gives conditions for a map to be linear. If the vector space is finite-dimensional and a basis $\{\mathbf{e}_i\}$ is selected then the familiar matrix picture is immediately recovered from the abstract definition. Let us first, for simplicity, consider a linear map $\hat{A} : V \to V$.

Statement 1: If $\hat{A}$ is a linear map $V \to V$ and $\{\mathbf{e}_j\}$ is a basis then there exist numbers A_{jk} $(j, k = 1, ..., n)$ such that the vector $\hat{A}\mathbf{v}$ has components $\sum_k A_{jk}v_k$ if a vector $\mathbf{v}$ has components v_k in the basis $\{\mathbf{e}_j\}$.

Proof: For any vector $\mathbf{v}$ we have a decomposition $\mathbf{v} = \sum_{k=1}^{n} v_k \mathbf{e}_k$ with some components v_k. By linearity, the result of application of the map $\hat{A}$ to the vector $\mathbf{v}$ is

$$\hat{A}\mathbf{v} = \hat{A}\Big(\sum_{k=1}^{n} v_k \mathbf{e}_k\Big) = \sum_{k=1}^{n} v_k (\hat{A}\mathbf{e}_k).$$

Therefore, it is sufficient to know how the map $\hat{A}$ transforms the basis vectors $\mathbf{e}_k$, $k = 1, ..., n$. Each of the vectors $\hat{A}\mathbf{e}_k$ has (in the basis $\{\mathbf{e}_i\}$) a decomposition

$$\hat{A}\mathbf{e}_k = \sum_{j=1}^{n} A_{jk}\mathbf{e}_j, \quad k = 1, ..., n,$$

where A_{jk} with $1 \leq j, k \leq n$ are some coefficients; these A_{jk} are just some numbers that we can calculate for a specific given linear transformation and a specific basis. It is convenient to arrange these numbers into a square table (matrix) A_{jk}. Finally, we compute $\hat{A}\mathbf{v}$ as

$$\hat{A}\mathbf{v} = \sum_{k=1}^{n} v_k \sum_{j=1}^{n} A_{jk}\mathbf{e}_j = \sum_{j=1}^{n} u_j \mathbf{e}_j,$$

where the components u_j of the vector $\mathbf{u} \equiv \hat{A}\mathbf{v}$ are

$$u_j \equiv \sum_{k=1}^{n} A_{jk} v_k.$$

This is exactly the law (1.12) of multiplication of the matrix A_{jk} by a column vector v_k. Therefore the formula of the matrix representation (1.12) is a necessary consequence of the linearity of a transformation. ■

The analogous matrix representation holds for linear maps $\hat{A} : V \to W$ between different vector spaces.

It is helpful to imagine that the linear transformation $\hat{A}$ somehow exists as a geometric object (an object that "knows how to transform vectors"), while the matrix representation A_{jk} is merely a set of coefficients needed to describe that transformation in a particular basis. The matrix A_{jk} depends on the choice of the basis, but there any many properties of the linear transformation $\hat{A}$ that *do not* depend on the basis; these properties can be thought of as the "geometric" properties of the transformation.[2] Below we will be concerned only with geometric properties of objects.

Definition: Two linear maps $\hat{A}$, $\hat{B}$ are **equal** if $\hat{A}\mathbf{v} = \hat{B}\mathbf{v}$ for all $\mathbf{v} \in V$. The **composition** of linear maps $\hat{A}$, $\hat{B}$ is the map $\hat{A}\hat{B}$ which acts on vectors $\mathbf{v}$ as $(\hat{A}\hat{B})\mathbf{v} \equiv \hat{A}(\hat{B}\mathbf{v})$.

Statement 2: The composition of two linear transformations is again a linear transformation.

Proof: I give two proofs to contrast the coordinate-free language with the language of matrices, and also to show the derivation of the matrix multiplication law.

(*Coordinate-free proof*:) We need to demonstrate the property (1.14). If $\hat{A}$ and $\hat{B}$ are linear transformations then we have, by definition,

$$\hat{A}\hat{B}\,(\mathbf{u} + \lambda\mathbf{v}) = \hat{A}(\hat{B}\mathbf{u} + \lambda\hat{B}\mathbf{v}) = \hat{A}\hat{B}\mathbf{u} + \lambda\hat{A}\hat{B}\mathbf{v}.$$

Therefore the composition $\hat{A}\hat{B}$ is a linear map.

(*Proof using matrices*:) We need to show that for any vector $\mathbf{v}$ with components v_i and for any two transformation matrices A_{ij} and B_{ij}, the result of first transforming with B_{ij} and then with A_{ij} is equivalent to transforming $\mathbf{v}$ with some other matrix. We calculate the components v'_i of the transformed vector,

$$v'_i = \sum_{j=1}^{n} A_{ij} \sum_{k=1}^{n} B_{jk} v_k = \sum_{k=1}^{n} \left(\sum_{j=1}^{n} A_{ij} B_{jk} \right) v_k \equiv \sum_{k=1}^{n} C_{ik} v_k,$$

where C_{ik} is the matrix of the new transformation. ■

Note that we need to work more in the second proof because matrices are *defined* through their components, as "tables of numbers." So we cannot prove linearity without also finding an *explicit formula* for the matrix product in terms of matrix components. The first proof does not use such a formula.

1.2.2 Examples of linear maps

The easiest example of a linear map is the **identity operator** $\hat{1}_V$. This is a map $V \to V$ defined by $\hat{1}_V \mathbf{v} = \mathbf{v}$. It is clear that this map is linear, and that its matrix elements in any basis are given by the **Kronecker delta** symbol

$$\delta_{ij} \equiv \begin{cases} 1, & i = j; \\ 0, & i \neq j. \end{cases}$$

We can also define a map which multiplies all vectors $\mathbf{v} \in V$ by a fixed number λ. This is also obviously a linear map, and we denote it by $\lambda\hat{1}_V$. If

[2]Example: the properties $A_{11} = 0$, $A_{11} > A_{12}$, and $A_{ij} = -2A_{ji}$ are not geometric properties of the linear transformation $\hat{A}$ because they may hold in one basis but not in another basis. However, the number $\sum_{i=1}^{n} A_{ii}$ turns out to be geometric (independent of the basis), as we will see below.

$\lambda = 0$, we may write $\hat{0}_V$ to denote the map that transforms all vectors into the zero vector.

Another example of a linear transformation is the following. Suppose that the set $\{\mathbf{e}_1, ..., \mathbf{e}_n\}$ is a basis in the space V; then any vector $\mathbf{v} \in V$ is uniquely expressed as a linear combination $\mathbf{v} = \sum_{j=1}^{n} v_j \mathbf{e}_j$. We denote by $\mathbf{e}_1^* (\mathbf{v})$ the function that gives the component v_1 of a vector $\mathbf{v}$ in the basis $\{\mathbf{e}_j\}$. Then we define the map $\hat{M}$ by the formula

$$\hat{M}\mathbf{v} \equiv v_1 \mathbf{e}_2 = \mathbf{e}_1^* (\mathbf{v})\, \mathbf{e}_2.$$

In other words, the new vector $\hat{M}\mathbf{v}$ is always parallel to $\mathbf{e}_2$ but has the coefficient v_1. It is easy to prove that this map is linear (you need to check that the first component of a sum of vectors is equal to the sum of their first components). The matrix corresponding to $\hat{M}$ in the basis $\{\mathbf{e}_j\}$ is

$$M_{ij} = \begin{pmatrix} 0 & 0 & 0 & ... \\ 1 & 0 & 0 & ... \\ 0 & 0 & 0 & ... \\ ... & ... & ... & ... \end{pmatrix}.$$

The map that shifts all vectors by a fixed vector, $\hat{S}_{\mathbf{a}}\mathbf{v} \equiv \mathbf{v} + \mathbf{a}$, is not linear because

$$\hat{S}_{\mathbf{a}} (\mathbf{u} + \mathbf{v}) = \mathbf{u} + \mathbf{v} + \mathbf{a} \neq \hat{S}_{\mathbf{a}} (\mathbf{u}) + \hat{S}_{\mathbf{a}} (\mathbf{v}) = \mathbf{u} + \mathbf{v} + 2\mathbf{a}.$$

Question: I understand how to work with a linear transformation specified by its matrix A_{jk}. But how can I work with an abstract "linear map" $\hat{A}$ if the only thing I know about $\hat{A}$ is that it is linear? It seems that I cannot specify linear transformations or perform calculations with them unless I use matrices.

Answer: It is true that the abstract definition of a linear map does not include a specification of a particular transformation, unlike the concrete definition in terms of a matrix. However, it does not mean that matrices are always needed. For a particular problem in linear algebra, a particular transformation is always specified either as a certain matrix in a given basis, or in a *geometric*, i.e. basis-free manner, e.g. "the transformation $\hat{B}$ multiplies a vector by $3/2$ and then projects onto the plane orthogonal to the fixed vector $\mathbf{a}$." In this book I concentrate on general properties of linear transformations, which are best formulated and studied in the geometric (coordinate-free) language rather than in the matrix language. Below we will see many coordinate-free calculations with linear maps. In Sec. 1.8 we will also see how to specify arbitrary linear transformations in a coordinate-free manner, although it will then be quite similar to the matrix notation.

Exercise 1: If V is a one-dimensional vector space over a field $\mathbb{K}$, prove that any linear operator $\hat{A}$ on V must act simply as a multiplication by a number.

Solution: Let $\mathbf{e} \neq 0$ be a basis vector; note that any nonzero vector $\mathbf{e}$ is a basis in V, and that every vector $\mathbf{v} \in V$ is proportional to $\mathbf{e}$. Consider the action of $\hat{A}$ on the vector $\mathbf{e}$: the vector $\hat{A}\mathbf{e}$ must also be proportional to $\mathbf{e}$, say $\hat{A}\mathbf{e} = a\mathbf{e}$ where $a \in \mathbb{K}$ is some constant. Then by linearity of $\hat{A}$, for any vector $\mathbf{v} = v\mathbf{e}$ we get $\hat{A}\mathbf{v} = \hat{A}v\mathbf{e} = av\mathbf{e} = a\mathbf{v}$, so the operator $\hat{A}$ multiplies all vectors by the same number a. ■

Exercise 2: If $\{\mathbf{e}_1, ..., \mathbf{e}_N\}$ is a basis in V and $\{\mathbf{v}_1, ..., \mathbf{v}_N\}$ is a set of N arbitrary vectors, does there exist a linear map $\hat{A}$ such that $\hat{A}\mathbf{e}_j = \mathbf{v}_j$ for $j = 1, ..., N$? If so, is this map unique?

Solution: For any $\mathbf{x} \in V$ there exists a unique set of N numbers x_1, ..., x_N such that $\mathbf{x} = \sum_{i=1}^{N} x_i \mathbf{e}_i$. Since $\hat{A}$ must be linear, the action of $\hat{A}$ on $\mathbf{x}$ *must* be given by the formula $\hat{A}\mathbf{x} = \sum_{i=1}^{N} x_i \mathbf{v}_i$. This formula defines $\hat{A}\mathbf{x}$ for all $\mathbf{x}$. Hence, the map $\hat{A}$ exists and is unique. ■

1.2.3 Vector space of all linear maps

Suppose that V and W are two vector spaces and consider *all* linear maps $\hat{A} : V \to W$. The set of all such maps is itself a vector space because we can add two linear maps and multiply linear maps by scalars, getting again a linear map. More formally, if $\hat{A}$ and $\hat{B}$ are linear maps from V to W and $\lambda \in \mathbb{K}$ is a number (a scalar) then we define $\lambda\hat{A}$ and $\hat{A} + \hat{B}$ in the natural way:

$$(\lambda\hat{A})\mathbf{v} \equiv \lambda(\hat{A}\mathbf{v}),$$
$$(\hat{A} + \hat{B})\mathbf{v} \equiv \hat{A}\mathbf{v} + \hat{B}\mathbf{v}, \quad \forall \mathbf{v} \in V.$$

In words: the map $\lambda\hat{A}$ acts on a vector $\mathbf{v}$ by first acting on it with $\hat{A}$ and then multiplying the result by the scalar λ; the map $\hat{A} + \hat{B}$ acts on a vector $\mathbf{v}$ by adding the vectors $\hat{A}\mathbf{v}$ and $\hat{B}\mathbf{v}$. It is straightforward to check that the maps $\lambda\hat{A}$ and $\hat{A} + \hat{B}$ defined in this way are *linear* maps $V \to W$. Therefore, the set of all linear maps $V \to W$ is a vector space. This vector space is denoted $\mathrm{Hom}\,(V, W)$, meaning the "space of **homomorphisms**" from V to W.

The space of linear maps from V to itself is called the space of **endomorphisms** of V and is denoted $\mathrm{End}\,V$. Endomorphisms of V are also called **linear operators** in the space V. (We have been talking about linear operators all along, but we did not call them endomorphisms until now.)

1.2.4 Eigenvectors and eigenvalues

Definition 1: Suppose $\hat{A} : V \to V$ is a linear operator, and a vector $\mathbf{v} \neq 0$ is such that $\hat{A}\mathbf{v} = \lambda\mathbf{v}$ where $\lambda \in \mathbb{K}$ is some number. Then $\mathbf{v}$ is called the **eigenvector of $\hat{A}$ with the eigenvalue λ**.

The geometric interpretation is that $\mathbf{v}$ is a special direction for the transformation $\hat{A}$ such that $\hat{A}$ acts simply as a scaling by a certain number λ in that direction.

Remark: Without the condition $\mathbf{v} \neq 0$ in the definition, it would follow that the zero vector is an eigenvector for any operator with any eigenvalue, which would not be very useful, so we exclude the trivial case $\mathbf{v} = 0$.

Example 1: Suppose $\hat{A}$ is the transformation that rotates vectors around some fixed axis by a fixed angle. Then any vector $\mathbf{v}$ parallel to the axis is unchanged by the rotation, so it is an eigenvector of $\hat{A}$ with eigenvalue 1.

Example 2: Suppose $\hat{A}$ is the operator of multiplication by a number α, i.e. we define $\hat{A}\mathbf{x} \equiv \alpha\mathbf{x}$ for all $\mathbf{x}$. Then *all* nonzero vectors $\mathbf{x} \neq 0$ are eigenvectors of $\hat{A}$ with eigenvalue α.

Exercise 1: Suppose $\mathbf{v}$ is an eigenvector of $\hat{A}$ with eigenvalue λ. Show that $c\mathbf{v}$ for any $c \in \mathbb{K}$, $c \neq 0$, is also an eigenvector with the same eigenvalue.

Solution: $\hat{A}(c\mathbf{v}) = c\hat{A}\mathbf{v} = c\lambda\mathbf{v} = \lambda(c\mathbf{v})$.

Example 3: Suppose that an operator $\hat{A} \in \operatorname{End} V$ is such that it has $N = \dim V$ eigenvectors $\mathbf{v}_1$, ..., $\mathbf{v}_N$ that constitute a basis in V. Suppose that λ_1, ..., λ_N are the corresponding eigenvalues (not necessarily different). Then the matrix representation of $\hat{A}$ in the basis $\{\mathbf{v}_j\}$ is a **diagonal** matrix

$$A_{ij} = \operatorname{diag}(\lambda_1, ..., \lambda_N) \equiv \begin{pmatrix} \lambda_1 & 0 & \dots & 0 \\ 0 & \lambda_2 & \dots & 0 \\ \vdots & \vdots & \ddots & \vdots \\ 0 & 0 & \dots & \lambda_N \end{pmatrix}.$$

Thus a basis consisting of eigenvectors (the **eigenbasis**), if it exists, is a particularly convenient choice of basis for a given operator.

Remark: The task of determining the eigenbasis (also called the **diagonalization of an operator**) is a standard, well-studied problem for which efficient numerical methods exist. (This book is not about these methods.) However, it is important to know that not all operators can be diagonalized. The simplest example of a non-diagonalizable operator is one with the matrix representation $\begin{pmatrix} 0 & 1 \\ 0 & 0 \end{pmatrix}$ in $\mathbb{R}^2$. This operator has *only one* eigenvector, $\binom{1}{0}$, so we have no hope of finding an eigenbasis. The theory of the "Jordan canonical form" (see Sec. 4.6) explains how to choose the basis for a non-diagonalizable operator so that its matrix in that basis becomes as simple as possible.

Definition 2: A map $\hat{A} : V \to W$ is **invertible** if there exists a map $\hat{A}^{-1} : W \to V$ such that $\hat{A}\hat{A}^{-1} = \hat{1}_W$ and $\hat{A}^{-1}\hat{A} = \hat{1}_V$. The map $\hat{A}^{-1}$ is called the **inverse** of $\hat{A}$.

Exercise 2: Suppose that an operator $\hat{A} \in \operatorname{End} V$ has an eigenvector with eigenvalue 0. Show that $\hat{A}$ describes a non-invertible transformation.

Outline of the solution: Show that the inverse of a linear operator (if the inverse exists) is again a linear operator. A linear operator must transform the zero vector into the zero vector. We have $\hat{A}\mathbf{v} = 0$ and yet we must have $\hat{A}^{-1}0 = 0$ if $\hat{A}^{-1}$ exists. ■

Exercise 3: Suppose that an operator $\hat{A} \in \operatorname{End} V$ in an n-dimensional vector space V describes a non-invertible transformation. Show that the operator $\hat{A}$ has *at least one* eigenvector $\mathbf{v}$ with eigenvalue 0.

Outline of the solution: Let $\{\mathbf{e}_1, ..., \mathbf{e}_n\}$ be a basis; consider the set of vectors $\{\hat{A}\mathbf{e}_1, ..., \hat{A}\mathbf{e}_n\}$ and show that it is not a basis, hence linearly *dependent* (otherwise $\hat{A}$ would be invertible). Then there exists a linear combination $\sum_j c_j(\hat{A}\mathbf{e}_j) = 0$ where not all c_j are zero; $\mathbf{v} \equiv \sum_j c_j \mathbf{e}_j$ is then nonzero, and is the desired eigenvector. ■

1.3 Subspaces

Definition: A **subspace** of a vector space V is a subset $S \subset V$ such that S is itself a vector space.

A subspace is not just any subset of V. For example, if $\mathbf{v} \in V$ is a nonzero vector then the subset S consisting of the single vector, $S = \{\mathbf{v}\}$, is not a subspace: for instance, $\mathbf{v} + \mathbf{v} = 2\mathbf{v}$, but $2\mathbf{v} \notin S$.

Example 1. The set $\{\lambda\mathbf{v} \,|\, \forall\lambda \in \mathbb{K}\}$ is called the subspace **spanned by** the vector $\mathbf{v}$. This set is a subspace because we can add vectors from this set to each other and obtain again vectors from the same set. More generally, if $\mathbf{v}_1, ..., \mathbf{v}_n \in V$ are some vectors, we define the **subspace spanned by** $\{\mathbf{v}_j\}$ as the set of all linear combinations

$$\text{Span}\,\{\mathbf{v}_1, ..., \mathbf{v}_n\} \equiv \{\lambda_1\mathbf{v}_1 + ... + \lambda_n\mathbf{v}_n \,|\, \forall\lambda_i \in \mathbb{K}\}.$$

It is obvious that $\text{Span}\,\{\mathbf{v}_1, ..., \mathbf{v}_n\}$ is a subspace of V.

If $\{\mathbf{e}_j\}$ is a basis in the space V then the subspace spanned by the vectors $\{\mathbf{e}_j\}$ is equal to V itself.

Exercise 1: Show that the intersection of two subspaces is also a subspace.

Example 2: Kernel of an operator. Suppose $\hat{A} \in \text{End}\, V$ is a linear operator. The set of all vectors $\mathbf{v}$ such that $\hat{A}\mathbf{v} = 0$ is called the **kernel** of the operator $\hat{A}$ and is denoted by $\ker \hat{A}$. In formal notation,

$$\ker \hat{A} \equiv \{\mathbf{u} \in V \,|\, \hat{A}\mathbf{u} = 0\}.$$

This set is a subspace of V because if $\mathbf{u}, \mathbf{v} \in \ker \hat{A}$ then

$$\hat{A}\,(\mathbf{u} + \lambda\mathbf{v}) = \hat{A}\mathbf{u} + \lambda\hat{A}\mathbf{v} = 0,$$

and so $\mathbf{u} + \lambda\mathbf{v} \in \ker \hat{A}$.

Example 3: Image of an operator. Suppose $\hat{A} : V \to V$ is a linear operator. The **image** of the operator $\hat{A}$, denoted $\text{im}\, A$, is by definition the set of all vectors $\mathbf{v}$ obtained by acting with $\hat{A}$ on some other vectors $\mathbf{u} \in V$. In formal notation,

$$\text{im}\, \hat{A} \equiv \{\hat{A}\mathbf{u} \,|\, \forall\mathbf{u} \in V\}.$$

This set is also a subspace of V (prove this!).

Exercise 2: In a vector space V, let us choose a vector $\mathbf{v} \neq 0$. Consider the set S_0 of all linear operators $\hat{A} \in \text{End}\, V$ such that $\hat{A}\mathbf{v} = 0$. Is S_0 a subspace? Same question for the set S_3 of operators $\hat{A}$ such that $\hat{A}\mathbf{v} = 3\mathbf{v}$. Same question for the set S' of all operators $\hat{A}$ for which there exists some $\lambda \in \mathbb{K}$ such that $\hat{A}\mathbf{v} = \lambda\mathbf{v}$, where λ may be different for each $\hat{A}$.

1.3.1 Projectors and subspaces

Definition: A linear operator $\hat{P} : V \to V$ is called a **projector** if $\hat{P}\hat{P} = \hat{P}$.

Projectors are useful for defining subspaces: The result of a projection remains invariant under further projections, $\hat{P}(\hat{P}\mathbf{v}) = \hat{P}\mathbf{v}$, so a projector $\hat{P}$ defines a subspace $\operatorname{im}\hat{P}$, which consists of all vectors invariant under $\hat{P}$.

As an example, consider the transformation of $\mathbb{R}^3$ given by the matrix

$$\hat{P} = \begin{pmatrix} 1 & 0 & a \\ 0 & 1 & b \\ 0 & 0 & 0 \end{pmatrix},$$

where a, b are arbitrary numbers. It is easy to check that $\hat{P}\hat{P} = \hat{P}$ for any a, b. This transformation is a projector onto the subspace spanned by the vectors $(1, 0, 0)$ and $(0, 1, 0)$. (Note that a and b can be chosen at will; there are many projectors onto the same subspace.)

Statement: Eigenvalues of a projector can be only the numbers 0 and 1.

Proof: If $\mathbf{v} \in V$ is an eigenvector of a projector $\hat{P}$ with the eigenvalue λ then

$$\lambda\mathbf{v} = \hat{P}\mathbf{v} = \hat{P}\hat{P}\mathbf{v} = \hat{P}\lambda\mathbf{v} = \lambda^2\mathbf{v} \Rightarrow \lambda\,(\lambda - 1)\,\mathbf{v} = 0.$$

Since $\mathbf{v} \neq 0$, we must have either $\lambda = 0$ or $\lambda = 1$. ■

1.3.2 Eigenspaces

Another way to specify a subspace is through eigenvectors of some operator.

Exercise 1: For a linear operator $\hat{A}$ and a fixed number $\lambda \in \mathbb{K}$, the set of all vectors $\mathbf{v} \in V$ such that $\hat{A}\mathbf{v} = \lambda\mathbf{v}$ is a *subspace* of V.

The subspace of all such vectors is called the **eigenspace** of $\hat{A}$ with the eigenvalue λ. Any nonzero vector from that subspace is an eigenvector of $\hat{A}$ with eigenvalue λ.

Example: If $\hat{P}$ is a projector then $\operatorname{im}\hat{P}$ is the eigenspace of $\hat{P}$ with eigenvalue 1.

Exercise 2: Show that eigenspaces V_λ and V_μ corresponding to different eigenvalues, $\lambda \neq \mu$, have only one common vector — the zero vector. ($V_\lambda \cap V_\mu = \{0\}$.)

By definition, a subspace $U \subset V$ is **invariant** under the action of some operator $\hat{A}$ if $\hat{A}\mathbf{u} \in U$ for all $\mathbf{u} \in U$.

Exercise 3: Show that the eigenspace of $\hat{A}$ with eigenvalue λ is invariant under $\hat{A}$.

Exercise 4: In a space of polynomials in the variable x of any (finite) degree, consider the subspace U of polynomials of degree not more than 2 and the operator $\hat{A} \equiv x\frac{d}{dx}$, that is,

$$\hat{A} : p(x) \mapsto x\frac{dp(x)}{dx}.$$

Show that U is invariant under $\hat{A}$.

1.4 Isomorphisms of vector spaces

Two vector spaces are **isomorphic** if there exists a one-to-one linear map between them. This linear map is called the **isomorphism**.

Exercise 1: If $\{\mathbf{v}_1, ..., \mathbf{v}_N\}$ is a linearly independent set of vectors ($\mathbf{v}_j \in V$) and $\hat{M} : V \to W$ is an isomorphism then the set $\{\hat{M}\mathbf{v}_1, ..., \hat{M}\mathbf{v}_N\}$ is also linearly independent. In particular, $\hat{M}$ maps a basis in V into a basis in W.

Hint: First show that $\hat{M}\mathbf{v} = 0$ if and only if $\mathbf{v} = 0$. Then consider the result of $\hat{M}(\lambda_1 \mathbf{v}_1 + ... + \lambda_N \mathbf{v}_N)$.

Statement 1: Any vector space V of dimension n is isomorphic to the space $\mathbb{K}^n$ of n-tuples.

Proof: To demonstrate this, it is sufficient to present *some* isomorphism. We can always choose a basis $\{\mathbf{e}_i\}$ in V, so that any vector $\mathbf{v} \in V$ is decomposed as $\mathbf{v} = \sum_{i=1}^{n} \lambda_i \mathbf{e}_i$. Then we define the isomorphism map $\hat{M}$ between V and the space $\mathbb{K}^n$ as

$$\hat{M}\mathbf{v} \equiv (\lambda_1, ..., \lambda_n).$$

It is easy to see that $\hat{M}$ is linear and one-to-one. ■

Vector spaces $\mathbb{K}^m$ and $\mathbb{K}^n$ are isomorphic only if they have equal dimension, $m = n$. The reason they are not isomorphic for $m \neq n$ is that they have different numbers of vectors in a basis, while one-to-one linear maps must preserve linear independence and map a basis to a basis. (For $m \neq n$, there are plenty of linear maps from $\mathbb{K}^m$ to $\mathbb{K}^n$ but none of them is a one-to-one map. It also follows that a one-to-one map between $\mathbb{K}^m$ and $\mathbb{K}^n$ cannot be linear.)

Note that the isomorphism $\hat{M}$ constructed in the proof of Statement 1 will depend on the choice of the basis: a different basis $\{\mathbf{e}'_i\}$ yields a different map $\hat{M}'$. For this reason, the isomorphism $\hat{M}$ is *not canonical*.

Definition: A linear map between two vector spaces V and W is **canonically defined** or **canonical** if it is defined independently of a choice of bases in V and W. (We are of course allowed to choose a basis *while* constructing a canonical map, but at the end we need to prove that the resulting map does not depend on that choice.) Vector spaces V and W are **canonically isomorphic** if there exists a canonically defined isomorphism between them; I write $V \cong W$ in this case.

Examples of canonical isomorphisms:

1. Any vector space V is canonically isomorphic to itself, $V \cong V$; the isomorphism is the identity map $\mathbf{v} \to \mathbf{v}$ which is defined regardless of any basis. (This is trivial but still, a valid example.)

2. If V is a one-dimensional vector space then $\text{End}\, V \cong \mathbb{K}$. You have seen the map $\text{End}\, V \to \mathbb{K}$ in the Exercise 1.2.2, where you had to show that any linear operator in V is a multiplication by a number; this number is the element of $\mathbb{K}$ corresponding to the given operator. Note that $V \ncong \mathbb{K}$ unless there is a "preferred" vector $\mathbf{e} \in V$, $\mathbf{e} \neq 0$ which would be mapped into the number $1 \in \mathbb{K}$. Usually vector spaces do not have any

special vectors, so there is no canonical isomorphism. (However, End V does have a special element — the identity $\hat{1}_V$.)

At this point I cannot give more interesting examples of canonical maps, but I will show many of them later. My intuitive picture is that canonically isomorphic spaces have a fundamental structural similarity. An isomorphism that depends on the choice of basis, as in the Statement 1 above, is unsatisfactory if we are interested in properties that can be formulated geometrically (independently of any basis).

1.5 Direct sum of vector spaces

If V and W are two given vector spaces over a field $\mathbb{K}$, we define a new vector space $V \oplus W$ as the space of pairs $(\mathbf{v}, \mathbf{w})$, where $\mathbf{v} \in V$ and $\mathbf{w} \in W$. The operations of vector sum and scalar multiplication are defined in the natural way,

$$(\mathbf{v}_1, \mathbf{w}_1) + (\mathbf{v}_2, \mathbf{w}_2) = (\mathbf{v}_1 + \mathbf{v}_2, \mathbf{w}_1 + \mathbf{w}_2),$$
$$\lambda (\mathbf{v}_1, \mathbf{w}_1) = (\lambda \mathbf{v}_1, \lambda \mathbf{w}_1).$$

The new vector space is called the **direct sum** of the spaces V and W.
Statement: The dimension of the direct sum is $\dim (V \oplus W) = \dim V + \dim W$.

Proof: If $\mathbf{v}_1$, ..., $\mathbf{v}_m$ and $\mathbf{w}_1$, ..., $\mathbf{w}_n$ are bases in V and W respectively, consider the set of $m + n$ vectors

$$(\mathbf{v}_1, 0), ..., (\mathbf{v}_m, 0), (0, \mathbf{w}_1), ..., (0, \mathbf{w}_n).$$

It is easy to prove that this set is linearly independent. Then it is clear that any vector $(\mathbf{v}, \mathbf{w}) \in V \oplus W$ can be represented as a linear combination of the vectors from the above set, therefore that set is a basis and the dimension of $V \oplus W$ is $m + n$. (This proof is sketchy but the material is standard and straightforward.) ■

Exercise 1: Complete the proof.

Hint: If $(\mathbf{v}, \mathbf{w}) = 0$ then $\mathbf{v} = 0$ and $\mathbf{w} = 0$ separately.

1.5.1 V and W as subspaces of $V \oplus W$; canonical projections

If V and W are two vector spaces then the space $V \oplus W$ has a certain subspace which is canonically isomorphic to V. This subspace is the set of all vectors from $V \oplus W$ of the form $(\mathbf{v}, 0)$, where $\mathbf{v} \in V$. It is obvious that this set forms a subspace (it is closed under linear operations) and is isomorphic to V. To demonstrate this, we present a canonical isomorphism which we denote $\hat{P}_V : V \oplus W \to V$. The isomorphism $\hat{P}_V$ is the **canonical projection** defined by

$$\hat{P}_V (\mathbf{v}, \mathbf{w}) \equiv \mathbf{v}.$$

It is easy to check that this is a linear and one-to-one map of the subspace $\{(\mathbf{v},0) \mid \mathbf{v} \in V\}$ to V, and that $\hat{P}$ is a projector. This projector is *canonical* because we have defined it without reference to any basis. The relation is so simple that it is convenient to write $\mathbf{v} \in V \oplus W$ instead of $(\mathbf{v},0) \in V \oplus W$.

Similarly, we define the subspace isomorphic to W and the corresponding canonical projection.

It is usually convenient to denote vectors from $V \oplus W$ by formal linear combinations, e.g. $\mathbf{v} + \mathbf{w}$, instead of the pair notation $(\mathbf{v}, \mathbf{w})$. A pair $(\mathbf{v}, 0)$ is denoted simply by $\mathbf{v} \in V \oplus W$.

Exercise 1: Show that the space $\mathbb{R}^n \oplus \mathbb{R}^m$ is isomorphic to $\mathbb{R}^{n+m}$, but not canonically.

Hint: The image of $\mathbb{R}^n \subset \mathbb{R}^n \oplus \mathbb{R}^m$ under the isomorphism is a subspace of $\mathbb{R}^{n+m}$, but there are no canonically defined subspaces in that space.

1.6 Dual (conjugate) vector space

Given a vector space V, we define another vector space V^* called the **dual** or the **conjugate** to V. The elements of V^* are **linear functions** on V, that is to say, maps $\mathbf{f}^* : V \to \mathbb{K}$ having the property

$$\mathbf{f}^*(\mathbf{u} + \lambda \mathbf{v}) = \mathbf{f}^*(\mathbf{u}) + \lambda \mathbf{f}^*(\mathbf{v}), \quad \forall \mathbf{u}, \mathbf{v} \in V, \forall \lambda \in \mathbb{K}.$$

The elements of V^* are called **dual vectors, covectors** or **linear forms**; I will say "covectors" to save space.

Definition: A **covector** is a linear map $V \to \mathbb{K}$. The set of all covectors is the **dual space** to the vector space V. The **zero covector** is the linear function that maps all vectors into zero. Covectors $\mathbf{f}^*$ and $\mathbf{g}^*$ are **equal** if

$$\mathbf{f}^*(\mathbf{v}) = \mathbf{g}^*(\mathbf{v}), \quad \forall \mathbf{v} \in V.$$

It is clear that the set of *all* linear functions is a vector space because e.g. the sum of linear functions is again a linear function. This "space of all linear functions" is the space we denote by V^*. In our earlier notation, this space is the same as $\text{Hom}(V, \mathbb{K})$.

Example 1: For the space $\mathbb{R}^2$ with vectors $\mathbf{v} \equiv (x, y)$, we may define the functions $\mathbf{f}^*(\mathbf{v}) \equiv 2x$, $\mathbf{g}^*(\mathbf{v}) \equiv y - x$. It is straightforward to check that these functions are linear.

Example 2: Let V be the space of polynomials of degree not more than 2 in the variable x with real coefficients. This space V is three-dimensional and contains elements such as $\mathbf{p} \equiv p(x) = a + bx + cx^2$. A linear function $\mathbf{f}^*$ on V could be defined in a way that might appear nontrivial, such as

$$\mathbf{f}^*(\mathbf{p}) = \int_0^\infty e^{-x} p(x) dx.$$

Nevertheless, it is clear that this is a *linear* function mapping V into $\mathbb{R}$. Similarly,

$$\mathbf{g}^*(\mathbf{p}) = \left.\frac{d}{dx}\right|_{x=1} p(x)$$

is a linear function. Hence, $\mathbf{f}^*$ and $\mathbf{g}^*$ belong to V^*.

Remark: One says that a covector $\mathbf{f}^*$ **is applied to** a vector $\mathbf{v}$ and yields a number $\mathbf{f}^*(\mathbf{v})$, or alternatively that a covector **acts on** a vector. This is similar to writing $\cos(0) = 1$ and saying that the cosine function is applied to the number 0, or "acts on the number 0," and then yields the number 1. Other notations for a covector acting on a vector are $\langle \mathbf{f}^*, \mathbf{v} \rangle$ and $\mathbf{f}^* \cdot \mathbf{v}$, and also $\iota_{\mathbf{v}}\mathbf{f}^*$ or $\iota_{\mathbf{f}^*}\mathbf{v}$ (here the symbol ι stands for "insert"). However, in this text I will always use the notation $\mathbf{f}^*(\mathbf{v})$ for clarity. The notation $\langle \mathbf{x}, \mathbf{y} \rangle$ will be used for scalar products.

Question: It is unclear how to visualize the dual space when it is defined in such abstract terms, as the set of *all* functions having some property. How do I know which functions are there, and how can I describe this space in more concrete terms?

Answer: Indeed, we need some work to characterize V^* more explicitly. We will do this in the next subsection by constructing a basis in V^*.

1.6.1 Dual basis

Suppose $\{\mathbf{e}_1, ..., \mathbf{e}_n\}$ is a basis in V; then any vector $\mathbf{v} \in V$ is uniquely expressed as a linear combination

$$\mathbf{v} = \sum_{j=1}^{n} v_j \mathbf{e}_j.$$

The coefficient v_1, understood *as a function of the vector* $\mathbf{v}$, is a linear function of $\mathbf{v}$ because

$$\mathbf{u} + \lambda\mathbf{v} = \sum_{j=1}^{n} u_j \mathbf{e}_j + \lambda \sum_{j=1}^{n} v_j \mathbf{e}_j = \sum_{j=1}^{n} (u_i + \lambda v_j)\, \mathbf{e}_j,$$

therefore the first coefficient of the vector $\mathbf{u}+\lambda\mathbf{v}$ is $u_1+\lambda v_1$. So the coefficients v_k, $1 \le k \le n$, are linear functions of the vector $\mathbf{v}$; therefore they are *covectors*, i.e. elements of V^*. Let us denote these covectors by $\mathbf{e}_1^*, ..., \mathbf{e}_n^*$. Please note that $\mathbf{e}_1^*$ depends on the *entire* basis $\{\mathbf{e}_j\}$ and not only on $\mathbf{e}_1$, as it might appear from the notation $\mathbf{e}_1^*$. In other words, $\mathbf{e}_1^*$ is not a result of some "star" operation applied only to $\mathbf{e}_1$. The covector $\mathbf{e}_1^*$ will change if we change $\mathbf{e}_2$ or any other basis vector. This is so because the component v_1 of a fixed vector $\mathbf{v}$ depends not only on $\mathbf{e}_1$ but also on every other basis vector $\mathbf{e}_j$.

Theorem: The set of n covectors $\mathbf{e}_1^*, ..., \mathbf{e}_n^*$ is a basis in V^*. Thus, the dimension of the dual space V^* is equal to that of V.

Proof: First, we show by an explicit calculation that any covector $\mathbf{f}^*$ is a linear combination of $\{\mathbf{e}_j^*\}$. Namely, for any $\mathbf{f}^* \in V^*$ and $\mathbf{v} \in V$ we have

$$\mathbf{f}^*(\mathbf{v}) = \mathbf{f}^*\Big(\sum_{j=1}^{n} v_j \mathbf{e}_j\Big) = \sum_{j=1}^{n} v_j \mathbf{f}^*(\mathbf{e}_j) = \sum_{j=1}^{n} \mathbf{e}_j^*(\mathbf{v})\, \mathbf{f}^*(\mathbf{e}_j).$$

Note that in the last line the quantities $\mathbf{f}^*(\mathbf{e}_j)$ are some numbers that do not depend on $\mathbf{v}$. Let us denote $\phi_j \equiv \mathbf{f}^*(\mathbf{e}_j)$ for brevity; then we obtain the following linear decomposition of $\mathbf{f}^*$ through the covectors $\{\mathbf{e}_j^*\}$,

$$\mathbf{f}^*(\mathbf{v}) = \sum_{j=1}^{n} \phi_j \mathbf{e}_j^*(\mathbf{v}) \Rightarrow \mathbf{f}^* = \sum_{j=1}^{n} \phi_j \mathbf{e}_j^*.$$

So indeed all covectors $\mathbf{f}^*$ are linear combinations of $\mathbf{e}_j^*$.

It remains to prove that the set $\{\mathbf{e}_j^*\}$ is linearly independent. If this were not so, we would have $\sum_i \lambda_i \mathbf{e}_i^* = 0$ where not all λ_i are zero. Act on a vector $\mathbf{e}_k$ ($k = 1, ..., n$) with this linear combination and get

$$0 \stackrel{!}{=} (\sum_{i=1}^{n} \lambda_i \mathbf{e}_i^*)(\mathbf{e}_k) = \lambda_k, \quad k = 1, ..., n.$$

Hence all λ_k are zero. ■

Remark: The theorem holds only for finite-dimensional spaces! For infinite-dimensional spaces V, the dual space V^* may be "larger" or "smaller" than V. Infinite-dimensional spaces are subtle, and one should not think that they are simply "spaces with infinitely many basis vectors." More detail (*much* more detail!) can be found in standard textbooks on functional analysis. ■

The set of covectors $\{\mathbf{e}_j^*\}$ is called the **dual basis** to the basis $\{\mathbf{e}_j\}$. The covectors $\mathbf{e}_j^*$ of the dual basis have the useful property

$$\mathbf{e}_i^*(\mathbf{e}_j) = \delta_{ij}$$

(please check this!). Here δ_{ij} is the **Kronecker symbol**: $\delta_{ij} = 0$ if $i \neq j$ and $\delta_{ii} = 1$. For instance, $\mathbf{e}_1^*(\mathbf{e}_1) = 1$ and $\mathbf{e}_1^*(\mathbf{e}_k) = 0$ for $k \geq 2$.

Question: I would like to see a concrete calculation. How do I compute $\mathbf{f}^*(\mathbf{v})$ if a vector $\mathbf{v} \in V$ and a covector $\mathbf{f}^* \in V^*$ are "given"?

Answer: Vectors are usually "given" by listing their components in some basis. Suppose $\{\mathbf{e}_1, ..., \mathbf{e}_N\}$ is a basis in V and $\{\mathbf{e}_1^*, ..., \mathbf{e}_N^*\}$ is its dual basis. If the vector $\mathbf{v}$ has components v_k in a basis $\{\mathbf{e}_k\}$ and the covector $\mathbf{f}^* \in V^*$ has components f_k^* in the dual basis $\{\mathbf{e}_k^*\}$, then

$$\mathbf{f}^*(\mathbf{v}) = \sum_{k=1}^{N} f_k^* \mathbf{e}_k^* (\sum_{l=1}^{N} v_l \mathbf{e}_l) = \sum_{k=1}^{N} f_k^* v_k. \tag{1.15}$$

Question: The formula (1.15) looks like the scalar product (1.4). How come?

Answer: Yes, it does look like that, but Eq. (1.15) does not describe a scalar product because for one thing, $\mathbf{f}^*$ and $\mathbf{v}$ are from *different* vector spaces. I would rather say that the scalar product resembles Eq. (1.15), and this happens only for a special choice of basis (an *orthonormal* basis) in V. This will be explained in more detail in Sec. 5.1.

Question: The dual basis still seems too abstract to me. Suppose V is the three-dimensional space of polynomials in the variable x with real coefficients and degree no more than 2. The three polynomials $\{1, x, x^2\}$ are a basis in V. How can I compute explicitly the dual basis to this basis?

Answer: An arbitrary vector from this space is a polynomial $a + bx + cx^2$. The basis dual to $\{1, x, x^2\}$ consists of three covectors. Let us denote the set of these covectors by $\{\mathbf{e}_1^*, \mathbf{e}_2^*, \mathbf{e}_3^*\}$. These covectors are linear functions defined like this:

$$\begin{aligned} \mathbf{e}_1^* \left(a + bx + cx^2\right) &= a, \\ \mathbf{e}_2^* \left(a + bx + cx^2\right) &= b, \\ \mathbf{e}_3^* \left(a + bx + cx^2\right) &= c. \end{aligned}$$

If you like, you can visualize them as differential operators acting on the polynomials $p(x)$ like this:

$$\mathbf{e}_1^*(p) = p(x)|_{x=0}; \quad \mathbf{e}_2^*(p) = \left.\frac{dp}{dx}\right|_{x=0}; \quad \mathbf{e}_3^*(p) = \left.\frac{1}{2}\frac{d^2p}{dx^2}\right|_{x=0}.$$

However, this is a bit too complicated; the covector $\mathbf{e}_3^*$ just extracts the coefficient of the polynomial $p(x)$ at x^2. To make it clear that, say, $\mathbf{e}_2^*$ and $\mathbf{e}_3^*$ can be evaluated without taking derivatives or limits, we may write the formulas for $\mathbf{e}_j^*(p)$ in another equivalent way, e.g.

$$\mathbf{e}_2^*(p) = \frac{p(1) - p(-1)}{2}, \quad \mathbf{e}_3^*(p) = \frac{p(1) - 2p(0) + p(-1)}{2}.$$

It is straightforward to check that these formulas are indeed equivalent by substituting $p(x) = a + bx + cx^2$.

Exercise 1: Compute $\mathbf{f}^*$ and $\mathbf{g}^*$ from Example 2 in terms of the basis $\{\mathbf{e}_i^*\}$ defined above.

Question: I'm still not sure what to do in the general case. For example, the set $\{1, 1 + x, 1 + x + \frac{1}{2}x^2\}$ is also a basis in the space V of quadratic polynomials. How do I explicitly compute the dual basis now? The previous trick with derivatives does not work.

Answer: Let's denote this basis by $\{\mathbf{f}_1, \mathbf{f}_2, \mathbf{f}_3\}$; we are looking for the dual basis $\{\mathbf{f}_1^*, \mathbf{f}_2^*, \mathbf{f}_3^*\}$. It will certainly be sufficiently explicit if we manage to express the covectors $\mathbf{f}_j^*$ through the covectors $\{\mathbf{e}_1^*, \mathbf{e}_2^*, \mathbf{e}_3^*\}$ that we just found previously. Since the set of covectors $\{\mathbf{e}_1^*, \mathbf{e}_2^*, \mathbf{e}_3^*\}$ is a basis in V^*, we expect that $\mathbf{f}_1^*$ is a linear combination of $\{\mathbf{e}_1^*, \mathbf{e}_2^*, \mathbf{e}_3^*\}$ with some constant coefficients, and similarly $\mathbf{f}_2^*$ and $\mathbf{f}_3^*$. Let us, for instance, determine $\mathbf{f}_1^*$. We write

$$\mathbf{f}_1^* = A\mathbf{e}_1^* + B\mathbf{e}_2^* + C\mathbf{e}_3^*$$

with unknown coefficients A, B, C. By definition, $\mathbf{f}_1^*$ acting on an arbitrary vector $\mathbf{v} = c_1\mathbf{f}_1 + c_2\mathbf{f}_2 + c_3\mathbf{f}_3$ must yield c_1. Recall that $\mathbf{e}_i^*$, $i = 1, 2, 3$ yield the coefficients of the polynomial at 1, x, and x^2. Therefore

$$\begin{aligned} c_1 &\overset{!}{=} \mathbf{f}_1^*(\mathbf{v}) = \mathbf{f}_1^*\left(c_1\mathbf{f}_1 + c_2\mathbf{f}_2 + c_3\mathbf{f}_3\right) \\ &= \left(A\mathbf{e}_1^* + B\mathbf{e}_2^* + C\mathbf{e}_3^*\right)\left(c_1\mathbf{f}_1 + c_2\mathbf{f}_2 + c_3\mathbf{f}_3\right) \\ &= \left(A\mathbf{e}_1^* + B\mathbf{e}_2^* + C\mathbf{e}_3^*\right)\left(c_1 + c_2\left(1 + x\right) + c_3\left(1 + x + \tfrac{1}{2}x^2\right)\right) \\ &= Ac_1 + Ac_2 + Ac_3 + Bc_2 + Bc_3 + \tfrac{1}{2}Cc_3. \end{aligned}$$

Since this must hold for every c_1, c_2, c_3, we obtain a system of equations for the unknown constants A, B, C:

$$A = 1;$$
$$A + B = 0;$$
$$A + B + \tfrac{1}{2}C = 0.$$

The solution is $A = 1$, $B = -1$, $C = 0$. Therefore $\mathbf{f}_1^* = \mathbf{e}_1^* - \mathbf{e}_2^*$. In the same way we can determine $\mathbf{f}_2^*$ and $\mathbf{f}_3^*$. ■

Here are some useful properties of covectors.

Statement: (1) If $\mathbf{f}^* \neq 0$ is a given covector, there exists a basis $\{\mathbf{v}_1, ..., \mathbf{v}_N\}$ of V such that $\mathbf{f}^*(\mathbf{v}_1) = 1$ while $\mathbf{f}^*(\mathbf{v}_i) = 0$ for $2 \leq i \leq N$.

(2) Once such a basis is found, the set $\{\mathbf{a}, \mathbf{v}_2, ..., \mathbf{v}_N\}$ will still be a basis in V for any vector $\mathbf{a}$ such that $\mathbf{f}^*(\mathbf{a}) \neq 0$.

Proof: **(1)** By definition, the property $\mathbf{f}^* \neq 0$ means that there exists at least one vector $\mathbf{u} \in V$ such that $\mathbf{f}^*(\mathbf{u}) \neq 0$. Given the vector $\mathbf{u}$, we define the vector $\mathbf{v}_1$ by

$$\mathbf{v}_1 \equiv \frac{1}{\mathbf{f}^*(\mathbf{u})}\mathbf{u}.$$

It follows (using the linearity of $\mathbf{f}^*$) that $\mathbf{f}^*(\mathbf{v}_1) = 1$. Then by Exercise 1 in Sec. 1.1.5 the vector $\mathbf{v}_1$ can be completed to *some* basis $\{\mathbf{v}_1, \mathbf{w}_2, ..., \mathbf{w}_N\}$. Thereafter we define the vectors $\mathbf{v}_2, ..., \mathbf{v}_N$ by the formula

$$\mathbf{v}_i \equiv \mathbf{w}_i - \mathbf{f}^*(\mathbf{w}_i)\,\mathbf{v}_1, \quad 2 \leq i \leq N,$$

and obtain a set of vectors $\{\mathbf{v}_1, ..., \mathbf{v}_N\}$ such that $\mathbf{f}^*(\mathbf{v}_1) = 1$ and $\mathbf{f}^*(\mathbf{v}_i) = 0$ for $2 \leq i \leq N$. This set is linearly independent because a linear dependence among $\{\mathbf{v}_j\}$,

$$0 = \sum_{i=1}^{N} \lambda_i \mathbf{v}_i = \big(\lambda_1 - \sum_{i=2}^{N} \lambda_i \mathbf{f}^*(\mathbf{w}_i)\big)\mathbf{v}_1 + \sum_{i=2}^{N} \lambda_i \mathbf{w}_i,$$

together with the linear independence of the basis $\{\mathbf{v}_1, \mathbf{w}_2, ..., \mathbf{w}_N\}$, forces $\lambda_i = 0$ for all $i \geq 2$ and hence also $\lambda_1 = 0$. Therefore, the set $\{\mathbf{v}_1, ..., \mathbf{v}_N\}$ is the required basis.

(2) If the set $\{\mathbf{a}, \mathbf{v}_2, ..., \mathbf{v}_N\}$ were linearly dependent,

$$\lambda \mathbf{a} + \sum_{j=2}^{N} \lambda_j \mathbf{v}_j = 0,$$

with λ_j, λ not all zero, then we would have

$$\mathbf{f}^*\big(\lambda \mathbf{a} + \sum_{j=2}^{N} \lambda_j \mathbf{v}_j\big) = \lambda \mathbf{f}^*(\mathbf{a}) = 0,$$

which forces $\lambda = 0$ since by assumption $\mathbf{f}^*(\mathbf{a}) \neq 0$. However, $\lambda = 0$ entails

$$\sum_{j=2}^{N} \lambda_j \mathbf{v}_j = 0,$$

with λ_j not all zero, which contradicts the linear independence of the set $\{\mathbf{v}_2, ..., \mathbf{v}_N\}$. ■

Exercise 2: Suppose that $\{\mathbf{v}_1, ..., \mathbf{v}_k\}$, $\mathbf{v}_j \in V$ is a linearly independent set (not necessarily a basis). Prove that there exists at least one covector $\mathbf{f}^* \in V^*$ such that

$$\mathbf{f}^*(\mathbf{v}_1) = 1, \text{ while } \mathbf{f}^*(\mathbf{v}_2) = ... = \mathbf{f}^*(\mathbf{v}_k) = 0.$$

Outline of proof: The set $\{\mathbf{v}_1, ..., \mathbf{v}_k\}$ can be completed to a basis in V, see Exercise 1 in Sec. 1.1.5. Then $\mathbf{f}^*$ is the covector dual to $\mathbf{v}_1$ in that basis.

Exercise 3: Prove that the space dual to V^* is canonically isomorphic to V, i.e. $V^{**} \cong V$ (for finite-dimensional V).

Hint: Vectors $\mathbf{v} \in V$ can be thought of as linear functions on V^*, defined by $\mathbf{v}(\mathbf{f}^*) \equiv \mathbf{f}^*(\mathbf{v})$. This provides a map $V \to V^{**}$, so the space V is a subspace of V^{**}. Show that this map is injective. The dimensions of the spaces V, V^*, and V^{**} are the same; deduce that V as a subspace of V^{**} coincides with the whole space V^{**}.

1.6.2 Hyperplanes

Covectors are convenient for characterizing hyperplanes.

Let us begin with a familiar example: In three dimensions, the set of points with coordinate $x = 0$ is a *plane.* The set of points whose coordinates satisfy the linear equation $x + 2y - z = 0$ is another plane.

Instead of writing a linear equation with coordinates, one can write a covector applied to the vector of coordinates. For example, the equation $x+2y-z = 0$ can be rewritten as $\mathbf{f}^*(\mathbf{x}) = 0$, where $\mathbf{x} \equiv \{x, y, z\} \in \mathbb{R}^3$, while the covector $\mathbf{f}^* \in (\mathbb{R}^3)^*$ is expressed through the dual basis $\{\mathbf{e}_j^*\}$ as

$$\mathbf{f}^* \equiv \mathbf{e}_1^* + 2\mathbf{e}_2^* - \mathbf{e}_3^*.$$

The generalization of this to N dimensions is as follows.

Definition 1: The **hyperplane** (i.e. subspace of **codimension** 1) **annihilated by** a covector $\mathbf{f}^* \in V^*$ is the set of all vectors $\mathbf{x} \in V$ such that $\mathbf{f}^*(\mathbf{x}) = 0$. (Note that the zero vector, $\mathbf{x} = 0$, belongs to the hyperplane.)

Statement: The hyperplane annihilated by a nonzero covector $\mathbf{f}^*$ is a subspace of V of dimension $N - 1$ (where $N \equiv \dim V$).

Proof: It is clear that the hyperplane is a subspace of V because for any $\mathbf{x}_1$ and $\mathbf{x}_2$ in the hyperplane we have

$$\mathbf{f}^*(\mathbf{x}_1 + \lambda\mathbf{x}_2) = \mathbf{f}^*(\mathbf{x}_1) + \lambda\mathbf{f}^*(\mathbf{x}_2) = 0.$$

Hence any linear combination of $\mathbf{x}_1$ and $\mathbf{x}_2$ also belongs to the hyperplane, so the hyperplane is a subspace.

To determine the dimension of this subspace, we would like to construct a basis for the hyperplane. Since $\mathbf{f}^* \in V^*$ is a nonzero covector, there exists some vector $\mathbf{u} \in V$ such that $\mathbf{f}^*(\mathbf{u}) \neq 0$. (This vector does not belong to the hyperplane.) The idea is to complete $\mathbf{u}$ to a basis $\{\mathbf{u}, \mathbf{v}_1, ..., \mathbf{v}_{N-1}\}$ in V, such that $\mathbf{f}^*(\mathbf{u}) \neq 0$ but $\mathbf{f}^*(\mathbf{v}_i) = 0$; then $\{\mathbf{v}_1, ..., \mathbf{v}_{N-1}\}$ will be a basis in the hyperplane. To find such a basis $\{\mathbf{u}, \mathbf{v}_1, ..., \mathbf{v}_{N-1}\}$, let us first complete $\mathbf{u}$ to *some* basis $\{\mathbf{u}, \mathbf{u}_1, ..., \mathbf{u}_{N-1}\}$. Then we define $\mathbf{v}_i = \mathbf{u}_i - c_i\mathbf{u}$ with appropriately chosen c_i. To achieve $\mathbf{f}^*(\mathbf{v}_i) = 0$, we set

$$c_i = \frac{\mathbf{f}^*(\mathbf{u}_i)}{\mathbf{f}^*(\mathbf{u})}.$$

It remains to prove that $\{\mathbf{u}, \mathbf{v}_1, ..., \mathbf{v}_{N-1}\}$ is again a basis. Applying $\mathbf{f}^*$ to a supposedly existing vanishing linear combination,

$$\lambda\mathbf{u} + \sum_{i=1}^{N-1} \lambda_i \mathbf{v}_i = 0,$$

we obtain $\lambda = 0$. Expressing $\mathbf{v}_i$ through $\mathbf{u}$ and $\mathbf{u}_i$, we obtain a vanishing linear combination of vectors $\{\mathbf{u}, \mathbf{u}_1, ..., \mathbf{u}_{N-1}\}$ with coefficients λ_i at $\mathbf{u}_i$. Hence, all λ_i are zero, and so the set $\{\mathbf{u}, \mathbf{v}_1, ..., \mathbf{v}_{N-1}\}$ is linearly independent and thus a basis in V.

Finally, we show that $\{\mathbf{v}_1, ..., \mathbf{v}_{N-1}\}$ is a basis in the hyperplane. By construction, every $\mathbf{v}_i$ belongs to the hyperplane, and so does every linear combination of the $\mathbf{v}_i$'s. It remains to show that every $\mathbf{x}$ such that $\mathbf{f}^*(\mathbf{x}) = 0$ can be expressed as a linear combination of the $\{\mathbf{v}_j\}$. For any such $\mathbf{x}$ we have the decomposition in the basis$\{\mathbf{u}, \mathbf{v}_1, ..., \mathbf{v}_{N-1}\}$,

$$\mathbf{x} = \lambda\mathbf{u} + \sum_{i=1}^{N-1} \lambda_i \mathbf{v}_i.$$

Applying $\mathbf{f}^*$ to this, we find $\lambda = 0$. Hence, $\mathbf{x}$ is a linear combination only of the $\{\mathbf{v}_j\}$. This shows that the set $\{\mathbf{v}_j\}$ spans the hyperplane. The set $\{\mathbf{v}_j\}$ is linearly independent since it is a subset of a basis in V. Hence, $\{\mathbf{v}_j\}$ is a basis in the hyperplane. Therefore, the hyperplane has dimension $N - 1$. ■

Hyperplanes considered so far always contain the zero vector. Another useful construction is that of an *affine* hyperplane: Geometrically speaking, this is a hyperplane that has been shifted away from the origin.

Definition 2: An **affine hyperplane** is the set of all vectors $\mathbf{x} \in V$ such that $\mathbf{f}^*(\mathbf{x}) = \alpha$, where $\mathbf{f}^* \in V^*$ is nonzero, and α is a number.

Remark: An affine hyperplane with $\alpha \neq 0$ is *not* a subspace of V and may be described more constructively as follows. We first obtain a basis $\{\mathbf{v}_1, ..., \mathbf{v}_{N-1}\}$ of the hyperplane $\mathbf{f}^*(\mathbf{x}) = 0$, as described above. We then choose some vector $\mathbf{u}$ such that $\mathbf{f}^*(\mathbf{u}) \neq 0$; such a vector exists since $\mathbf{f}^* \neq 0$. We can then multiply $\mathbf{u}$ by a constant λ such that $\mathbf{f}^*(\lambda\mathbf{u}) = \alpha$, that is, the vector $\lambda\mathbf{u}$ belongs to the affine hyperplane. Now, every vector $\mathbf{x}$ of the form

$$\mathbf{x} = \lambda\mathbf{u} + \sum_{i=1}^{N-1} \lambda_i \mathbf{v}_i,$$

with arbitrary λ_i, belongs to the hyperplane since $\mathbf{f}^*(\mathbf{x}) = \alpha$ by construction. Thus, the set $\{\mathbf{x} \,|\, \mathbf{f}^*(\mathbf{x}) = \alpha\}$ is a hyperplane drawn through $\lambda\mathbf{u}$ parallel to the vectors $\{\mathbf{v}_i\}$. Affine hyperplanes described by the same covector $\mathbf{f}^*$ but with different values of α will differ only in the choice of the initial vector $\lambda\mathbf{u}$ and thus are parallel to each other, in the geometric sense.

Exercise: Intersection of many hyperplanes. a) Suppose $\mathbf{f}_1^*, ..., \mathbf{f}_k^* \in V$. Show that the set of all vectors $\mathbf{x} \in V$ such that $\mathbf{f}_i^*(\mathbf{x}) = 0$ $(i = 1, ...k)$ is a subspace of V.

b)* Show that the dimension of that subspace is equal to $N - k$ (where $N \equiv \dim V$) if the set $\{\mathbf{f}_1^*, ..., \mathbf{f}_k^*\}$ is linearly independent.

1.7 Tensor product of vector spaces

The tensor product is an abstract construction which is important in many applications. The motivation is that we would like to define a product of vectors, $\mathbf{u} \otimes \mathbf{v}$, which behaves as we expect a product to behave, e.g.

$$(\mathbf{a} + \lambda\mathbf{b}) \otimes \mathbf{c} = \mathbf{a} \otimes \mathbf{c} + \lambda\mathbf{b} \otimes \mathbf{c}, \quad \forall\lambda \in \mathbb{K}, \ \forall\mathbf{a}, \mathbf{b}, \mathbf{c} \in V,$$

and the same with respect to the second vector. This property is called **bilinearity**. A "trivial" product would be $\mathbf{a} \otimes \mathbf{b} = 0$ for all $\mathbf{a}, \mathbf{b}$; of course, this product has the bilinearity property but is useless. It turns out to be impossible to define a nontrivial product of vectors in a general vector space, such that the result is again a vector in the same space.[3] The solution is to define a product of vectors so that the resulting object $\mathbf{u} \otimes \mathbf{v}$ is not a vector from V but an element of *another space*. This space is constructed in the following definition.

Definition: Suppose V and W are two vector spaces over a field $\mathbb{K}$; then one defines a new vector space, which is called the **tensor product** of V and W and denoted by $V \otimes W$. This is the space of *expressions* of the form

$$\mathbf{v}_1 \otimes \mathbf{w}_1 + ... + \mathbf{v}_n \otimes \mathbf{w}_n, \tag{1.16}$$

where $\mathbf{v}_i \in V$, $\mathbf{w}_i \in W$. The plus sign behaves as usual (commutative and associative). The symbol $\otimes$ is a special separator symbol. Further, we postulate that the following combinations are equal,

$$\lambda(\mathbf{v} \otimes \mathbf{w}) = (\lambda\mathbf{v}) \otimes \mathbf{w} = \mathbf{v} \otimes (\lambda\mathbf{w}), \tag{1.17}$$

$$(\mathbf{v}_1 + \mathbf{v}_2) \otimes \mathbf{w} = \mathbf{v}_1 \otimes \mathbf{w} + \mathbf{v}_2 \otimes \mathbf{w}, \tag{1.18}$$

$$\mathbf{v} \otimes (\mathbf{w}_1 + \mathbf{w}_2) = \mathbf{v} \otimes \mathbf{w}_1 + \mathbf{v} \otimes \mathbf{w}_2, \tag{1.19}$$

for any vectors $\mathbf{v}, \mathbf{w}, \mathbf{v}_{1,2}, \mathbf{w}_{1,2}$ and for any constant λ. (One could say that the symbol $\otimes$ "behaves as a noncommutative product sign".) The expression $\mathbf{v} \otimes \mathbf{w}$, which is by definition an element of $V \otimes W$, is called the **tensor product** of vectors $\mathbf{v}$ and $\mathbf{w}$. In the space $V \otimes W$, the operations of addition and multiplication by scalars are defined in the natural way. Elements of the tensor product space are called **tensors**.

[3]The impossibility of this is proved in abstract algebra but I do not know the proof.

Question: The set $V \otimes W$ is a vector space. What is the zero vector in that space?

Answer: Since $V \otimes W$ is a vector space, the zero element $0 \in V \otimes W$ can be obtained by multiplying any other element of $V \otimes W$ by the number 0. So, according to Eq. (1.17), we have $0 = 0\,(\mathbf{v} \otimes \mathbf{w}) = (0\mathbf{v}) \otimes \mathbf{w} = 0 \otimes \mathbf{w} = 0 \otimes (0\mathbf{w}) = 0 \otimes 0$. In other words, the zero element is represented by the tensor $0 \otimes 0$. It will not cause confusion if we simply write 0 for this zero tensor. ■

Generally, one calls something a **tensor** if it belongs to a space that was previously defined as a tensor product of some other vector spaces.

According to the above definition, we may perform calculations with the tensor product expressions by expanding brackets or moving scalar factors, as if $\otimes$ is a kind of multiplication. For example, if $\mathbf{v}_i \in V$ and $\mathbf{w}_i \in W$ then

$$\begin{aligned}\frac{1}{3}(\mathbf{v}_1 - \mathbf{v}_2) \otimes (\mathbf{w}_1 - 2\mathbf{w}_2) &= \frac{1}{3}\mathbf{v}_1 \otimes \mathbf{w}_1 - \frac{1}{3}\mathbf{v}_2 \otimes \mathbf{w}_1 \\ &\quad - \frac{2}{3}\mathbf{v}_1 \otimes \mathbf{w}_2 + \frac{2}{3}\mathbf{v}_2 \otimes \mathbf{w}_2.\end{aligned}$$

Note that we cannot simplify this expression any further, because by definition *no other combinations* of tensor products are equal *except* those specified in Eqs. (1.17)–(1.19). This calculation illustrates that $\otimes$ is a formal symbol, so in particular $\mathbf{v} \otimes \mathbf{w}$ is not a new vector from V or from W but is a new entity, an element of a new vector space that we just defined.

Question: The logic behind the operation $\otimes$ is still unclear. How could we write the properties (1.17)–(1.19) if the operation $\otimes$ was not yet defined?

Answer: We actually *define* the operation $\otimes$ through these properties. In other words, the object $\mathbf{a} \otimes \mathbf{b}$ is defined as an expression with which one may perform certain manipulations. Here is a more formal definition of the tensor product space. We first consider the space of *all* formal linear combinations

$$\lambda_1 \mathbf{v}_1 \otimes \mathbf{w}_1 + ... + \lambda_n \mathbf{v}_n \otimes \mathbf{w}_n,$$

which is a very large vector space. Then we introduce equivalence relations expressed by Eqs. (1.17)–(1.19). The space $V \otimes W$ is, by definition, the set of equivalence classes of linear combinations with respect to these relations. Representatives of these equivalence classes may be written in the form (1.16) and calculations can be performed using only the axioms (1.17)–(1.19). ■

Note that $\mathbf{v} \otimes \mathbf{w}$ is generally different from $\mathbf{w} \otimes \mathbf{v}$ because the vectors $\mathbf{v}$ and $\mathbf{w}$ can belong to different vector spaces. Pedantically, one can also define the tensor product space $W \otimes V$ and then demonstrate a canonical isomorphism $V \otimes W \cong W \otimes V$.

Exercise: Prove that the spaces $V \otimes W$ and $W \otimes V$ are canonically isomorphic.

Answer: A canonical isomorphism will map the expression $\mathbf{v} \otimes \mathbf{w} \in V \otimes W$ into $\mathbf{w} \otimes \mathbf{v} \in W \otimes V$. ■

The representation of a tensor $A \in V \otimes W$ in the form (1.16) is *not unique*, i.e. there may be many possible choices of the vectors $\mathbf{v}_j$ and $\mathbf{w}_j$ that give the same tensor A. For example,

$$A \equiv \mathbf{v}_1 \otimes \mathbf{w}_1 + \mathbf{v}_2 \otimes \mathbf{w}_2 = (\mathbf{v}_1 - \mathbf{v}_2) \otimes \mathbf{w}_1 + \mathbf{v}_2 \otimes (\mathbf{w}_1 + \mathbf{w}_2).$$

This is quite similar to the identity $2 + 3 = (2 - 1) + (3 + 1)$, except that in this case we can simplify $2 + 3 = 5$ while in the tensor product space no such simplification is possible. I stress that two tensor expressions $\sum_k \mathbf{v}_k \otimes \mathbf{w}_k$ and $\sum_k \mathbf{v}'_k \otimes \mathbf{w}'_k$ are equal *only if* they can be related by a chain of identities of the form (1.17)–(1.19); such are the axioms of the tensor product.

1.7.1 First examples

Example 1: polynomials. Let V be the space of polynomials having a degree ≤ 2 in the variable x, and let W be the space of polynomials of degree ≤ 2 in the variable y. We consider the tensor product of the elements $p(x) = 1 + x$ and $q(y) = y^2 - 2y$. Expanding the tensor product according to the axioms, we find

$$(1 + x) \otimes (y^2 - 2y) = 1 \otimes y^2 - 1 \otimes 2y + x \otimes y^2 - x \otimes 2y.$$

Let us compare this with the formula we would obtain by multiplying the polynomials in the conventional way,

$$(1 + x)(y^2 - 2y) = y^2 - 2y + xy^2 - 2xy.$$

Note that $1 \otimes 2y = 2 \otimes y$ and $x \otimes 2y = 2x \otimes y$ according to the axioms of the tensor product. So we can see that the tensor product space $V \otimes W$ has a natural interpretation through the algebra of polynomials. The space $V \otimes W$ can be visualized as the space of polynomials in both x and y of degree at most 2 in each variable. To make this interpretation precise, we can construct a canonical isomorphism between the space $V \otimes W$ and the space of polynomials in x and y of degree at most 2 in each variable. The isomorphism maps the tensor $p(x) \otimes q(y)$ to the polynomial $p(x)q(y)$.

Example 2: $\mathbb{R}^3 \otimes \mathbb{C}$. Let V be the three-dimensional space $\mathbb{R}^3$, and let W be the set of all complex numbers $\mathbb{C}$ considered as a vector space over $\mathbb{R}$. Then the tensor product of V and W is, by definition, the space of combinations of the form

$$(x_1, y_1, z_1) \otimes (a_1 + b_1 \mathrm{i}) + (x_2, y_2, z_2) \otimes (a_2 + b_2 \mathrm{i}) + ...$$

Here "i" can be treated as a formal symbol; of course we know that $\mathrm{i}^2 = -1$, but our vector spaces are over $\mathbb{R}$ and so we will not need to *multiply* complex numbers when we perform calculations in these spaces. Since

$$(x, y, z) \otimes (a + b\mathrm{i}) = (ax, ay, az) \otimes 1 + (bx, by, bz) \otimes \mathrm{i},$$

any element of $\mathbb{R}^3 \otimes \mathbb{C}$ can be represented by the expression $\mathbf{v}_1 \otimes 1 + \mathbf{v}_2 \otimes \mathrm{i}$, where $\mathbf{v}_{1,2} \in \mathbb{R}^3$. For brevity one can write such expressions as $\mathbf{v}_1 + \mathbf{v}_2 \mathrm{i}$. One also writes $\mathbb{R}^3 \otimes_{\mathbb{R}} \mathbb{C}$ to emphasize the fact that it is a space over $\mathbb{R}$. In other words, $\mathbb{R}^3 \otimes_{\mathbb{R}} \mathbb{C}$ is the space of three-dimensional vectors "with complex coefficients." This space is six-dimensional.

Exercise: We can consider $\mathbb{R}^3 \otimes_{\mathbb{R}} \mathbb{C}$ as a vector space over $\mathbb{C}$ if we define the multiplication by a complex number λ by $\lambda(\mathbf{v} \otimes z) \equiv \mathbf{v} \otimes (\lambda z)$ for $\mathbf{v} \in V$ and $\lambda, z \in \mathbb{C}$. Compute explicitly

$$\lambda \left(\mathbf{v}_1 \otimes 1 + \mathbf{v}_2 \otimes \mathrm{i} \right) = ?$$

Determine the dimension of the space $\mathbb{R}^3 \otimes_{\mathbb{R}} \mathbb{C}$ when viewed as a vector space over $\mathbb{C}$ in this way.

Example 3: $V \otimes \mathbb{K}$ is isomorphic to V. Since $\mathbb{K}$ is a vector space over itself, we can consider the tensor product of V and $\mathbb{K}$. However, nothing is gained: the space $V \otimes \mathbb{K}$ is canonically isomorphic to V. This can be easily verified: an element $\mathbf{x}$ of $V \otimes \mathbb{K}$ is by definition an expression of the form $\mathbf{x} = \mathbf{v}_1 \otimes \lambda_1 + ... + \mathbf{v}_n \otimes \lambda_n$, however, it follows from the axiom (1.17) that $\mathbf{v}_1 \otimes \lambda_1 = (\lambda_1 \mathbf{v}_1) \otimes 1$, therefore $\mathbf{x} = (\lambda_1 \mathbf{v}_1 + ... + \lambda_n \mathbf{v}_n) \otimes 1$. Thus for any $\mathbf{x} \in V \otimes \mathbb{K}$ there exists a unique $\mathbf{v} \in V$ such that $\mathbf{x} = \mathbf{v} \otimes 1$. In other words, there is a canonical isomorphism $V \to V \otimes \mathbb{K}$ which maps $\mathbf{v}$ into $\mathbf{v} \otimes 1$.

1.7.2 Example: $\mathbb{R}^m \otimes \mathbb{R}^n$

Let $\{\mathbf{e}_1, ..., \mathbf{e}_m\}$ and $\{\mathbf{f}_1, ..., \mathbf{f}_n\}$ be the standard bases in $\mathbb{R}^m$ and $\mathbb{R}^n$ respectively. The vector space $\mathbb{R}^m \otimes \mathbb{R}^n$ consists, by definition, of expressions of the form

$$\mathbf{v}_1 \otimes \mathbf{w}_1 + ... + \mathbf{v}_k \otimes \mathbf{w}_k = \sum_{i=1}^{k} \mathbf{v}_i \otimes \mathbf{w}_i, \quad \mathbf{v}_i \in \mathbb{R}^m, \mathbf{w}_i \in \mathbb{R}^n.$$

The vectors $\mathbf{v}_i, \mathbf{w}_i$ can be decomposed as follows,

$$\mathbf{v}_i = \sum_{j=1}^{m} \lambda_{ij} \mathbf{e}_j, \quad \mathbf{w}_i = \sum_{l=1}^{n} \mu_{il} \mathbf{f}_l, \tag{1.20}$$

where λ_{ij} and μ_{ij} are some coefficients. Then

$$\begin{aligned}\sum_{i=1}^{k} \mathbf{v}_i \otimes \mathbf{w}_i &= \sum_{i=1}^{k} \left(\sum_{j=1}^{m} \lambda_{ij} \mathbf{e}_j \right) \otimes \left(\sum_{l=1}^{n} \mu_{il} \mathbf{f}_l \right) \\ &= \sum_{j=1}^{m} \sum_{l=1}^{n} \left(\sum_{i=1}^{k} \lambda_{ij} \mu_{il} \right) (\mathbf{e}_j \otimes \mathbf{f}_l) \\ &= \sum_{j=1}^{m} \sum_{l=1}^{n} C_{jl} \mathbf{e}_j \otimes \mathbf{f}_l,\end{aligned}$$

where $C_{jl} \equiv \sum_{i=1}^{k} \lambda_{ij} \mu_{il}$ is a certain set of numbers. In other words, an arbitrary element of $\mathbb{R}^m \otimes \mathbb{R}^n$ can be expressed as a linear combination of $\mathbf{e}_j \otimes \mathbf{f}_l$. In Sec. 1.7.3 (after some preparatory work) we will prove that the the set of tensors

$$\{\mathbf{e}_j \otimes \mathbf{f}_l \,|\, 1 \leq j \leq m, 1 \leq l \leq n\}$$

is linearly independent and therefore is a basis in the space $\mathbb{R}^m \otimes \mathbb{R}^n$. It follows that the space $\mathbb{R}^m \otimes \mathbb{R}^n$ has dimension mn and that elements of $\mathbb{R}^m \otimes \mathbb{R}^n$ can be represented by *rectangular tables* of components C_{jl}, where $1 \leq j \leq m$, $1 \leq l \leq n$. In other words, the space $\mathbb{R}^m \otimes \mathbb{R}^n$ is isomorphic to the linear space of rectangular $m \times n$ matrices with coefficients from $\mathbb{K}$. This isomorphism is *not canonical* because the components C_{jl} depend on the choice of the bases $\{\mathbf{e}_j\}$ and $\{\mathbf{f}_j\}$.

1.7.3 Dimension of tensor product is the product of dimensions

We have seen above that the dimension of a direct sum $V \oplus W$ is the sum of dimensions of V and of W. Now the analogous statement: The dimension of a tensor product space $V \otimes W$ is equal to $\dim V \cdot \dim W$.

To prove this statement, we will explicitly construct a basis in $V \otimes W$ out of two given bases in V and in W. Throughout this section, we consider finite-dimensional vector spaces V and W and vectors $\mathbf{v}_j \in V$, $\mathbf{w}_j \in W$.

Lemma 1: a) If $\{\mathbf{v}_1, ..., \mathbf{v}_m\}$ and $\{\mathbf{w}_1, ..., \mathbf{w}_n\}$ are two bases in their respective spaces then any element $A \in V \otimes W$ can be expressed as a linear combination of the form

$$A = \sum_{j=1}^{m} \sum_{k=1}^{n} \lambda_{jk} \mathbf{v}_j \otimes \mathbf{w}_k$$

with some coefficients λ_{jk}.

b) Any tensor $A \in V \otimes W$ can be written as a linear combination $A = \sum_k \mathbf{a}_k \otimes \mathbf{b}_k$, where $\mathbf{a}_k \in V$ and $\mathbf{b}_k \in W$, with at most $\min(m, n)$ terms in the sum.

***Proof:* a)** The required decomposition was given in Example 1.7.2.

b) We can group the n terms $\lambda_{jk} \mathbf{w}_k$ into new vectors $\mathbf{b}_j$ and obtain the required formula with m terms:

$$A = \sum_{j=1}^{m} \sum_{k=1}^{n} \lambda_{jk} \mathbf{v}_j \otimes \mathbf{w}_k = \sum_{j=1}^{m} \mathbf{v}_j \otimes \mathbf{b}_j, \quad \mathbf{b}_j \equiv \sum_{k=1}^{n} \lambda_{jk} \mathbf{w}_k.$$

I will call this formula the **decomposition** of the tensor A in the basis $\{\mathbf{v}_j\}$. Since a similar decomposition with n terms exists for the basis $\{\mathbf{w}_k\}$, it follows that A has a decomposition with at most $\min(m, n)$ terms (not all terms in the decomposition need to be nonzero). ■

We have proved that the set $\{\mathbf{v}_j \otimes \mathbf{w}_k\}$ allows us to express any tensor A as a linear combination; in other words, the set

$$\{\mathbf{v}_j \otimes \mathbf{w}_k \,|\, 1 \leq j \leq m, \, 1 \leq k \leq n\}$$

spans the space $V \otimes W$. This set will be a basis in $V \otimes W$ if it is linearly independent, which we have not yet proved. This is a somewhat subtle point; indeed, how do we show that there exists no linear dependence, say, of the form

$$\lambda_1 \mathbf{v}_1 \otimes \mathbf{w}_1 + \lambda_2 \mathbf{v}_2 \otimes \mathbf{w}_2 = 0$$

with some nonzero coefficients λ_i? Is it perhaps possible to juggle tensor products to obtain such a relation? The answer is negative, but the proof is a bit circumspect. We will use covectors from V^* in a nontraditional way, namely not as linear maps $V \to \mathbb{K}$ but as maps $V \otimes W \to W$.

Lemma 2: If $\mathbf{f}^* \in V^*$ is any covector, we define the map $\mathbf{f}^* : V \otimes W \to W$ (tensors into vectors) by the formula

$$\mathbf{f}^*\big(\sum_k \mathbf{v}_k \otimes \mathbf{w}_k\big) \equiv \sum_k \mathbf{f}^*(\mathbf{v}_k)\,\mathbf{w}_k. \tag{1.21}$$

Then this map is a linear map $V \otimes W \to W$.

Proof: The formula (1.21) defines the map explicitly (and canonically!). It is easy to see that any linear combinations of tensors are mapped into the corresponding linear combinations of vectors,

$$\mathbf{f}^*(\mathbf{v}_k \otimes \mathbf{w}_k + \lambda \mathbf{v}'_k \otimes \mathbf{w}'_k) = \mathbf{f}^*(\mathbf{v}_k)\,\mathbf{w}_k + \lambda \mathbf{f}^*(\mathbf{v}'_k)\,\mathbf{w}'_k.$$

This follows from the definition (1.21) and the linearity of the map $\mathbf{f}^*$. However, there is one potential problem: there exist *many* representations of an element $A \in V \otimes W$ as an expression of the form $\sum_k \mathbf{v}_k \otimes \mathbf{w}_k$ with different choices of $\mathbf{v}_k, \mathbf{w}_k$. Thus we need to show that the map $\mathbf{f}^*$ is well-defined by Eq. (1.21), i.e. that $\mathbf{f}^*(A)$ is always the same vector regardless of the choice of the vectors $\mathbf{v}_k$ and $\mathbf{w}_k$ used to represent A as $A = \sum_k \mathbf{v}_k \otimes \mathbf{w}_k$. Recall that different expressions of the form $\sum_k \mathbf{v}_k \otimes \mathbf{w}_k$ can be equal as a consequence of the axioms (1.17)–(1.19).

In other words, we need to prove that a tensor equality

$$\sum_k \mathbf{v}_k \otimes \mathbf{w}_k = \sum_k \mathbf{v}'_k \otimes \mathbf{w}'_k \tag{1.22}$$

entails

$$\mathbf{f}^*\big(\sum_k \mathbf{v}_k \otimes \mathbf{w}_k\big) = \mathbf{f}^*\big(\sum_k \mathbf{v}'_k \otimes \mathbf{w}'_k\big).$$

To prove this, we need to use the definition of the tensor product. Two expressions in Eq. (1.22) can be equal *only* if they are related by a chain of identities of the form (1.17)–(1.19), therefore it is sufficient to prove that the map $\mathbf{f}^*$ transforms both sides of each of those identities into the same vector. This is verified by explicit calculations, for example we need to check that

$$\begin{aligned}
\mathbf{f}^*(\lambda \mathbf{v} \otimes \mathbf{w}) &= \lambda \mathbf{f}^*(\mathbf{v} \otimes \mathbf{w}),\\
\mathbf{f}^*\left[(\mathbf{v}_1 + \mathbf{v}_2) \otimes \mathbf{w}\right] &= \mathbf{f}^*(\mathbf{v}_1 \otimes \mathbf{w}) + \mathbf{f}^*(\mathbf{v}_2 \otimes \mathbf{w}),\\
\mathbf{f}^*\left[\mathbf{v} \otimes (\mathbf{w}_1 + \mathbf{w}_2)\right] &= \mathbf{f}^*(\mathbf{v} \otimes \mathbf{w}_1) + \mathbf{f}^*(\mathbf{v} \otimes \mathbf{w}_2).
\end{aligned}$$

These simple calculations look tautological, so please check that you can do them and explain why they are necessary for this proof. ■

Lemma 3: If $\{\mathbf{v}_1, ..., \mathbf{v}_m\}$ and $\{\mathbf{u}_1, ..., \mathbf{u}_n\}$ are two linearly independent sets in their respective spaces then the set

$$\{\mathbf{v}_j \otimes \mathbf{w}_k\} \equiv \{\mathbf{v}_1 \otimes \mathbf{w}_1, \mathbf{v}_1 \otimes \mathbf{w}_2, ..., \mathbf{v}_m \otimes \mathbf{w}_{n-1}, \mathbf{v}_m \otimes \mathbf{w}_n\}$$

is linearly independent in the space $V \otimes W$.

Proof: We need to prove that a vanishing linear combination

$$\sum_{j=1}^{m}\sum_{k=1}^{n}\lambda_{jk}\mathbf{v}_j\otimes\mathbf{w}_k=0 \tag{1.23}$$

is possible only if all $\lambda_{jk}=0$. Let us choose some fixed value j_1; we will now prove that $\lambda_{j_1 k}=0$ for all k. By the result of Exercise 1 in Sec. 1.6 there exists a covector $\mathbf{f}^*\in V^*$ such that $\mathbf{f}^*(\mathbf{v}_j)=\delta_{j_1 j}$ for $j=1,...,n$. Then we apply the map $\mathbf{f}^*:V\otimes W\to W$ defined in Lemma 1 to Eq. (1.23). On the one hand, it follows from Eq. (1.23) that

$$\mathbf{f}^*\big[\sum_{j=1}^{m}\sum_{k=1}^{n}\lambda_{jk}\mathbf{v}_j\otimes\mathbf{w}_k\big]=\mathbf{f}^*(0)=0.$$

On the other hand, by definition of the map $\mathbf{f}^*$ we have

$$\begin{aligned}\mathbf{f}^*\big[\sum_{j=1}^{m}\sum_{k=1}^{n}\lambda_{jk}\mathbf{v}_j\otimes\mathbf{w}_k\big]&=\sum_{j=1}^{m}\sum_{k=1}^{n}\lambda_{jk}\mathbf{f}^*(\mathbf{v}_j)\,\mathbf{w}_k\\&=\sum_{j=1}^{m}\sum_{k=1}^{n}\lambda_{jk}\delta_{j_1 j}\mathbf{w}_k=\sum_{k=1}^{n}\lambda_{j_1 k}\mathbf{w}_k.\end{aligned}$$

Therefore $\sum_k\lambda_{j_1 k}\mathbf{w}_k=0$. Since the set $\{\mathbf{w}_k\}$ is linearly independent, we must have $\lambda_{j_1 k}=0$ for all $k=1,...,n$. ■

Now we are ready to prove the main statement of this section.

Theorem: If V and W are finite-dimensional vector spaces then

$$\dim(V\otimes W)=\dim V\cdot\dim W.$$

Proof: By definition of dimension, there exist linearly independent sets of $m\equiv\dim V$ vectors in V and of $n\equiv\dim W$ vectors in W, and by the basis theorem these sets are bases in V and W respectively. By Lemma 1 the set of mn elements $\{\mathbf{v}_j\otimes\mathbf{w}_k\}$ spans the space $V\otimes W$, and by Lemma 3 this set is linearly independent. Therefore this set is a basis. Hence, there are no linearly independent sets of $mn+1$ elements in $V\otimes W$, so $\dim(V\otimes W)=mn$. ■

1.7.4 Higher-rank tensor products

The tensor product of several spaces is defined similarly, e.g. $U\otimes V\otimes W$ is the space of expressions of the form

$$\mathbf{u}_1\otimes\mathbf{v}_1\otimes\mathbf{w}_1+...+\mathbf{u}_n\otimes\mathbf{v}_n\otimes\mathbf{w}_n,\quad\mathbf{u}_i,\mathbf{v}_i,\mathbf{w}_i\in V.$$

Alternatively (and equivalently) one can define the space $U\otimes V\otimes W$ as the tensor product of the spaces $U\otimes V$ and W.

Exercise*: Prove that $(U\otimes V)\otimes W\cong U\otimes(V\otimes W)$.

Definition: If we only work with one space V and if all other spaces are constructed out of V and V^* using the tensor product, then we only need spaces of the form

$$\underbrace{V \otimes ... \otimes V}_{m} \otimes \underbrace{V^* \otimes ... \otimes V^*}_{n}.$$

Elements of such spaces are called **tensors of rank** (m, n). For example, vectors $\mathbf{v} \in V$ have rank $(1, 0)$, covectors $\mathbf{f}^* \in V^*$ have rank $(0, 1)$, tensors from $V \otimes V^*$ have rank $(1, 1)$, tensors from $V \otimes V$ have rank $(2, 0)$, and so on. Scalars from $\mathbb{K}$ have rank $(0, 0)$.

In many applications, the spaces V and V^* are identified (e.g. using a scalar product; see below). In that case, the rank is reduced to a single number — the sum of m and n. Thus, in this simplified counting, tensors from $V \otimes V^*$ as well as tensors from $V \otimes V$ have rank 2.

1.7.5 * Distributivity of tensor product

We have two operations that build new vector spaces out of old ones: the direct sum $V \oplus W$ and the tensor product $V \otimes W$. Is there something like the formula $(U \oplus V) \otimes W \cong (U \otimes W) \oplus (V \otimes W)$? The answer is positive. I will not need this construction below; this is just another example of how different spaces are related by a canonical isomorphism.

Statement: The spaces $(U \oplus V) \otimes W$ and $(U \otimes W) \oplus (V \otimes W)$ are canonically isomorphic.

Proof: An element $(\mathbf{u}, \mathbf{v}) \otimes \mathbf{w} \in (U \oplus V) \otimes W$ is mapped into the pair $(\mathbf{u} \otimes \mathbf{w}, \mathbf{v} \otimes \mathbf{w}) \in (U \otimes W) \oplus (V \otimes W)$. It is easy to see that this map is a canonical isomorphism. I leave the details to you. ■

Exercise: Let U, V, and W be some vector spaces. Demonstrate the following canonical isomorphisms:

$$(U \oplus V)^* \cong U^* \oplus V^*,$$
$$(U \otimes V)^* \cong U^* \otimes V^*.$$

1.8 Linear maps and tensors

The tensor product construction may appear an abstract plaything at this point, but in fact it is a universal tool to describe linear maps.

We have seen that the set of all linear operators $\hat{A} : V \to V$ is a vector space because one can naturally define the sum of two operators and the product of a number and an operator. This vector space is called the space of **endomorphisms** of V and denoted by End V.

In this section I will show that linear operators can be thought of as elements of the space $V \otimes V^*$. This gives a convenient way to represent a linear operator by a coordinate-free formula. Later we will see that the space Hom (V, W) of linear maps $V \to W$ is canonically isomorphic to $W \otimes V^*$.

1.8.1 Tensors as linear operators

First, we will show that any tensor from the space $V \otimes V^*$ acts as a linear map $V \to V$.

Lemma: A tensor $A \in V \otimes V^*$ expressed as

$$A \equiv \sum_{j=1}^{k} \mathbf{v}_j \otimes \mathbf{f}_j^*$$

defines a linear operator $\hat{A} : V \to V$ according to the formula

$$\hat{A}\mathbf{x} \equiv \sum_{j=1}^{k} \mathbf{f}_j^*(\mathbf{x})\, \mathbf{v}_j. \tag{1.24}$$

Proof: Compare this linear map with the linear map defined in Eq. (1.21), Lemma 2 of Sec. 1.7.3. We need to prove two statements:

(1) The transformation is linear, $\hat{A}(\mathbf{x} + \lambda \mathbf{y}) = \hat{A}\mathbf{x} + \lambda \hat{A}\mathbf{y}$.

(2) The operator $\hat{A}$ does not depend on the decomposition of the tensor A using particular vectors $\mathbf{v}_j$ and covectors $\mathbf{f}_j^*$: two decompositions of the tensor A,

$$A = \sum_{j=1}^{k} \mathbf{v}_j \otimes \mathbf{f}_j^* = \sum_{j=1}^{l} \mathbf{w}_j \otimes \mathbf{g}_j^*,$$

yield the same operator,

$$\hat{A}\mathbf{x} = \sum_{j=1}^{k} \mathbf{f}_j^*(\mathbf{x})\, \mathbf{v}_j = \sum_{j=1}^{l} \mathbf{g}_j^*(\mathbf{x})\, \mathbf{w}_j, \quad \forall \mathbf{x}.$$

The first statement, $\hat{A}\,(\mathbf{x} + \lambda \mathbf{y}) = \hat{A}\mathbf{x} + \lambda \hat{A}\mathbf{y}$, follows from the linearity of $\mathbf{f}_j^*$ as a map $V \to \mathbb{K}$ and is easy to verify by explicit calculation:

$$\begin{aligned}
\hat{A}(\mathbf{x} + \lambda \mathbf{y}) &= \sum_{j=1}^{k} \mathbf{f}_j^*(\mathbf{x} + \lambda \mathbf{y})\, \mathbf{v}_j \\
&= \sum_{j=1}^{k} \mathbf{f}_j^*(\mathbf{x})\, \mathbf{v}_j + \lambda \sum_{j=1}^{k} \mathbf{f}_j^*(\mathbf{y})\, \mathbf{v}_j \\
&= \hat{A}\mathbf{x} + \lambda \hat{A}\mathbf{y}.
\end{aligned}$$

The second statement is proved using the axioms (1.17)–(1.19) of the tensor product. Two different expressions for the tensor A can be equal only if they are related through the axioms (1.17)–(1.19). So it suffices to check that the operator $\hat{A}$ remains unchanged when we use each of the three axioms to replace $\sum_{j=1}^{k} \mathbf{v}_j \otimes \mathbf{f}_j^*$ by an equivalent tensor expression. Let us check the first axiom: We need to compare the action of $\sum_j (\mathbf{u}_j + \mathbf{v}_j) \otimes \mathbf{f}_j^*$ on a vector $\mathbf{x} \in V$ and the action of the sum of $\sum_j \mathbf{u}_j \otimes \mathbf{f}_j^*$ and $\sum_j \mathbf{v}_j \otimes \mathbf{f}_j^*$ on the same vector:

$$\begin{aligned}\hat{A}\mathbf{x} &= \Big[\sum_j (\mathbf{u}_j + \mathbf{v}_j) \otimes \mathbf{f}_j^*\Big]\mathbf{x} \\ &= \sum_j \mathbf{f}_j^*(\mathbf{x})(\mathbf{u}_j + \mathbf{v}_j) \\ &= \Big[\sum_j \mathbf{u}_j \otimes \mathbf{f}_j^*\Big]\mathbf{x} + \Big[\sum_j \mathbf{v}_j \otimes \mathbf{f}_j^*\Big]\mathbf{x}.\end{aligned}$$

The action of $\hat{A}$ on $\mathbf{x}$ remains unchanged for every $\mathbf{x}$, which means that the operator $\hat{A}$ itself is unchanged. Similarly, we (more precisely, *you*) can check directly that the other two axioms also leave $\hat{A}$ unchanged. It follows that the action of $\hat{A}$ on a vector $\mathbf{x}$, as defined by Eq. (1.24), is independent of the choice of representation of the tensor A through vectors $\mathbf{v}_j$ and covectors $\mathbf{f}_j^*$. ■

Question: I am wondering what kind of operators correspond to tensor expressions. For example, take the single-term tensor $A = \mathbf{v} \otimes \mathbf{w}^*$. What is the geometric meaning of the corresponding operator $\hat{A}$?

Answer: Let us calculate: $\hat{A}\mathbf{x} = \mathbf{w}^*(\mathbf{x})\,\mathbf{v}$, i.e. the operator $\hat{A}$ acts on any vector $\mathbf{x} \in V$ and produces a vector that is always proportional to the fixed vector $\mathbf{v}$. Hence, the image of the operator $\hat{A}$ is the one-dimensional subspace spanned by $\mathbf{v}$. However, $\hat{A}$ is not necessarily a projector because in general $\hat{A}\hat{A} \neq \hat{A}$:

$$\hat{A}(\hat{A}\mathbf{x}) = \mathbf{w}^*(\mathbf{v})\,\mathbf{w}^*(\mathbf{x})\,\mathbf{v} \neq \mathbf{w}^*(\mathbf{x})\,\mathbf{v}, \text{ unless } \mathbf{w}^*(\mathbf{v}) = 1.$$

Exercise 1: An operator $\hat{A}$ is given by the formula

$$\hat{A} = \hat{1}_V + \lambda \mathbf{v} \otimes \mathbf{w}^*,$$

where $\lambda \in \mathbb{K}$, $\mathbf{v} \in V$, $\mathbf{w}^* \in V^*$. Compute $\hat{A}\mathbf{x}$ for any $\mathbf{x} \in V$.

Answer: $\hat{A}\mathbf{x} = \mathbf{x} + \lambda \mathbf{w}^*(\mathbf{x})\,\mathbf{v}$.

Exercise 2: Let $\mathbf{n} \in V$ and $\mathbf{f}^* \in V^*$ such that $\mathbf{f}^*(\mathbf{n}) = 1$. Show that the operator $\hat{P} \equiv \hat{1}_V - \mathbf{n} \otimes \mathbf{f}^*$ is a projector onto the subspace annihilated by $\mathbf{f}^*$.

Hint: You need to show that $\hat{P}\hat{P} = \hat{P}$; that any vector $\mathbf{x}$ annihilated by $\mathbf{f}^*$ is invariant under $\hat{P}$ (i.e. if $\mathbf{f}^*(\mathbf{x}) = 0$ then $\hat{P}\mathbf{x} = \mathbf{x}$); and that for any vector $\mathbf{x}$, $\mathbf{f}^*(\hat{P}\mathbf{x}) = 0$.

1.8.2 Linear operators as tensors

We have seen that any tensor $A \in V \otimes V^*$ has a corresponding linear map in $\text{End}\, V$. Now conversely, let $\hat{A} \in \text{End}\, V$ be a linear operator and let $\{\mathbf{v}_1, ..., \mathbf{v}_n\}$ be a basis in V. We will now find such covectors $\mathbf{f}_k^* \in V^*$ that the tensor $\sum_k \mathbf{v}_k \otimes \mathbf{f}_k^*$ corresponds to $\hat{A}$. The required covectors $\mathbf{f}_k^* \in V^*$ can be defined by the formula

$$\mathbf{f}_k^*(\mathbf{x}) \equiv \mathbf{v}_k^*(\hat{A}\mathbf{x}), \quad \forall \mathbf{x} \in V,$$

where $\{\mathbf{v}_k^*\}$ is the dual basis. With this definition, we have

$$\left[\sum_{k=1}^{n} \mathbf{v}_k \otimes \mathbf{f}_k^*\right]\mathbf{x} = \sum_{k=1}^{n} \mathbf{f}_k^*(\mathbf{x})\, \mathbf{v}_k = \sum_{k=1}^{n} \mathbf{v}_k^*(\hat{A}\mathbf{x})\mathbf{v}_k = \hat{A}\mathbf{x}.$$

The last equality is based on the formula

$$\sum_{k=1}^{n} \mathbf{v}_k^*(\mathbf{y})\, \mathbf{v}_k = \mathbf{y},$$

which holds because the components of a vector $\mathbf{y}$ in the basis $\{\mathbf{v}_k\}$ are $\mathbf{v}_k^*(\mathbf{y})$. Then it follows from the definition (1.24) that $\left[\sum_k \mathbf{v}_k \otimes \mathbf{f}_k^*\right]\mathbf{x} = \hat{A}\mathbf{x}$.

Let us look at this construction in another way: we have defined a map $\hat{} : V \otimes V^* \to \text{End}\, V$ whereby any tensor $A \in V \otimes V^*$ is transformed into a linear operator $\hat{A} \in \text{End}\, V$.

Theorem: (1) There is a canonical isomorphism $A \to \hat{A}$ between the spaces $V \otimes V^*$ and $\text{End}\, V$. In other words, linear operators are canonically (without choosing a basis) and uniquely mapped into tensors of the form

$$\mathbf{v}_1 \otimes \mathbf{f}_1^* + ... + \mathbf{v}_n \otimes \mathbf{f}_n^*.$$

Conversely, a tensor $\sum_{k=1}^{n} \mathbf{v}_k \otimes \mathbf{f}_k^*$ is mapped into the operator $\hat{A}$ defined by Eq. (1.24).

(2) It is possible to write a tensor A as a sum of not more than $N \equiv \dim V$ terms,

$$A = \sum_{k=1}^{n} \mathbf{v}_k \otimes \mathbf{f}_k^*, \quad n \leq N.$$

Proof: **(1)** To prove that a map is an isomorphism of vector spaces, we need to show that this map is linear and **bijective** (one-to-one). Linearity easily follows from the definition of the map $\hat{}$: if $A, B \in V \otimes V^*$ are two tensors then $A + \lambda B \in V \otimes V^*$ is mapped into $\hat{A} + \lambda \hat{B}$. To prove the bijectivity, we need to show that for any operator $\hat{A}$ there exists a corresponding tensor $A = \sum_k \mathbf{v}_k \otimes \mathbf{f}_k^*$ (this we have already shown above), and that two different tensors $A \neq B$ cannot be mapped into the same operator $\hat{A} = \hat{B}$. If two different tensors $A \neq B$ were mapped into the same operator $\hat{A} = \hat{B}$, it would follow from the linearity of $\hat{}$ that $\widehat{A - B} = \hat{A} - \hat{B} = 0$, in other words, that a nonzero tensor $C \equiv A - B \neq 0$ is mapped into the zero operator, $\hat{C} = 0$. We will now arrive to a contradiction. The tensor C has a decomposition $C = \sum_k \mathbf{v}_k \otimes \mathbf{c}_k^*$ in the basis $\{\mathbf{v}_k\}$. Since $C \neq 0$, it follows that at least one covector $\mathbf{c}_k^*$ is nonzero. Suppose $\mathbf{c}_1^* \neq 0$; then there exists at least one vector $\mathbf{x} \in V$ such that $\mathbf{c}_1^*(\mathbf{x}) \neq 0$. We now act on $\mathbf{x}$ with the operator $\hat{C}$: by assumption, $\hat{C} = \hat{A} - \hat{B} = 0$, but at the same time

$$0 = \hat{C}\mathbf{x} \equiv \sum_k \mathbf{v}_k \mathbf{c}_k^*(\mathbf{x}) = \mathbf{v}_1 \mathbf{c}_1(\mathbf{x}) + ...$$

This is a contradiction because a linear combination of vectors $\mathbf{v}_k$ with at least one nonzero coefficient cannot vanish (the vectors $\{\mathbf{v}_k\}$ are a basis).

Note that we *did* use a basis $\{\mathbf{v}_k\}$ in the construction of the map $\text{End}\, V \to V \otimes V^*$, when we defined the covectors $\mathbf{f}_k^*$. However, this map is canonical because it is the same map for all choices of the basis. Indeed, if we choose another basis $\{\mathbf{v}'_k\}$ then of course the covectors $\mathbf{f}'^*_k$ will be different from $\mathbf{f}_k^*$, but the tensor A will remain the same,

$$A = \sum_{k=1}^{n} \mathbf{v}_k \otimes \mathbf{f}_k^* = A' = \sum_{k=1}^{n} \mathbf{v}'_k \otimes \mathbf{f}'^*_k \in V \otimes V^*,$$

because (as we just proved) different tensors are always mapped into different operators.

(2) This follows from Lemma 1 of Sec. 1.7.3. ■

From now on, I will not use the map ˆ explicitly. Rather, I will simply not distinguish between the spaces $\text{End}\, V$ and $V \otimes V^*$. I will write things like $\mathbf{v} \otimes \mathbf{w}^* \in \text{End}\, V$ or $\hat{A} = \mathbf{x} \otimes \mathbf{y}^*$. The space implied in each case will be clear from the context.

1.8.3 Examples and exercises

Example 1: The identity operator. How to represent the identity operator $\hat{1}_V$ by a tensor $A \in V \otimes V^*$?

Choose a basis $\{\mathbf{v}_k\}$ in V; this choice defines the dual basis $\{\mathbf{v}_k^*\}$ in V^* (see Sec. 1.6) such that $\mathbf{v}_j^*(\mathbf{v}_k) = \delta_{jk}$. Now apply the construction of Sec. 1.8.2 to find

$$A = \sum_{k=1}^{n} \mathbf{v}_k \otimes \mathbf{f}_k^*, \quad \mathbf{f}_k^*(\mathbf{x}) = \mathbf{v}_k^*(\hat{1}_V \mathbf{x}) = \mathbf{v}_k^*(\mathbf{x}) \Rightarrow \mathbf{f}_k^* = \mathbf{v}_k^*.$$

Therefore

$$\hat{1}_V = \sum_{k=1}^{n} \mathbf{v}_k \otimes \mathbf{v}_k^*. \tag{1.25}$$

Question: The identity operator $\hat{1}_V$ is defined **canonically**, i.e. independently of a basis in V; it is simply the transformation that does not change any vectors. However, the tensor representation (1.25) seems to depend on the choice of a basis $\{\mathbf{v}_k\}$. What is going on? Is the tensor $\hat{1} \in V \otimes V^*$ defined canonically?

Answer: Yes. The tensor $\sum_k \mathbf{v}_k \otimes \mathbf{v}_k^*$ is *the same tensor* regardless of which basis $\{\mathbf{v}_k\}$ we choose; of course the correct dual basis $\{\mathbf{v}_k^*\}$ must be used. In other words, for any two bases $\{\mathbf{v}_k\}$ and $\{\tilde{\mathbf{v}}_k\}$, and with $\{\mathbf{v}_k^*\}$ and $\{\tilde{\mathbf{v}}_k^*\}$ being the corresponding dual bases, we have the tensor equality

$$\sum_k \mathbf{v}_k \otimes \mathbf{v}_k^* = \sum_k \tilde{\mathbf{v}}_k \otimes \tilde{\mathbf{v}}_k^*.$$

We have proved this in Theorem 1.8.2 when we established that two different tensors are always mapped into different operators by the map ˆ. One can say that $\sum_k \mathbf{v}_k \otimes \mathbf{v}_k^*$ is a *canonically defined tensor* in $V \otimes V^*$ since it is the unique tensor corresponding to the canonically defined identity operator $\hat{1}_V$. Recall that a given tensor can be written as a linear combination of tensor products in many different ways! Here is a worked-out example:

Let $\{\mathbf{v}_1, \mathbf{v}_2\}$ be a basis in a two-dimensional space; let $\{\mathbf{v}_1^*, \mathbf{v}_2^*\}$ be the corresponding dual basis. We can choose another basis, e.g.

$$\{\mathbf{w}_1, \mathbf{w}_2\} \equiv \{\mathbf{v}_1 + \mathbf{v}_2, \mathbf{v}_1 - \mathbf{v}_2\}.$$

Its dual basis is (verify this!)

$$\mathbf{w}_1^* = \frac{1}{2}(\mathbf{v}_1^* + \mathbf{v}_2^*), \quad \mathbf{w}_2^* = \frac{1}{2}(\mathbf{v}_1^* - \mathbf{v}_2^*).$$

Then we compute the identity tensor:

$$\begin{aligned}\hat{1} = \mathbf{w}_1 \otimes \mathbf{w}_1^* + \mathbf{w}_2 \otimes \mathbf{w}_2^* &= (\mathbf{v}_1 + \mathbf{v}_2) \otimes \frac{1}{2}(\mathbf{v}_1^* + \mathbf{v}_2^*) \\ &\quad + (\mathbf{v}_1 - \mathbf{v}_2) \otimes \frac{1}{2}(\mathbf{v}_1^* - \mathbf{v}_2^*) \\ &= \mathbf{v}_1 \otimes \mathbf{v}_1^* + \mathbf{v}_2 \otimes \mathbf{v}_2^*.\end{aligned}$$

The tensor expressions $\mathbf{w}_1 \otimes \mathbf{w}_1^* + \mathbf{w}_2 \otimes \mathbf{w}_2^*$ and $\mathbf{v}_1 \otimes \mathbf{v}_1^* + \mathbf{v}_2 \otimes \mathbf{v}_2^*$ are *equal* because of distributivity and linearity of tensor product, i.e. due to the axioms of the tensor product.

Exercise 1: Matrices as tensors. Now suppose we have a matrix A_{jk} that specifies the linear operator $\hat{A}$ in a basis $\{\mathbf{e}_k\}$. Which tensor $A \in V \otimes V^*$ corresponds to this operator?

Answer: $A = \sum_{j,k=1}^n A_{jk} \mathbf{e}_j \otimes \mathbf{e}_k^*$.

Exercise 2: Product of linear operators. Suppose $\hat{A} = \sum_{k=1}^n \mathbf{v}_k \otimes \mathbf{f}_k^*$ and $\hat{B} = \sum_{l=1}^n \mathbf{w}_l \otimes \mathbf{g}_l^*$ are two operators. Obtain the tensor representation of the product $\hat{A}\hat{B}$.

Answer: $\hat{A}\hat{B} = \sum_{k=1}^n \sum_{l=1}^n \mathbf{f}_k^*(\mathbf{w}_l)\, \mathbf{v}_k \otimes \mathbf{g}_l^*$.

Exercise 3: Verify that $\hat{1}_V \hat{1}_V = \hat{1}_V$ by explicit computation using the tensor representation (1.25).

Hint: Use the formula $\mathbf{v}_j^*(\mathbf{v}_k) = \delta_{jk}$.

Exercise 4: Eigenvalues. Suppose $\hat{A} = \alpha \hat{1}_V + \mathbf{u} \otimes \mathbf{f}^*$ and $\hat{B} = \mathbf{u} \otimes \mathbf{f}^* + \mathbf{v} \otimes \mathbf{g}^*$, where $\mathbf{u}, \mathbf{v} \in V$ are a linearly independent set, $\alpha \in \mathbb{K}$, and $\mathbf{f}^*, \mathbf{g}^* \in V^*$ are nonzero but such that $\mathbf{f}^*(\mathbf{v}) = 0$ and $\mathbf{g}^*(\mathbf{u}) = 0$ while $\mathbf{f}^*(\mathbf{u}) \neq 0$ and $\mathbf{g}^*(\mathbf{v}) \neq 0$. Determine the eigenvalues and eigenvectors of the operators $\hat{A}$ and $\hat{B}$.

Solution: (I give a solution because it is an instructive calculation showing how to handle tensors in the index-free approach. Note that the vectors $\mathbf{u}, \mathbf{v}$ and the covectors $\mathbf{f}^*, \mathbf{g}^*$ are "given," which means that numbers such as $\mathbf{f}^*(\mathbf{u})$ are known constants.)

For the operator $\hat{A}$, the eigenvalue equation $\hat{A}\mathbf{x} = \lambda \mathbf{x}$ yields

$$\alpha \mathbf{x} + \mathbf{u}\mathbf{f}^*(\mathbf{x}) = \lambda \mathbf{x}.$$

Either $\lambda = \alpha$ and then $\mathbf{f}^*(\mathbf{x}) = 0$, or $\lambda \neq \alpha$ and then $\mathbf{x}$ is proportional to $\mathbf{u}$; substituting $\mathbf{x} = \mathbf{u}$ into the above equation, we find $\lambda = \alpha + \mathbf{f}^*(\mathbf{u})$. Therefore the operator $\hat{A}$ has two eigenvalues, $\lambda = \alpha$ and $\lambda = \alpha + \mathbf{f}^*(\mathbf{u})$. The eigenspace with the eigenvalue $\lambda = \alpha$ is the set of all $\mathbf{x} \in V$ such that $\mathbf{f}^*(\mathbf{x}) = 0$. The eigenspace with the eigenvalue $\lambda = \alpha + \mathbf{f}^*(\mathbf{u})$ is the set of vectors proportional to $\mathbf{u}$. (It might happen that $\mathbf{f}^*(\mathbf{u}) = 0$; then there is only one eigenvalue, $\lambda = \alpha$, and no second eigenspace.)

For the operator $\hat{B}$, the calculations are longer. Since $\{\mathbf{u}, \mathbf{v}\}$ is a linearly independent set, we may add some vectors $\mathbf{e}_k$ to that set in order to complete it to a basis $\{\mathbf{u}, \mathbf{v}, \mathbf{e}_3, ..., \mathbf{e}_N\}$. It is convenient to adapt this basis to the given covectors $\mathbf{f}^*$ and $\mathbf{g}^*$; namely, it is possible to choose this basis such that $\mathbf{f}^*(\mathbf{e}_k) = 0$ and $\mathbf{g}^*(\mathbf{e}_k) = 0$ for $k = 3, ..., N$. (We may replace $\mathbf{e}_k \mapsto \mathbf{e}_k - a_k\mathbf{u} - b_k\mathbf{v}$ with some suitable constants a_k, b_k to achieve this, using the given properties $\mathbf{f}^*(\mathbf{v}) = 0$, $\mathbf{g}^*(\mathbf{u}) = 0$, $\mathbf{f}^*(\mathbf{u}) \neq 0$, and $\mathbf{g}^*(\mathbf{v}) \neq 0$.) Suppose $\mathbf{x}$ is an unknown eigenvector with the eigenvalue λ; then $\mathbf{x}$ can be expressed as $\mathbf{x} = \alpha\mathbf{u} + \beta\mathbf{v} + \sum_{k=3}^{N} y_k\mathbf{e}_k$ in this basis, where α, β, and y_k are unknown constants. Our goal is therefore to determine α, β, y_k, and λ. Denote $\mathbf{y} \equiv \sum_{k=3}^{N} y_k\mathbf{e}_k$ and transform the eigenvalue equation using the given conditions $\mathbf{f}^*(\mathbf{v}) = \mathbf{g}^*(\mathbf{u}) = 0$ as well as the properties $\mathbf{f}^*(\mathbf{y}) = \mathbf{g}^*(\mathbf{y}) = 0$,

$$\begin{aligned}\hat{B}\mathbf{x} - \lambda\mathbf{x} =& \mathbf{u}\left(\alpha\mathbf{f}^*(\mathbf{u}) + \beta\mathbf{f}^*(\mathbf{v}) + \mathbf{f}^*(\mathbf{y}) - \alpha\lambda\right) \\ &+ \mathbf{v}\left(\alpha\mathbf{g}^*(\mathbf{u}) + \beta\mathbf{g}^*(\mathbf{v}) + \mathbf{g}^*(\mathbf{y}) - \beta\lambda\right) - \lambda\mathbf{y} \\ =& \mathbf{u}\left(\alpha\mathbf{f}^*(\mathbf{u}) - \alpha\lambda\right) + \mathbf{v}\left(\beta\mathbf{g}^*(\mathbf{v}) - \beta\lambda\right) - \lambda\mathbf{y} = 0.\end{aligned}$$

The above equation says that a certain linear combination of the vectors $\mathbf{u}$, $\mathbf{v}$, and $\mathbf{y}$ is zero. If $\mathbf{y} \neq 0$, the set $\{\mathbf{u}, \mathbf{v}, \mathbf{y}\}$ is linearly independent since $\{\mathbf{u}, \mathbf{v}, \mathbf{e}_3, ..., \mathbf{e}_N\}$ is a basis (see Exercise 1 in Sec. 1.1.4). Then the linear combination of the three vectors $\mathbf{u}$, $\mathbf{v}$, and $\mathbf{y}$ can be zero only if all three coefficients are zero. On the other hand, if $\mathbf{y} = 0$ then we are left only with two coefficients that must vanish. Thus, we can proceed by considering separately the two possible cases, $\mathbf{y} \neq 0$ and $\mathbf{y} = 0$.

We begin with the case $\mathbf{y} = 0$. In this case, $\hat{B}\mathbf{x} - \lambda\mathbf{x} = 0$ is equivalent to the vanishing of the linear combination

$$\mathbf{u}\left(\alpha\mathbf{f}^*(\mathbf{u}) - \alpha\lambda\right) + \mathbf{v}\left(\beta\mathbf{g}^*(\mathbf{v}) - \beta\lambda\right) = 0.$$

Since $\{\mathbf{u}, \mathbf{v}\}$ is linearly independent, this linear combination can vanish only when both coefficients vanish:

$$\begin{aligned}\alpha\left(\mathbf{f}^*(\mathbf{u}) - \lambda\right) &= 0, \\ \beta\left(\mathbf{g}^*(\mathbf{v}) - \lambda\right) &= 0.\end{aligned}$$

This is a system of two linear equations for the two unknowns α and β; when we solve it, we will determine the possible eigenvectors $\mathbf{x} = \alpha\mathbf{u} + \beta\mathbf{v}$ and the corresponding eigenvalues λ. Note that we are looking for *nonzero* solutions, so α and β cannot be both zero. If $\alpha \neq 0$, we must have $\lambda = \mathbf{f}^*(\mathbf{u})$. If $\mathbf{f}^*(\mathbf{u}) \neq \mathbf{g}^*(\mathbf{v})$, the second equation forces $\beta = 0$. Otherwise, any β is a solution.

Likewise, if $\beta \neq 0$ then we must have $\lambda = \mathbf{g}^*(\mathbf{v})$. Therefore we obtain the following possibilities:

a) $\mathbf{f}^*(\mathbf{u}) \neq \mathbf{g}^*(\mathbf{v})$, two nonzero eigenvalues $\lambda_1 = \mathbf{f}^*(\mathbf{u})$ with eigenvector $\mathbf{x}_1 = \alpha\mathbf{u}$ (with any $\alpha \neq 0$) and $\lambda_2 = \mathbf{g}^*(\mathbf{v})$ with eigenvector $\mathbf{x}_2 = \beta\mathbf{v}$ (with any $\beta \neq 0$).

b) $\mathbf{f}^*(\mathbf{u}) = \mathbf{g}^*(\mathbf{v})$, one nonzero eigenvalue $\lambda = \mathbf{f}^*(\mathbf{u}) = \mathbf{g}^*(\mathbf{v})$, two-dimensional eigenspace with eigenvectors $\mathbf{x} = \alpha\mathbf{u} + \beta\mathbf{v}$ where at least one of α, β is nonzero.

Now we consider the case $\mathbf{y} \neq 0$ (recall that $\mathbf{y}$ is an unknown vector from the subspace Span $\{\mathbf{e}_3, ..., \mathbf{e}_N\}$). In this case, we obtain a system of linear equations for the set of unknowns $(\alpha, \beta, \lambda, \mathbf{y})$:

$$\begin{aligned} \alpha\mathbf{f}^*(\mathbf{u}) - \alpha\lambda &= 0, \\ \beta\mathbf{g}^*(\mathbf{v}) - \beta\lambda &= 0, \\ -\lambda &= 0. \end{aligned}$$

This system is simplified, using $\lambda = 0$, to

$$\begin{aligned} \alpha\mathbf{f}^*(\mathbf{u}) &= 0, \\ \beta\mathbf{g}^*(\mathbf{v}) &= 0. \end{aligned}$$

Since $\mathbf{f}^*(\mathbf{u}) \neq 0$ and $\mathbf{g}^*(\mathbf{v}) \neq 0$, the only solution is $\alpha = \beta = 0$. Hence, the eigenvector is $\mathbf{x} = \mathbf{y}$ for any nonzero $\mathbf{y} \in$ Span $\{\mathbf{e}_3, ..., \mathbf{e}_N\}$. In other words, there is an $(N-2)$-dimensional eigenspace corresponding to the eigenvalue $\lambda = 0$. ■

Remark: The preceding exercise serves to show that calculations in the coordinate-free approach are not always short! (I even specified some additional constraints on $\mathbf{u}, \mathbf{v}, \mathbf{f}^*, \mathbf{g}^*$ in order to make the solution shorter. Without these constraints, there are many more cases to be considered.) The coordinate-free approach does not necessarily provide a shorter way to find eigenvalues of matrices than the usual methods based on the evaluation of determinants. However, the coordinate-free method is efficient for the operator $\hat{A}$. The end result is that we are able to determine eigenvalues and eigenspaces of operators such as $\hat{A}$ and $\hat{B}$, regardless of the number of dimensions in the space, by using the special structure of these operators, which is specified in a purely geometric way.

Exercise 5: Find the inverse operator to $\hat{A} = \hat{1}_V + \mathbf{u} \otimes \mathbf{f}^*$, where $\mathbf{u} \in V$, $\mathbf{f}^* \in V^*$. Determine when $\hat{A}^{-1}$ exists.

Answer: The inverse operator exists only if $\mathbf{f}^*(\mathbf{u}) \neq -1$: then

$$\hat{A}^{-1} = \hat{1}_V - \frac{1}{1 + \mathbf{f}^*(\mathbf{u})}\mathbf{u} \otimes \mathbf{f}^*.$$

When $\mathbf{f}^*(\mathbf{u}) = -1$, the operator $\hat{A}$ has an eigenvector $\mathbf{u}$ with eigenvalue 0, so $\hat{A}^{-1}$ cannot exist.

1.8.4 Linear maps between *different* spaces

So far we have been dealing with linear operators that map a space V into itself; what about linear maps $V \to W$ between *different* spaces? If we replace V^* by W^* in many of our definitions and proofs, we will obtain a parallel set of results for linear maps $V \to W$.

Theorem 1: Any tensor $A \equiv \sum_{j=1}^{k} \mathbf{w}_j \otimes \mathbf{f}_j^* \in W \otimes V^*$ acts as a linear map $V \to W$ according to the formula

$$A\mathbf{x} \equiv \sum_{j=1}^{k} \mathbf{f}_j^*(\mathbf{x})\,\mathbf{w}_j.$$

The space $\mathrm{Hom}(V, W)$ of all linear operators $V \to W$ is canonically isomorphic to the space $W \otimes V^*$.

Proof: Left as an exercise since it is fully analogous to previous proofs.

Example 1: Covectors as tensors. We know that the number field $\mathbb{K}$ is a vector space over itself and $V \cong V \otimes \mathbb{K}$. Therefore linear maps $V \to \mathbb{K}$ are tensors from $V^* \otimes \mathbb{K} \cong V^*$, i.e. covectors, in agreement with the definition of V^*.

Example 2: If V and W are vector spaces, what are tensors from $V^* \otimes W^*$?

They can be viewed as (1) linear maps from V into W^*, (2) linear maps from W into V^*, (3) linear maps from $V \otimes W$ into $\mathbb{K}$. These possibilities can be written as canonical isomorphisms:

$$V^* \otimes W^* \cong \mathrm{Hom}(V, W^*) \cong \mathrm{Hom}(W, V^*) \cong \mathrm{Hom}(V \otimes W, \mathbb{K}).$$

Exercise 1: How can we interpret the space $V \otimes V \otimes V^*$? Same question for the space $V^* \otimes V^* \otimes V \otimes V$.

Answer: In many different ways:

$$\begin{aligned}
&V \otimes V \otimes V^* \cong \mathrm{Hom}(V, V \otimes V)\\
&\cong \mathrm{Hom}(\mathrm{End}\, V, V) \cong \mathrm{Hom}(V^*, \mathrm{End}\, V) \cong \ldots \text{ and}\\
&V^* \otimes V^* \otimes V \otimes V \cong \mathrm{Hom}(V, V^* \otimes V \otimes V)\\
&\cong \mathrm{Hom}(V \otimes V, V \otimes V) \cong \mathrm{Hom}(\mathrm{End}\, V, \mathrm{End}\, V) \cong \ldots
\end{aligned}$$

For example, $V \otimes V \otimes V^*$ can be visualized as the space of linear maps from V^* to linear operators in V. The action of a tensor $\mathbf{u} \otimes \mathbf{v} \otimes \mathbf{w}^* \in V \otimes V \otimes V^*$ on a covector $\mathbf{f}^* \in V^*$ may be defined either as $\mathbf{f}^*(\mathbf{u})\,\mathbf{v} \otimes \mathbf{w}^* \in V \otimes V^*$ or alternatively as $\mathbf{f}^*(\mathbf{v})\,\mathbf{u} \otimes \mathbf{w}^* \in V \otimes V^*$. Note that these two definitions are *not* equivalent, i.e. the same tensors are mapped to *different* operators. In each case, one of the copies of V (from $V \otimes V \otimes V^*$) is "paired up" with V^*.

Question: We have seen in the proof of Lemma 1 in Sec. 1.7.3 that covectors $\mathbf{f}^* \in V^*$ act as linear maps $V \otimes W \to W$. However, I am now sufficiently illuminated to know that linear maps $V \otimes W \to W$ are elements of the space $W \otimes W^* \otimes V^*$ and not elements of V^*. How can this be reconciled?

Answer: There is an injection map $V^* \to W \otimes W^* \otimes V^*$ defined by the formula $\mathbf{f}^* \to \hat{1}_W \otimes \mathbf{f}^*$, where $\hat{1}_W \in W \otimes W^*$ is the identity operator. Since $\hat{1}_W$ is a canonically defined element of $W \otimes W^*$, the map is canonical (defined without choice of basis, i.e. *geometrically*). Thus covectors $\mathbf{f}^* \in V^*$ can be naturally considered as elements of the space $\mathrm{Hom}(V \otimes W, W)$.

Question: The space $V \otimes V^*$ can be interpreted as End V, as End V^*, or as Hom $(V \otimes V^*, \mathbb{K})$. This means that one tensor $A \in V \otimes V^*$ represents an operator in V, an operator in V^*, or a map from operators into numbers. What is the relation between all these different interpretations of the tensor A? For example, what is the interpretation of the identity operator $\hat{1}_V \in V \otimes V^*$ as an element of Hom $(V \otimes V^*, \mathbb{K})$?

Answer: The identity tensor $\hat{1}_V$ represents the identity operator in V and in V^*. It also represents the following map $V \otimes V^* \to \mathbb{K}$,

$$\hat{1}_V : \mathbf{v} \otimes \mathbf{f}^* \mapsto \mathbf{f}^* (\mathbf{v}) .$$

This map applied to an operator $\hat{A} \in V \otimes V^*$ yields the **trace** of that operator (see Sec. 3.8).

The definition below explains the relation between operators in V and operators in V^* represented by the same tensor.

Definition: If $\hat{A} : V \to W$ is a linear map then the **transposed operator** $\hat{A}^T : W^* \to V^*$ is the map defined by

$$(\hat{A}^T \mathbf{f}^*) (\mathbf{v}) \equiv \mathbf{f}^* (\hat{A}\mathbf{v}), \quad \forall \mathbf{v} \in V, \forall \mathbf{f}^* \in W^*. \qquad (1.26)$$

In particular, this defines the transposed operator $\hat{A}^T : V^* \to V^*$ given an operator $\hat{A} : V \to V$.

Remark: The above definition is an example of "mathematical style": I just wrote formula (1.26) and left it for you to digest. In case you have trouble with this formula, let me translate: The operator $\hat{A}^T$ is by definition such that it will transform an arbitrary covector $\mathbf{f}^* \in W^*$ into a new covector $(\hat{A}^T \mathbf{f}^*) \in V^*$, which is a linear function defined by its action on vectors $\mathbf{v} \in V$. The formula says that the value of that linear function applied to an arbitrary vector $\mathbf{v}$ should be equal to the number $\mathbf{f}^* (\hat{A}\mathbf{v})$; thus we defined the action of the covector $\hat{A}^T \mathbf{f}^*$ on any vector $\mathbf{v}$. Note how in the formula $(\hat{A}^T \mathbf{f}^*) (\mathbf{v})$ the parentheses are used to show that the first object is acting on the second.

Since we have defined the covector $\hat{A}^T \mathbf{f}^*$ for any $\mathbf{f}^* \in W^*$, it follows that we have thereby defined the operator $\hat{A}^T$ acting in the space W^* and yielding a covector from V^*. Please read the formula again and check that you can understand it. The difficulty of understanding equations such as Eq. (1.26) is that one needs to keep in mind all the mathematical notations introduced previously and used here, and one also needs to guess the argument implied by the formula. In this case, the implied argument is that we will *define a new operator* $\hat{A}^T$ if we show, for any $\mathbf{f}^* \in W^*$, how the new covector $(\hat{A}^T \mathbf{f}^*) \in V^*$ works on any vector $\mathbf{v} \in V$. Only after some practice with such arguments will it become easier to read mathematical definitions. ■

Note that the transpose map $\hat{A}^T$ is defined **canonically** (i.e. without choosing a basis) through the original map $\hat{A}$.

Question: How to use this definition when the operator $\hat{A}$ is given? Eq. (1.26) is not a formula that gives $\hat{A}^T \mathbf{f}^*$ directly; rather, it is an identity connecting some values for arbitrary $\mathbf{v}$ and $\mathbf{f}^*$.

Answer: In order to use this definition, we need to apply $\hat{A}^T \mathbf{f}^*$ to an arbitrary vector $\mathbf{v}$ and transform the resulting expression. We could also compute the coefficients of the operator $\hat{A}^T$ in some basis.

Exercise 2: If $A = \sum_k \mathbf{w}_k \otimes \mathbf{f}_k^* \in W \otimes V^*$ is a linear map $V \to W$, what is the tensor representation of its transpose A^T? What is its matrix representation in a suitable basis?

Answer: The transpose operator A^T maps $W^* \to V^*$, so the corresponding tensor is $A^T = \sum_k \mathbf{f}_k^* \otimes \mathbf{w}_k \in V^* \otimes W$. Its tensor representation consists of the same vectors $\mathbf{w}_k \in W$ and covectors $\mathbf{f}_k^* \in V^*$ as the tensor representation of A. The matrix representation of A^T is the transposed matrix of A if we use the same basis $\{\mathbf{e}_j\}$ and its dual basis $\{\mathbf{e}_j^*\}$. ■

An important characteristic of linear operators is the rank. (Note that we have already used the word "rank" to denote the degree of a tensor product; the following definition presents a *different* meaning of the word "rank.")

Definition: The **rank** of a linear map $\hat{A} : V \to W$ is the dimension of the image subspace $\text{im}\,\hat{A} \subset W$. (Recall that $\text{im}\,\hat{A}$ is a linear subspace of W that contains all vectors $\mathbf{w} \in W$ expressed as $\mathbf{w} = \hat{A}\mathbf{v}$ with some $\mathbf{v} \in V$.) The rank may be denoted by $\text{rank}\,\hat{A} \equiv \dim(\text{im}\,\hat{A})$.

Theorem 2: The rank of $\hat{A}$ is the smallest number of terms necessary to write an operator $\hat{A} : V \to W$ as a sum of single-term tensor products. In other words, the operator $\hat{A}$ can be expressed as

$$\hat{A} = \sum_{k=1}^{\text{rank}\,\hat{A}} \mathbf{w}_k \otimes \mathbf{f}_k^* \in W \otimes V^*,$$

with suitably chosen $\mathbf{w}_k \in W$ and $\mathbf{f}_k^* \in V^*$, but not as a sum of fewer terms.

Proof: We know that $\hat{A}$ can be written as a sum of tensor product terms,

$$\hat{A} = \sum_{k=1}^{n} \mathbf{w}_k \otimes \mathbf{f}_k^*, \tag{1.27}$$

where $\mathbf{w}_k \in W$, $\mathbf{f}_k^* \in V^*$ are *some* vectors and covectors, and n is *some* integer. There are many possible choices of these vectors and the covectors. Let us suppose that Eq. (1.27) represents a choice such that n is the smallest possible number of terms. We will first show that n is not smaller than the rank of $\hat{A}$; then we will show that n is not larger than the rank of $\hat{A}$.

If n is the smallest number of terms, the set $\{\mathbf{w}_1, ..., \mathbf{w}_n\}$ must be linearly independent, or else we can reduce the number of terms in the sum (1.27). To show this, suppose that $\mathbf{w}_1$ is equal to a linear combination of other $\mathbf{w}_k$,

$$\mathbf{w}_1 = \sum_{k=2}^{n} \lambda_k \mathbf{w}_k,$$

then we can rewrite $\hat{A}$ as

$$\hat{A} = \mathbf{w}_1 \otimes \mathbf{f}_1^* + \sum_{k=2}^{n} \mathbf{w}_k \otimes \mathbf{f}_k^* = \sum_{k=2}^{n} \mathbf{w}_k \otimes (\mathbf{f}_k^* + \lambda_k \mathbf{f}_1^*),$$

reducing the number of terms from n to $n-1$. Since by assumption the number of terms cannot be made less than n, the set $\{\mathbf{w}_k\}$ must be linearly independent. In particular, the subspace spanned by $\{\mathbf{w}_k\}$ is n-dimensional. (The same reasoning shows that the set $\{\mathbf{f}_k^*\}$ must be also linearly independent, but we will not need to use this.)

The rank of $\hat{A}$ is the dimension of the image of $\hat{A}$; let us denote $m \equiv \text{rank}\,\hat{A}$. It follows from the definition of the map $\hat{A}$ that for any $\mathbf{v} \in V$, the image $\hat{A}\mathbf{v}$ is a linear combination of the vectors $\mathbf{w}_k$,

$$\hat{A}\mathbf{v} = \sum_{k=1}^{n} \mathbf{f}_k^* (\mathbf{v})\, \mathbf{w}_k.$$

Therefore, the m-dimensional subspace $\text{im}\hat{A}$ is contained within the n-dimensional subspace $\text{Span}\{\mathbf{w}_1, ..., \mathbf{w}_n\}$, so $m \leq n$.

Now, we may choose a basis $\{\mathbf{b}_1, ..., \mathbf{b}_m\}$ in the subspace $\text{im}\hat{A}$; then for every $\mathbf{v} \in V$ we have

$$\hat{A}\mathbf{v} = \sum_{i=1}^{m} \beta_i \mathbf{b}_i$$

with some coefficients β_i that are uniquely determined for each vector $\mathbf{v}$; in other words, β_i are *functions* of $\mathbf{v}$. It is easy to see that the coefficients β_i are *linear* functions of the vector $\mathbf{v}$ since

$$\hat{A}(\mathbf{v} + \lambda\mathbf{u}) = \sum_{i=1}^{m} (\beta_i + \lambda\alpha_i)\mathbf{b}_i$$

if $\hat{A}\mathbf{u} = \sum_{i=1}^{m} \alpha_i \mathbf{b}_i$. Hence there exist some covectors $\mathbf{g}_i^*$ such that $\beta_i = \mathbf{g}_i^*(\mathbf{v})$. It follows that we are able to express $\hat{A}$ as the tensor $\sum_{i=1}^{m} \mathbf{b}_i \otimes \mathbf{g}_i^*$ using m terms. Since the smallest possible number of terms is n, we must have $m \geq n$.

We have shown that $m \leq n$ and $m \geq n$, therefore $n = m = \text{rank}\,\hat{A}$. ■

Corollary: The rank of a map $\hat{A} : V \to W$ is equal to the rank of its transpose $\hat{A}^T : W^* \to V^*$.

Proof: The maps $\hat{A}$ and $\hat{A}^T$ are represented by the same tensor from the space $W \otimes V^*$. Since the rank is equal to the minimum number of terms necessary to express that tensor, the ranks of $\hat{A}$ and $\hat{A}^T$ always coincide. ■

We conclude that tensor product is a general construction that represents the space of linear maps between various previously defined spaces. For example, matrices are representations of linear maps from vectors to vectors; tensors from $V^* \otimes V \otimes V$ can be viewed as linear maps from matrices to vectors, etc.

Exercise 3: Prove that the tensor equality $\mathbf{a} \otimes \mathbf{a} + \mathbf{b} \otimes \mathbf{b} = \mathbf{v} \otimes \mathbf{w}$ where $\mathbf{a} \neq 0$ and $\mathbf{b} \neq 0$ can hold only when $\mathbf{a} = \lambda\mathbf{b}$ for some scalar λ.

Hint: If $\mathbf{a} \neq \lambda\mathbf{b}$ then there exists a covector $\mathbf{f}^*$ such that $\mathbf{f}^*(\mathbf{a}) = 1$ and $\mathbf{f}^*(\mathbf{b}) = 0$. Define the map $\mathbf{f}^* : V \otimes V \to V$ as $\mathbf{f}^*(\mathbf{x} \otimes \mathbf{y}) = \mathbf{f}^*(\mathbf{x})\mathbf{y}$. Compute

$$\mathbf{f}^*(\mathbf{a} \otimes \mathbf{a} + \mathbf{b} \otimes \mathbf{b}) = \mathbf{a} = \mathbf{f}^*(\mathbf{v})\mathbf{w},$$

hence $\mathbf{w}$ is proportional to $\mathbf{a}$. Similarly you can show that $\mathbf{w}$ is proportional to $\mathbf{b}$.

1.9 Index notation for tensors

So far we have used a purely coordinate-free formalism to define and describe tensors from spaces such as $V \otimes V^*$. However, in many calculations a basis in V is fixed, and one needs to compute the components of tensors in that basis. Also, the coordinate-free notation becomes cumbersome for computations in higher-rank tensor spaces such as $V \otimes V \otimes V^*$ because there is no direct means of referring to an individual component in the tensor product. The **index notation** makes such calculations easier.

Suppose a basis $\{\mathbf{e}_1, ..., \mathbf{e}_N\}$ in V is fixed; then the dual basis $\{\mathbf{e}_k^*\}$ is also fixed. Any vector $\mathbf{v} \in V$ is decomposed as $\mathbf{v} = \sum_k v_k \mathbf{e}_k$ and any covector as $\mathbf{f}^* = \sum_k f_k \mathbf{e}_k^*$. Any tensor from $V \otimes V$ is decomposed as

$$A = \sum_{j,k} A_{jk} \mathbf{e}_j \otimes \mathbf{e}_k \in V \otimes V$$

and so on. The action of a covector on a vector is $\mathbf{f}^*(\mathbf{v}) = \sum_k f_k v_k$, and the action of an operator on a vector is $\sum_{j,k} A_{jk} v_k \mathbf{e}_k$. However, it is cumbersome to keep writing these sums. In the index notation, one writes *only* the components v_k or A_{jk} of vectors and tensors.

1.9.1 Definition of index notation

The rules are as follows:

- Basis vectors $\mathbf{e}_k$ and basis tensors $\mathbf{e}_k \otimes \mathbf{e}_l^*$ are never written explicitly. (It is assumed that the basis is fixed and known.)
- Instead of a vector $\mathbf{v} \in V$, one writes its array of components v^k with the *superscript* index. Covectors $\mathbf{f}^* \in V^*$ are written f_k with the *subscript* index. The index k runs over integers from 1 to N. Components of vectors and tensors may be thought of as numbers (e.g. elements of the number field $\mathbb{K}$).
- Tensors are written as multidimensional arrays of components with superscript or subscript indices as necessary, for example $A_{jk} \in V^* \otimes V^*$ or $B_k^{lm} \in V \otimes V \otimes V^*$. Thus e.g. the Kronecker delta symbol is written as δ_k^j when it represents the identity operator $\hat{1}_V$.
- The choice of indices must be consistent; each index corresponds to a particular copy of V or V^*. Thus it is wrong to write $v_j = u_k$ or $v_i + u^i = 0$. Correct equations are $v_j = u_j$ and $v^i + u^i = 0$. This disallows meaningless expressions such as $\mathbf{v}^* + \mathbf{u}$ (one cannot add vectors from different spaces).
- Sums over indices such as $\sum_{k=1}^N a_k b_k$ are not written explicitly, the $\sum$ symbol is omitted, and the **Einstein summation convention** is used instead: Summation over all values of an index is *always implied* when that index letter appears once as a subscript and once as a superscript.

In this case the letter is called a **dummy** (or **mute**) **index**. Thus one writes $f_k v^k$ instead of $\sum_k f_k v_k$ and $A^j_k v^k$ instead of $\sum_k A_{jk} v_k$.

- Summation is allowed *only* over one subscript and one superscript but never over two subscripts or two superscripts and never over three or more coincident indices. This corresponds to requiring that we are only allowed to compute the canonical pairing of V and V^* [see Eq. (1.15)] but no other pairing. The expression $v^k v^k$ is not allowed because there is no canonical pairing of V and V, so, for instance, the sum $\sum_{k=1}^N v^k v^k$ depends on the choice of the basis. For the same reason (dependence on the basis), expressions such as $u^i v^i w^i$ or $A_{ii} B^{ii}$ are not allowed. Correct expressions are $u_i v^i w_k$ and $A_{ik} B^{ik}$.

- One needs to pay close attention to the choice and the position of the letters such as j, k, l,... used as indices. Indices that are not repeated are **free** indices. The rank of a tensor expression is equal to the number of free subscript and superscript indices. Thus $A^j_k v^k$ is a rank 1 tensor (i.e. a vector) because the expression $A^j_k v^k$ has a single free index, j, and a summation over k is implied.

- The tensor product symbol $\otimes$ is never written. For example, if $\mathbf{v} \otimes \mathbf{f}^* = \sum_{jk} v_j f^*_k \mathbf{e}_j \otimes \mathbf{e}^*_k$, one writes $v^k f_j$ to represent the tensor $\mathbf{v} \otimes \mathbf{f}^*$. The index letters in the expression $v^k f_j$ are intentionally chosen to be *different* (in this case, k and j) so that no summation would be implied. In other words, a tensor product is written simply as a product of components, and the index letters are chosen appropriately. Then one can interpret $v^k f_j$ as simply the product of *numbers*. In particular, it makes no difference whether one writes $f_j v^k$ or $v^k f_j$. The *position of the indices* (rather than the ordering of vectors) shows in every case how the tensor product is formed. Note that it is not possible to distinguish $V \otimes V^*$ from $V^* \otimes V$ in the index notation.

Example 1: It follows from the definition of δ^i_j that $\delta^i_j v^j = v^i$. This is the index representation of $\hat{1}\mathbf{v} = \mathbf{v}$.

Example 2: Suppose $\mathbf{w}, \mathbf{x}, \mathbf{y}$, and $\mathbf{z}$ are vectors from V whose components are w^i, x^i, y^i, z^i. What are the components of the tensor $\mathbf{w} \otimes \mathbf{x} + 2\mathbf{y} \otimes \mathbf{z} \in V \otimes V$?

Answer: $w^i x^k + 2y^i z^k$. (We need to choose another letter for the second free index, k, which corresponds to the second copy of V in $V \otimes V$.)

Example 3: The operator $\hat{A} \equiv \hat{1}_V + \lambda \mathbf{v} \otimes \mathbf{u}^* \in V \otimes V^*$ acts on a vector $\mathbf{x} \in V$. Calculate the resulting vector $\mathbf{y} \equiv \hat{A}\mathbf{x}$.

In the index-free notation, the calculation is

$$\mathbf{y} = \hat{A}\mathbf{x} = \left(\hat{1}_V + \lambda \mathbf{v} \otimes \mathbf{u}^*\right) \mathbf{x} = \mathbf{x} + \lambda \mathbf{u}^* (\mathbf{x})\, \mathbf{v}.$$

In the index notation, the calculation looks like this:

$$y^k = \left(\delta^k_j + \lambda v^k u_j\right) x^j = x^k + \lambda v^k u_j x^j.$$

In this formula, j is a dummy index and k is a free index. We could have also written $\lambda x^j v^k u_j$ instead of $\lambda v^k u_j x^j$ since the ordering of components makes no difference in the index notation.

Exercise: In a physics book you find the following formula,

$$H^{\alpha}_{\mu\nu} = \frac{1}{2}\left(h_{\beta\mu\nu} + h_{\beta\nu\mu} - h_{\mu\nu\beta}\right) g^{\alpha\beta}.$$

To what spaces do the tensors H, g, h belong (assuming these quantities represent tensors)? Rewrite this formula in the coordinate-free notation.

Answer: $H \in V \otimes V^* \otimes V^*$, $h \in V^* \otimes V^* \otimes V^*$, $g \in V \otimes V$. Assuming the simplest case,

$$h = \mathbf{h}_1^* \otimes \mathbf{h}_2^* \otimes \mathbf{h}_3^*,\ g = \mathbf{g}_1 \otimes \mathbf{g}_2,$$

the coordinate-free formula is

$$H = \frac{1}{2}\mathbf{g}_1 \otimes \left(\mathbf{h}_1^*(\mathbf{g}_2)\,\mathbf{h}_2^* \otimes \mathbf{h}_3^* + \mathbf{h}_1^*(\mathbf{g}_2)\,\mathbf{h}_3^* \otimes \mathbf{h}_2^* - \mathbf{h}_3^*(\mathbf{g}_2)\,\mathbf{h}_1^* \otimes \mathbf{h}_2^*\right).$$

Question: I would like to decompose a vector $\mathbf{v}$ in the basis $\{\mathbf{e}_j\}$ using the index notation, $\mathbf{v} = v^j \mathbf{e}_j$. Is it okay to write the *lower* index j on the basis vectors $\mathbf{e}_j$? I also want to write $v^j = \mathbf{e}_j^*(\mathbf{v})$ using the dual basis $\{\mathbf{e}_j^*\}$, but then the index j is not correctly matched at both sides.

Answer: The index notation is designed so that you never use the basis vectors $\mathbf{e}_j$ or $\mathbf{e}_j^*$ — you only use components such as v^j or f_j. The only way to keep the upper and the lower indices consistent (i.e. having the summation always over one upper and one lower index) when you want to use both the components v^j and the basis vectors $\mathbf{e}_j$ is to use *upper* indices on the dual basis, i.e. writing $\{\mathbf{e}^{*j}\}$. Then a covector will have components with lower indices, $\mathbf{f}^* = f_j \mathbf{e}^{*j}$, and the index notation remains consistent. A further problem occurs when you have a scalar product and you would like to express the component v^j as $v^j = \langle \mathbf{v}, \mathbf{e}_j \rangle$. In this case, the only way to keep the notation consistent is to use explicitly a suitable matrix, say g^{ij}, in order to represent the scalar product. Then one would be able to write $v^j = g^{jk}\langle \mathbf{v}, \mathbf{e}_k \rangle$ and keep the index notation consistent.

1.9.2 Advantages and disadvantages of index notation

Index notation is conceptually easier than the index-free notation because one can imagine manipulating "merely" some tables of numbers, rather than "abstract vectors." In other words, we are working with less abstract objects. The price is that we obscure the geometric interpretation of what we are doing, and proofs of general theorems become more difficult to understand.

The main advantage of the index notation is that it makes computations with complicated tensors quicker. Consider, for example, the space $V \otimes V \otimes V^* \otimes V^*$ whose elements can be interpreted as operators from $\mathrm{Hom}\,(V \otimes V, V \otimes V)$. The action of such an operator on a tensor $a^{jk} \in V \otimes V$ is expressed in the index notation as

$$b^{lm} = A^{lm}_{jk} a^{jk},$$

where a^{lm} and b^{lm} represent tensors from $V \otimes V$ and A^{lm}_{jk} is a tensor from $V \otimes V \otimes V^* \otimes V^*$, while the summation over the indices j and k is implied. Each index letter refers unambiguously to one tensor product factor. Note that the formula

$$b^{lm} = A^{lm}_{kj} a^{jk}$$

describes another (*inequivalent*) way to define the isomorphism between the spaces $V \otimes V \otimes V^* \otimes V^*$ and $\mathrm{Hom}\,(V \otimes V, V \otimes V)$. The index notation expresses this difference in a concise way; of course, one needs to pay close attention to the position and the order of indices.

Note that in the coordinate-free notation it is much more cumbersome to describe and manipulate such tensors. Without the index notation, it is cumbersome to perform calculations with a tensor such as

$$B^{ik}_{jl} \equiv \delta^i_j \delta^k_l - \delta^k_j \delta^i_l \in V \otimes V \otimes V^* \otimes V^*$$

which acts as an operator in $V \otimes V$, exchanging the two vector factors:

$$\left(\delta^i_j \delta^k_l - \delta^k_j \delta^i_l\right) a^{jl} = a^{ik} - a^{ki}.$$

The index-free definition of this operator is simple with single-term tensor products,

$$\hat{B}\,(\mathbf{u} \otimes \mathbf{v}) \equiv \mathbf{u} \otimes \mathbf{v} - \mathbf{v} \otimes \mathbf{u}.$$

Having defined $\hat{B}$ on single-term tensor products, we require linearity and so define the operator $\hat{B}$ on the entire space $V \otimes V$. However, practical calculations are cumbersome if we are applying $\hat{B}$ to a complicated tensor $X \in V \otimes V$ rather than to a single-term product $\mathbf{u} \otimes \mathbf{v}$, because, in particular, we are obliged to decompose X into single-term tensor products in order to perform such a calculation.

Some *disadvantages* of the index notation are as follows: (1) If the basis is changed, all components need to be recomputed. In textbooks that use the index notation, quite some time is spent studying the transformation laws of tensor components under a change of basis. If different bases are used simultaneously, confusion may result as to which basis is implied in a particular formula. (2) If we are using unrelated vector spaces V and W, we need to choose a basis in each of them and always remember which index belongs to which space. The index notation does not show this explicitly. To alleviate this problem, one may use e.g. Greek and Latin indices to distinguish different spaces, but this is not always convenient or sufficient. (3) The geometrical meaning of many calculations appears hidden behind a mass of indices. It is sometimes unclear whether a long expression with indices can be simplified and how to proceed with calculations. (Do we need to try all possible relabellings of indices and see what happens?)

Despite these disadvantages, the index notation enables one to perform practical calculations with high-rank tensor spaces, such as those required in field theory and in general relativity. For this reason, and also for historical reasons (Einstein used the index notation when developing the theory of relativity), most physics textbooks use the index notation. In some cases, calculations can be performed equally quickly using index and index-free notations. In other cases, especially when deriving general properties of tensors, the index-free notation is superior.[4] I use the index-free notation in this book because calculations in coordinates are not essential for this book's central topics. However, I will occasionally show how to do some calculations also in the index notation.

1.10 Dirac notation for vectors and covectors

The Dirac notation was developed for quantum mechanics where one needs to perform many computations with operators, vectors and covectors (but *not* with higher-rank tensors!). The Dirac notation is index-free.

1.10.1 Definition of Dirac notation

The rules are as follows:

- One writes the symbol $|v\rangle$ for a vector $\mathbf{v} \in V$ and $\langle f|$ for a covector $\mathbf{f}^* \in V^*$. The labels inside the special brackets $|\,\rangle$ and $\langle\,|$ are chosen according to the problem at hand, e.g. one can denote specific vectors by $|0\rangle$, $|1\rangle$, $|x\rangle$, $|v_1\rangle$, or even $\langle {}^{(0)}\tilde{a}_{ij}; l, m|$ if that helps. (Note that $|0\rangle$ is normally *not* the zero vector; the latter is denoted simply by 0, as usual.)

- Linear combinations of vectors are written like this: $2\,|v\rangle - 3\,|u\rangle$ instead of $2\mathbf{v} - 3\mathbf{u}$.

- The action of a covector on a vector is written as $\langle f|v\rangle$; the result is a number. The mnemonic for this is "bra-ket", so $\langle f|$ is a "bra vector" and $|v\rangle$ is a "ket vector." The action of an operator $\hat{A}$ on a vector $|v\rangle$ is written $\hat{A}\,|v\rangle$.

- The action of the transposed operator $\hat{A}^T$ on a covector $\langle f|$ is written $\langle f|\,\hat{A}$. Note that the transposition label (^{T}) is *not* used. This is consistent within the Dirac notation: The covector $\langle f|\,\hat{A}$ acts on a vector $|v\rangle$ as $\langle f|\,\hat{A}\,|v\rangle$, which is the same (by definition of $\hat{A}^T$) as the covector $\langle f|$ acting on $\hat{A}\,|v\rangle$.

- The tensor product symbol $\otimes$ is omitted. Instead of $\mathbf{v} \otimes \mathbf{f}^* \in V \otimes V^*$ or $\mathbf{a} \otimes \mathbf{b} \in V \otimes V$, one writes $|v\rangle\,\langle f|$ and $|a\rangle\,|b\rangle$ respectively. The tensor space to which a tensor belongs will be clear from the notation or from explanations in the text. Note that one cannot write $\mathbf{f}^* \otimes \mathbf{v}$ as $\langle f|\,|v\rangle$ since $\langle f|\,|v\rangle$ already means $\mathbf{f}^*(\mathbf{v})$ in the Dirac notation. Instead, one always writes $|v\rangle\,\langle f|$ and does not distinguish between $\mathbf{f}^* \otimes \mathbf{v}$ and $\mathbf{v} \otimes \mathbf{f}^*$.

[4] I have developed an advanced textbook on general relativity entirely in the index-free notation and displayed the infrequent cases where the index notation is easier to use.

Example 1: The action of an operator $\mathbf{a} \otimes \mathbf{b}^* \in V \otimes V^*$ on a vector $\mathbf{v} \in V$ has been defined by $(\mathbf{a} \otimes \mathbf{b}^*)\,\mathbf{v} = \mathbf{b}^*(\mathbf{v})\,\mathbf{a}$. In the Dirac notation, this is very easy to express: one acts with $|a\rangle\,\langle b|$ on a vector $|v\rangle$ by writing

$$(|a\rangle\,\langle b|)\,|v\rangle = |a\rangle\,\langle b|\,|v\rangle = |a\rangle\,\langle b|v\rangle\,.$$

In other words, we mentally remove one vertical line and get the vector $|a\rangle$ times the number $\langle b|v\rangle$. This is entirely consistent with the definition of the operator $\mathbf{a} \otimes \mathbf{b}^* \in \operatorname{End} V$.

Example 2: The action of $\hat{A} \equiv \hat{1}_V + \frac{1}{2}\mathbf{v} \otimes \mathbf{u}^* \in V \otimes V^*$ on a vector $\mathbf{x} \in V$ is written as follows:

$$\begin{aligned} |y\rangle &= \hat{A}\,|x\rangle = \left(\hat{1} + \tfrac{1}{2}\,|v\rangle\,\langle u|\right)|x\rangle = |x\rangle + \tfrac{1}{2}\,|v\rangle\,\langle u|\,|x\rangle \\ &= |x\rangle + \frac{\langle u|x\rangle}{2}\,|v\rangle\,. \end{aligned}$$

Note that we have again "simplified" $\langle u|\,|x\rangle$ to $\langle u|x\rangle$, and the result is correct. Compare this notation with the same calculation written in the index-free notation:

$$\mathbf{y} = \hat{A}\mathbf{x} = \left(\hat{1} + \tfrac{1}{2}\mathbf{v} \otimes \mathbf{u}^*\right)\mathbf{x} = \mathbf{x} + \frac{\mathbf{u}^*(\mathbf{x})}{2}\mathbf{v}.$$

Example 3: If $|e_1\rangle$, ..., $|e_N\rangle$ is a basis, we denote by $\langle e_k|$ the covectors from the dual basis, so that $\langle e_j|e_k\rangle = \delta_{jk}$. A vector $|v\rangle$ is expressed through the basis vectors as

$$|v\rangle = \sum_k v_k\,|e_k\rangle\,,$$

where the coefficients v_k can be computed as $v_k = \langle e_k|v\rangle$. An arbitrary operator $\hat{A}$ is decomposed as

$$\hat{A} = \sum_{j,k} A_{jk}\,|e_j\rangle\,\langle e_k|\,.$$

The **matrix elements** A_{jk} of the operator $\hat{A}$ in this basis are found as

$$A_{jk} = \langle e_j|\,\hat{A}\,|e_k\rangle\,.$$

The identity operator is decomposed as follows,

$$\hat{1} = \sum_k |e_k\rangle\,\langle e_k|\,.$$

Expressions of this sort abound in quantum mechanics textbooks.

1.10.2 Advantages and disadvantages of Dirac notation

The Dirac notation is convenient when many calculations with vectors and covectors are required. But calculations become cumbersome if we need many tensor powers. For example, suppose we would like to apply a covector $\langle f|$ to the *second* vector in the tensor product $|a\rangle\,|b\rangle\,|c\rangle$, so that the answer is $|a\rangle\,\langle f\,|b\rangle\,|c\rangle$. Now one cannot simply write $\langle f|\,X$ with $X = |a\rangle\,|b\rangle\,|c\rangle$ because $\langle f|\,X$ is ambiguous in this case. The desired kind of action of covectors on tensors is difficult to express using the Dirac notation. Only the index notation allows one to write and to carry out arbitrary operations with this kind of tensor product. In the example just mentioned, one writes $f_j a^i b^j c^k$ to indicate that the covector f_j acts on the vector b^j but not on the other vectors. Of course, the resulting expression is harder to read because one needs to pay close attention to every index.

2 Exterior product

In this chapter I introduce one of the most useful constructions in basic linear algebra — the exterior product, denoted by $\mathbf{a} \wedge \mathbf{b}$, where $\mathbf{a}$ and $\mathbf{b}$ are vectors from a space V. The basic idea of the exterior product is that we would like to define an *antisymmetric* and bilinear product of vectors. In other words, we would like to have the properties $\mathbf{a}\wedge\mathbf{b} = -\mathbf{b}\wedge\mathbf{a}$ and $\mathbf{a}\wedge(\mathbf{b}+\lambda\mathbf{c}) = \mathbf{a}\wedge\mathbf{b}+\lambda\mathbf{a}\wedge\mathbf{c}$.

2.1 Motivation

Here I discuss, at some length, the motivation for introducing the exterior product. The motivation is geometrical and comes from considering the properties of areas and volumes in the framework of elementary Euclidean geometry. I will proceed with a formal definition of the exterior product in Sec. 2.2. In order to understand the definition explained there, it is not necessary to use this geometric motivation because the definition will be purely algebraic. Nevertheless, I feel that this motivation will be helpful for some readers.

2.1.1 Two-dimensional oriented area

We work in a two-dimensional Euclidean space, such as that considered in elementary geometry. We assume that the usual geometrical definition of the area of a parallelogram is known.

Consider the area $Ar(\mathbf{a}, \mathbf{b})$ of a parallelogram spanned by vectors $\mathbf{a}$ and $\mathbf{b}$. It is known from elementary geometry that $Ar(\mathbf{a}, \mathbf{b}) = |\mathbf{a}| \cdot |\mathbf{b}| \cdot \sin\alpha$ where α is the angle between the two vectors, which is always between 0 and π (we do not take into account the orientation of this angle). Thus defined, the area Ar is always non-negative.

Let us investigate $Ar(\mathbf{a}, \mathbf{b})$ as a function of the vectors $\mathbf{a}$ and $\mathbf{b}$. If we stretch the vector $\mathbf{a}$, say, by factor 2, the area is also increased by factor 2. However, if we multiply $\mathbf{a}$ by the number -2, the area will be multiplied by 2 rather than by -2:

$$Ar(\mathbf{a}, 2\mathbf{b}) = Ar(\mathbf{a}, -2\mathbf{b}) = 2Ar(\mathbf{a}, \mathbf{b}).$$

Similarly, for some vectors $\mathbf{a}, \mathbf{b}, \mathbf{c}$ such as shown in Fig. 2.2, we have $Ar(\mathbf{a}, \mathbf{b}+\mathbf{c}) = Ar(\mathbf{a}, \mathbf{b}) + Ar(\mathbf{a}, \mathbf{c})$. However, if we consider $\mathbf{b} = -\mathbf{c}$ then we obtain

$$\begin{aligned} Ar(\mathbf{a}, \mathbf{b}+\mathbf{c}) &= Ar(\mathbf{a}, 0) = 0 \\ &\neq Ar(\mathbf{a}, \mathbf{b}) + Ar(\mathbf{a}, -\mathbf{b}) = 2Ar(\mathbf{a}, \mathbf{b}). \end{aligned}$$

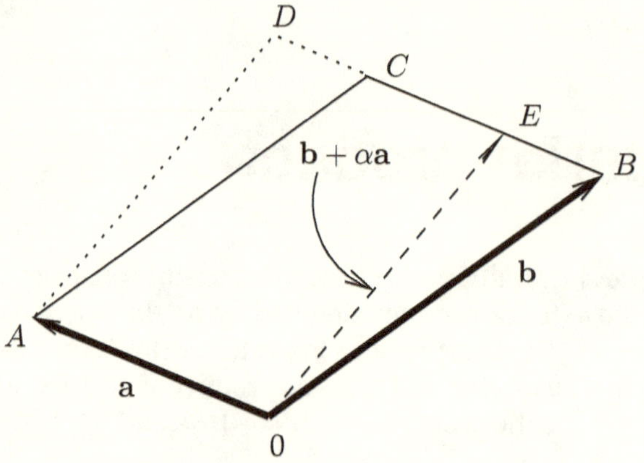

Figure 2.1: The area of the parallelogram $0ACB$ spanned by $\mathbf{a}$ and $\mathbf{b}$ is equal to the area of the parallelogram $0ADE$ spanned by $\mathbf{a}$ and $\mathbf{b} + \alpha\mathbf{a}$ due to the equality of areas ACD and $0BE$.

Hence, the area $Ar(\mathbf{a}, \mathbf{b})$ is, strictly speaking, *not* a linear function of the vectors $\mathbf{a}$ and $\mathbf{b}$:

$$Ar(\lambda\mathbf{a}, \mathbf{b}) = |\lambda|\, Ar(\mathbf{a}, \mathbf{b}) \neq \lambda\, Ar(\mathbf{a}, \mathbf{b}),$$
$$Ar(\mathbf{a}, \mathbf{b} + \mathbf{c}) \neq Ar(\mathbf{a}, \mathbf{b}) + Ar(\mathbf{a}, \mathbf{c}).$$

Nevertheless, as we have seen, the properties of linearity hold in *some* cases. If we look closely at those cases, we find that linearly holds precisely when we do not change the orientation of the vectors. It would be more convenient if the linearity properties held in all cases.

The trick is to replace the area function Ar with the **oriented area** function $A(\mathbf{a}, \mathbf{b})$. Namely, we define the function $A(\mathbf{a}, \mathbf{b})$ by

$$A(\mathbf{a}, \mathbf{b}) = \pm\, |\mathbf{a}| \cdot |\mathbf{b}| \cdot \sin\alpha,$$

where the sign is chosen positive when the angle α is measured from the vector $\mathbf{a}$ to the vector $\mathbf{b}$ in the counterclockwise direction, and negative otherwise.

Statement: The oriented area $A(\mathbf{a}, \mathbf{b})$ of a parallelogram spanned by the vectors $\mathbf{a}$ and $\mathbf{b}$ in the two-dimensional Euclidean space is an antisymmetric and bilinear function of the vectors $\mathbf{a}$ and $\mathbf{b}$:

$$A(\mathbf{a}, \mathbf{b}) = -A(\mathbf{b}, \mathbf{a}),$$
$$A(\lambda\mathbf{a}, \mathbf{b}) = \lambda\, A(\mathbf{a}, \mathbf{b}),$$
$$A(\mathbf{a}, \mathbf{b} + \mathbf{c}) = A(\mathbf{a}, \mathbf{b}) + A(\mathbf{a}, \mathbf{c}). \qquad \text{(the sum law)}$$

Proof: The first property is a straightforward consequence of the sign rule in the definition of A.

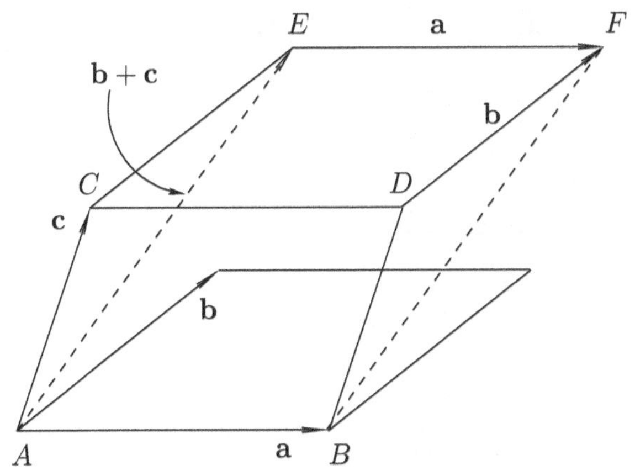

Figure 2.2: The area of the parallelogram spanned by **a** and **b** (equal to the area of $CEFD$) plus the area of the parallelogram spanned by **a** and **c** (the area of $ACDB$) equals the area of the parallelogram spanned by **a** and **b**+**c** (the area of $AEFB$) because of the equality of the areas of ACE and BDF.

Proving the second property requires considering the cases $\lambda > 0$ and $\lambda < 0$ separately. If $\lambda > 0$ then the orientation of the pair $(\mathbf{a}, \mathbf{b})$ remains the same and then it is clear that the property holds: When we rescale **a** by λ, the parallelogram is stretched and its area increases by factor λ. If $\lambda < 0$ then the orientation of the parallelogram is reversed and the oriented area changes sign.

To prove the sum law, we consider two cases: either **c** is parallel to **a** or it is not. If **c** is parallel to **a**, say $\mathbf{c} = \alpha\mathbf{a}$, we use Fig. 2.1 to show that $A(\mathbf{a}, \mathbf{b}+\lambda\mathbf{a}) = A(\mathbf{a}, \mathbf{b})$, which yields the desired statement since $A(\mathbf{a}, \lambda\mathbf{a}) = 0$. If **c** is not parallel to **a**, we use Fig. 2.2 to show that $A(\mathbf{a}, \mathbf{b}+\mathbf{c}) = A(\mathbf{a}, \mathbf{b}) + A(\mathbf{a}, \mathbf{c})$. Analogous geometric constructions can be made for different possible orientations of the vectors **a**, **b**, **c**. ■

It is relatively easy to compute the oriented area because of its algebraic properties. Suppose the vectors **a** and **b** are given through their components in a standard basis $\{\mathbf{e}_1, \mathbf{e}_2\}$, for instance

$$\mathbf{a} = \alpha_1\mathbf{e}_1 + \alpha_2\mathbf{e}_2, \quad \mathbf{b} = \beta_1\mathbf{e}_1 + \beta_2\mathbf{e}_2.$$

We assume, of course, that the vectors $\mathbf{e}_1$ and $\mathbf{e}_2$ are orthogonal to each other and have unit length, as is appropriate in a Euclidean space. We also assume that the right angle is measured from $\mathbf{e}_1$ to $\mathbf{e}_2$ in the counter-clockwise direction, so that $A(\mathbf{e}_1, \mathbf{e}_2) = +1$. Then we use the Statement and the properties $A(\mathbf{e}_1, \mathbf{e}_1) = 0$, $A(\mathbf{e}_1, \mathbf{e}_2) = 1$, $A(\mathbf{e}_2, \mathbf{e}_2) = 0$ to compute

$$\begin{aligned} A(\mathbf{a}, \mathbf{b}) &= A(\alpha_1\mathbf{e}_1 + \alpha_2\mathbf{e}_2, \beta_1\mathbf{e}_1 + \beta_2\mathbf{e}_2) \\ &= \alpha_1\beta_2 A(\mathbf{e}_1, \mathbf{e}_2) + \alpha_2\beta_1 A(\mathbf{e}_2, \mathbf{e}_1) \\ &= \alpha_1\beta_2 - \alpha_2\beta_1. \end{aligned}$$

The ordinary (unoriented) area is then obtained as the absolute value of the oriented area, $Ar(\mathbf{a},\mathbf{b}) = |A(\mathbf{a},\mathbf{b})|$. It turns out that the oriented area, due to its strict linearity properties, is a much more convenient and powerful construction than the unoriented area.

2.1.2 Parallelograms in $\mathbb{R}^3$ and in $\mathbb{R}^n$

Let us now work in the Euclidean space $\mathbb{R}^3$ with a standard basis $\{\mathbf{e}_1, \mathbf{e}_2, \mathbf{e}_3\}$. We can similarly try to characterize the area of a parallelogram spanned by two vectors $\mathbf{a}$, $\mathbf{b}$. It is, however, not possible to characterize the orientation of the area simply by a sign. We also cannot use a geometric construction such as that in Fig. 2.2; in fact it is *not true* in three dimensions that the area spanned by $\mathbf{a}$ and $\mathbf{b}+\mathbf{c}$ is equal to the sum of $Ar(\mathbf{a},\mathbf{b})$ and $Ar(\mathbf{a},\mathbf{c})$. Can we still define some kind of "oriented area" that obeys the sum law?

Let us consider Fig. 2.2 as a figure showing the *projection* of the areas of the three parallelograms onto some coordinate plane, say, the plane of the basis vectors $\{\mathbf{e}_1, \mathbf{e}_2\}$. It is straightforward to see that the projections of the areas obey the sum law as oriented areas.

Statement: Let $\mathbf{a}$, $\mathbf{b}$ be two vectors in $\mathbb{R}^3$, and let $P(\mathbf{a},\mathbf{b})$ be the parallelogram spanned by these vectors. Denote by $P(\mathbf{a},\mathbf{b})_{\mathbf{e}_1,\mathbf{e}_2}$ the parallelogram within the coordinate plane Span $\{\mathbf{e}_1, \mathbf{e}_2\}$ obtained by projecting $P(\mathbf{a},\mathbf{b})$ onto that coordinate plane, and similarly for the other two coordinate planes. Denote by $A(\mathbf{a},\mathbf{b})_{\mathbf{e}_1,\mathbf{e}_2}$ the oriented area of $P(\mathbf{a},\mathbf{b})_{\mathbf{e}_1,\mathbf{e}_2}$. Then $A(\mathbf{a},\mathbf{b})_{\mathbf{e}_1,\mathbf{e}_2}$ is a bilinear, antisymmetric function of $\mathbf{a}$ and $\mathbf{b}$.

Proof: The projection onto the coordinate plane of $\mathbf{e}_1, \mathbf{e}_2$ is a linear transformation. Hence, the vector $\mathbf{a}+\lambda\mathbf{b}$ is projected onto the sum of the projections of $\mathbf{a}$ and $\lambda\mathbf{b}$. Then we apply the arguments in the proof of Statement 2.1.1 to the *projections* of the vectors; in particular, Figs. 2.1 and 2.2 are interpreted as showing the projections of all vectors onto the coordinate plane $\mathbf{e}_1, \mathbf{e}_2$. It is then straightforward to see that all the properties of the oriented area hold for the projected oriented areas. Details left as exercise. ■

It is therefore convenient to consider the oriented areas of the three projections — $A(\mathbf{a},\mathbf{b})_{\mathbf{e}_1,\mathbf{e}_2}$, $A(\mathbf{a},\mathbf{b})_{\mathbf{e}_2,\mathbf{e}_3}$, $A(\mathbf{a},\mathbf{b})_{\mathbf{e}_3,\mathbf{e}_1}$ — as three components of a *vector-valued* area $A(\mathbf{a},\mathbf{b})$ of the parallelogram spanned by $\mathbf{a}$, $\mathbf{b}$. Indeed, it can be shown that these three projected areas coincide with the three Euclidean components of the vector product $\mathbf{a}\times\mathbf{b}$. The vector product is the traditional way such areas are represented in geometry: the vector $\mathbf{a}\times\mathbf{b}$ represents at once the magnitude of the area and the orientation of the parallelogram. One computes the unoriented area of a parallelogram as the length of the vector $\mathbf{a}\times\mathbf{b}$ representing the oriented area,

$$Ar(\mathbf{a},\mathbf{b}) = \left[A(\mathbf{a},\mathbf{b})^2_{\mathbf{e}_1,\mathbf{e}_2} + A(\mathbf{a},\mathbf{b})^2_{\mathbf{e}_2,\mathbf{e}_3} + A(\mathbf{a},\mathbf{b})^2_{\mathbf{e}_3,\mathbf{e}_1}\right]^{\frac{1}{2}}.$$

However, the vector product cannot be generalized to all higher-dimensional spaces. Luckily, the vector product does not play an essential role in the construction of the oriented area.

Instead of working with the vector product, we will generalize the idea of projecting the parallelogram onto coordinate planes. Consider a parallelogram spanned by vectors $\mathbf{a}, \mathbf{b}$ in an n-dimensional Euclidean space V with the standard basis $\{\mathbf{e}_1, ..., \mathbf{e}_n\}$. While in three-dimensional space we had just three projections (onto the coordinate planes xy, xz, yz), in an n-dimensional space we have $\frac{1}{2}n(n-1)$ coordinate planes, which can be denoted by $\text{Span}\{\mathbf{e}_i, \mathbf{e}_j\}$ (with $1 \leq i < j \leq n$). We may construct the $\frac{1}{2}n(n-1)$ projections of the parallelogram onto these coordinate planes. Each of these projections has an oriented area; that area is a bilinear, antisymmetric number-valued function of the vectors $\mathbf{a}, \mathbf{b}$. (The proof of the Statement above does not use the fact that the space is *three*-dimensional!) We may then regard these $\frac{1}{2}n(n-1)$ numbers as the components of a vector representing the oriented area of the parallelogram. It is clear that all these components are needed in order to describe the actual geometric *orientation* of the parallelogram in the n-dimensional space.

We arrived at the idea that the oriented area of the parallelogram spanned by $\mathbf{a}, \mathbf{b}$ is an antisymmetric, bilinear function $A(\mathbf{a}, \mathbf{b})$ whose value is a vector with $\frac{1}{2}n(n-1)$ components, i.e. a vector *in a new space* — the "space of oriented areas," as it were. This space is $\frac{1}{2}n(n-1)$-dimensional. We will construct this space explicitly below; it is the space of bivectors, to be denoted by $\wedge^2 V$.

We will see that the unoriented area of the parallelogram is computed as the *length* of the vector $A(\mathbf{a}, \mathbf{b})$, i.e. as the square root of the sum of squares of the areas of the projections of the parallelogram onto the coordinate planes. This is a generalization of the Pythagoras theorem to areas in higher-dimensional spaces.

The analogy between ordinary vectors and vector-valued areas can be understood visually as follows. A straight line segment in an n-dimensional space is represented by a vector whose n components (in an orthonormal basis) are the signed lengths of the n projections of the line segment onto the coordinate axes. (The components are *signed,* or *oriented,* i.e. taken with a negative sign if the orientation of the vector is opposite to the orientation of the axis.) The length of a straight line segment, i.e. the length of the vector $\mathbf{v}$, is then computed as $\sqrt{\langle \mathbf{v}, \mathbf{v} \rangle}$. The scalar product $\langle \mathbf{v}, \mathbf{v} \rangle$ is equal to the sum of squared lengths of the projections because we are using an orthonormal basis. A parallelogram in space is represented by a vector ψ whose $\binom{n}{2}$ components are the *oriented* areas of the $\binom{n}{2}$ projections of the parallelogram onto the coordinate planes. (The vector ψ belongs to the space of oriented areas, not to the original n-dimensional space.) The numerical value of the area of the parallelogram is then computed as $\sqrt{\langle \psi, \psi \rangle}$. The scalar product $\langle \psi, \psi \rangle$ in the space of oriented areas is equal to the sum of squared areas of the projections because the $\binom{n}{2}$ unit areas in the coordinate planes are an orthonormal basis (according to the definition of the scalar product in the space of oriented areas).

The generalization of the Pythagoras theorem holds not only for areas but also for higher-dimensional volumes. A general proof of this theorem will be given in Sec. 5.5.2, using the exterior product and several other constructions to be developed below.

2.2 Exterior product

In the previous section I motivated the introduction of the antisymmetric product by showing its connection to areas and volumes. In this section I will give the definition and work out the properties of the exterior product in a purely algebraic manner, without using any geometric intuition. This will enable us to work with vectors in arbitrary dimensions, to obtain many useful results, and eventually also to appreciate more fully the geometric significance of the exterior product.

As explained in Sec. 2.1.2, it is possible to represent the oriented area of a parallelogram by a vector in some auxiliary space. The oriented area is much more convenient to work with because it is a *bilinear* function of the vectors $\mathbf{a}$ and $\mathbf{b}$ (this is explained in detail in Sec. 2.1). "Product" is another word for "bilinear function." We have also seen that the oriented area is an *antisymmetric* function of the vectors $\mathbf{a}$ and $\mathbf{b}$.

In three dimensions, an oriented area is represented by the cross product $\mathbf{a} \times \mathbf{b}$, which is indeed an antisymmetric and bilinear product. So we expect that the oriented area in higher dimensions can be represented by some kind of new antisymmetric product of $\mathbf{a}$ and $\mathbf{b}$; let us denote this product (to be defined below) by $\mathbf{a} \wedge \mathbf{b}$, pronounced "a wedge b." The value of $\mathbf{a} \wedge \mathbf{b}$ will be a vector in a *new* vector space. We will also construct this new space explicitly.

2.2.1 Definition of exterior product

Like the tensor product space, the space of exterior products can be defined solely by its algebraic properties. We can consider the space of *formal expressions* like $\mathbf{a} \wedge \mathbf{b}$, $3\mathbf{a} \wedge \mathbf{b} + 2\mathbf{c} \wedge \mathbf{d}$, etc., and *require* the properties of an antisymmetric, bilinear product to hold.

Here is a more formal definition of the exterior product space: We will construct an antisymmetric product "by hand," using the tensor product space.

Definition 1: Given a vector space V, we define a new vector space $V \wedge V$ called the **exterior product** (or antisymmetric tensor product, or alternating product, or **wedge product**) of two copies of V. The space $V \wedge V$ is the subspace in $V \otimes V$ consisting of all **antisymmetric** tensors, i.e. tensors of the form

$$\mathbf{v}_1 \otimes \mathbf{v}_2 - \mathbf{v}_2 \otimes \mathbf{v}_1, \quad \mathbf{v}_{1,2} \in V,$$

and all linear combinations of such tensors. The exterior product of two vectors $\mathbf{v}_1$ and $\mathbf{v}_2$ is the expression shown above; it is obviously an antisymmetric and bilinear function of $\mathbf{v}_1$ and $\mathbf{v}_2$.

For example, here is one particular element from $V \wedge V$, which we write in two different ways using the axioms of the tensor product:

$$\begin{aligned}(\mathbf{u}+\mathbf{v}) \otimes (\mathbf{v}+\mathbf{w}) - (\mathbf{v}+\mathbf{w}) \otimes (\mathbf{u}+\mathbf{v}) &= \mathbf{u} \otimes \mathbf{v} - \mathbf{v} \otimes \mathbf{u} \\ &+ \mathbf{u} \otimes \mathbf{w} - \mathbf{w} \otimes \mathbf{u} + \mathbf{v} \otimes \mathbf{w} - \mathbf{w} \otimes \mathbf{v} \in V \wedge V. \end{aligned} \qquad (2.1)$$

Remark: A tensor $\mathbf{v}_1 \otimes \mathbf{v}_2 \in V \otimes V$ is not equal to the tensor $\mathbf{v}_2 \otimes \mathbf{v}_1$ if $\mathbf{v}_1 \neq \mathbf{v}_2$. This is so because there is no identity among the axioms of the tensor product that would allow us to exchange the factors $\mathbf{v}_1$ and $\mathbf{v}_2$ in the expression $\mathbf{v}_1 \otimes \mathbf{v}_2$.

Exercise 1: Prove that the "exchange map" $\hat{T}(\mathbf{v}_1 \otimes \mathbf{v}_2) \equiv \mathbf{v}_2 \otimes \mathbf{v}_1$ is a canonically defined, linear map of $V \otimes V$ into itself. Show that $\hat{T}$ has only two eigenvalues which are ± 1. Give examples of eigenvectors with eigenvalues $+1$ and -1. Show that the subspace $V \wedge V \subset V \otimes V$ is the eigenspace of the exchange operator $\hat{T}$ with eigenvalue -1

Hint: $\hat{T}\hat{T} = \hat{1}_{V \otimes V}$. Consider tensors of the form $\mathbf{u} \otimes \mathbf{v} \pm \mathbf{v} \otimes \mathbf{u}$ as candidate eigenvectors of $\hat{T}$. ∎

It is quite cumbersome to perform calculations in the tensor product notation as we did in Eq. (2.1). So let us write the exterior product as $\mathbf{u} \wedge \mathbf{v}$ instead of $\mathbf{u} \otimes \mathbf{v} - \mathbf{v} \otimes \mathbf{u}$. It is then straightforward to see that the "wedge" symbol $\wedge$ indeed works like an anti-commutative multiplication, as we intended. The rules of computation are summarized in the following statement.

Statement 1: One may save time and write $\mathbf{u} \otimes \mathbf{v} - \mathbf{v} \otimes \mathbf{u} \equiv \mathbf{u} \wedge \mathbf{v} \in V \wedge V$, and the result of any calculation will be correct, as long as one follows the rules:

$$\mathbf{u} \wedge \mathbf{v} = -\mathbf{v} \wedge \mathbf{u}, \tag{2.2}$$

$$(\lambda \mathbf{u}) \wedge \mathbf{v} = \lambda (\mathbf{u} \wedge \mathbf{v}), \tag{2.3}$$

$$(\mathbf{u} + \mathbf{v}) \wedge \mathbf{x} = \mathbf{u} \wedge \mathbf{x} + \mathbf{v} \wedge \mathbf{x}. \tag{2.4}$$

It follows also that $\mathbf{u} \wedge (\lambda \mathbf{v}) = \lambda (\mathbf{u} \wedge \mathbf{v})$ and that $\mathbf{v} \wedge \mathbf{v} = 0$. (These identities hold for any vectors $\mathbf{u}, \mathbf{v} \in V$ and any scalars $\lambda \in \mathbb{K}$.)

Proof: These properties are direct consequences of the axioms of the tensor product when applied to antisymmetric tensors. For example, the calculation (2.1) now requires a simple expansion of brackets,

$$(\mathbf{u} + \mathbf{v}) \wedge (\mathbf{v} + \mathbf{w}) = \mathbf{u} \wedge \mathbf{v} + \mathbf{u} \wedge \mathbf{w} + \mathbf{v} \wedge \mathbf{w}.$$

Here we removed the term $\mathbf{v} \wedge \mathbf{v}$ which vanishes due to the antisymmetry of $\wedge$. Details left as exercise. ∎

Elements of the space $V \wedge V$, such as $\mathbf{a} \wedge \mathbf{b} + \mathbf{c} \wedge \mathbf{d}$, are sometimes called **bivectors.**[1] We will also want to define the exterior product of more than two vectors. To define the exterior product of *three* vectors, we consider the subspace of $V \otimes V \otimes V$ that consists of antisymmetric tensors of the form

$$\begin{aligned} \mathbf{a} \otimes \mathbf{b} \otimes \mathbf{c} - \mathbf{b} \otimes \mathbf{a} \otimes \mathbf{c} + \mathbf{c} \otimes \mathbf{a} \otimes \mathbf{b} - \mathbf{c} \otimes \mathbf{b} \otimes \mathbf{a} \\ + \mathbf{b} \otimes \mathbf{c} \otimes \mathbf{a} - \mathbf{a} \otimes \mathbf{c} \otimes \mathbf{b} \end{aligned} \tag{2.5}$$

and linear combinations of such tensors. These tensors are called **totally antisymmetric** because they can be viewed as (tensor-valued) functions of the vectors $\mathbf{a}, \mathbf{b}, \mathbf{c}$ that change sign under exchange of any two vectors. The expression in Eq. (2.5) will be denoted for brevity by $\mathbf{a} \wedge \mathbf{b} \wedge \mathbf{c}$, similarly to the exterior product of two vectors, $\mathbf{a} \otimes \mathbf{b} - \mathbf{b} \otimes \mathbf{a}$, which is denoted for brevity by $\mathbf{a} \wedge \mathbf{b}$. Here is a general definition.

[1] It is important to note that a bivector is not necessarily expressible as a single-term product of two vectors; see the Exercise at the end of Sec. 2.3.2.

Definition 2: The **exterior product of k copies** of V (also called the **k-th exterior power** of V) is denoted by $\wedge^k V$ and is defined as the subspace of totally antisymmetric tensors within $V \otimes ... \otimes V$. In the concise notation, this is the space spanned by expressions of the form

$$\mathbf{v}_1 \wedge \mathbf{v}_2 \wedge ... \wedge \mathbf{v}_k, \quad \mathbf{v}_j \in V,$$

assuming that the properties of the wedge product (linearity and antisymmetry) hold as given by Statement 1. For instance,

$$\mathbf{u} \wedge \mathbf{v}_1 \wedge ... \wedge \mathbf{v}_k = (-1)^k \mathbf{v}_1 \wedge ... \wedge \mathbf{v}_k \wedge \mathbf{u} \tag{2.6}$$

("pulling a vector through k other vectors changes sign k times"). ■

The previously defined space of bivectors is in this notation $V \wedge V \equiv \wedge^2 V$. A natural extension of this notation is $\wedge^0 V = \mathbb{K}$ and $\wedge^1 V = V$. I will also use the following "wedge product" notation,

$$\bigwedge_{k=1}^{n} \mathbf{v}_k \equiv \mathbf{v}_1 \wedge \mathbf{v}_2 \wedge ... \wedge \mathbf{v}_n.$$

Tensors from the space $\wedge^n V$ are also called n-**vectors** or **antisymmetric tensors** of rank n.

Question: How to compute expressions containing multiple products such as $\mathbf{a} \wedge \mathbf{b} \wedge \mathbf{c}$?

Answer: Apply the rules shown in Statement 1. For example, one can permute adjacent vectors and change sign,

$$\mathbf{a} \wedge \mathbf{b} \wedge \mathbf{c} = -\mathbf{b} \wedge \mathbf{a} \wedge \mathbf{c} = \mathbf{b} \wedge \mathbf{c} \wedge \mathbf{a},$$

one can expand brackets,

$$\mathbf{a} \wedge (\mathbf{x} + 4\mathbf{y}) \wedge \mathbf{b} = \mathbf{a} \wedge \mathbf{x} \wedge \mathbf{b} + 4\mathbf{a} \wedge \mathbf{y} \wedge \mathbf{b},$$

and so on. If the vectors $\mathbf{a}, \mathbf{b}, \mathbf{c}$ are given as linear combinations of some basis vectors $\{\mathbf{e}_j\}$, we can thus reduce $\mathbf{a} \wedge \mathbf{b} \wedge \mathbf{c}$ to a linear combination of exterior products of basis vectors, such as $\mathbf{e}_1 \wedge \mathbf{e}_2 \wedge \mathbf{e}_3$, $\mathbf{e}_1 \wedge \mathbf{e}_2 \wedge \mathbf{e}_4$, etc.

Question: The notation $\mathbf{a} \wedge \mathbf{b} \wedge \mathbf{c}$ suggests that the exterior product is associative,

$$\mathbf{a} \wedge \mathbf{b} \wedge \mathbf{c} = (\mathbf{a} \wedge \mathbf{b}) \wedge \mathbf{c} = \mathbf{a} \wedge (\mathbf{b} \wedge \mathbf{c}).$$

How can we make sense of this?

Answer: If we want to be pedantic, we need to define the exterior product operation $\wedge$ between a single-term bivector $\mathbf{a} \wedge \mathbf{b}$ and a vector $\mathbf{c}$, such that the result is *by definition* the 3-vector $\mathbf{a} \wedge \mathbf{b} \wedge \mathbf{c}$. We then define the same operation on linear combinations of single-term bivectors,

$$(\mathbf{a} \wedge \mathbf{b} + \mathbf{x} \wedge \mathbf{y}) \wedge \mathbf{c} \equiv \mathbf{a} \wedge \mathbf{b} \wedge \mathbf{c} + \mathbf{x} \wedge \mathbf{y} \wedge \mathbf{c}.$$

Thus we have defined the exterior product between $\wedge^2 V$ and V, the result being a 3-vector from $\wedge^3 V$. We then need to verify that the results do not depend on the choice of the vectors such as $\mathbf{a}, \mathbf{b}, \mathbf{x}, \mathbf{y}$ in the representation of a bivector: A different representation can be achieved only by using the properties of the exterior product (i.e. the axioms of the tensor product), e.g. we may replace $\mathbf{a} \wedge \mathbf{b}$ by $-\mathbf{b} \wedge (\mathbf{a} + \lambda\mathbf{b})$. It is easy to verify that any such replacements will not modify the resulting 3-vector, e.g.

$$\mathbf{a} \wedge \mathbf{b} \wedge \mathbf{c} = -\mathbf{b} \wedge (\mathbf{a} + \lambda\mathbf{b}) \wedge \mathbf{c},$$

again due to the properties of the exterior product. This consideration shows that calculations with exterior products are consistent with our algebraic intuition. We may indeed compute $\mathbf{a} \wedge \mathbf{b} \wedge \mathbf{c}$ as $(\mathbf{a} \wedge \mathbf{b}) \wedge \mathbf{c}$ or as $\mathbf{a} \wedge (\mathbf{b} \wedge \mathbf{c})$.

Example 1: Suppose we work in $\mathbb{R}^3$ and have vectors $\mathbf{a} = (0, \frac{1}{2}, -\frac{1}{2})$, $\mathbf{b} = (2, -2, 0)$, $\mathbf{c} = (-2, 5, -3)$. Let us compute various exterior products. Calculations are easier if we introduce the basis $\{\mathbf{e}_1, \mathbf{e}_2, \mathbf{e}_3\}$ explicitly:

$$\mathbf{a} = \frac{1}{2}(\mathbf{e}_2 - \mathbf{e}_3), \quad \mathbf{b} = 2(\mathbf{e}_1 - \mathbf{e}_2), \quad \mathbf{c} = -2\mathbf{e}_1 + 5\mathbf{e}_2 - 3\mathbf{e}_3.$$

We compute the 2-vector $\mathbf{a} \wedge \mathbf{b}$ by using the properties of the exterior product, such as $\mathbf{x} \wedge \mathbf{x} = 0$ and $\mathbf{x} \wedge \mathbf{y} = -\mathbf{y} \wedge \mathbf{x}$, and simply expanding the brackets as usual in algebra:

$$\begin{aligned}
\mathbf{a} \wedge \mathbf{b} &= \frac{1}{2}(\mathbf{e}_2 - \mathbf{e}_3) \wedge 2(\mathbf{e}_1 - \mathbf{e}_2) \\
&= (\mathbf{e}_2 - \mathbf{e}_3) \wedge (\mathbf{e}_1 - \mathbf{e}_2) \\
&= \mathbf{e}_2 \wedge \mathbf{e}_1 - \mathbf{e}_3 \wedge \mathbf{e}_1 - \mathbf{e}_2 \wedge \mathbf{e}_2 + \mathbf{e}_3 \wedge \mathbf{e}_2 \\
&= -\mathbf{e}_1 \wedge \mathbf{e}_2 + \mathbf{e}_1 \wedge \mathbf{e}_3 - \mathbf{e}_2 \wedge \mathbf{e}_3.
\end{aligned}$$

The last expression is the result; note that now there is nothing more to compute or to simplify. The expressions such as $\mathbf{e}_1 \wedge \mathbf{e}_2$ are the basic expressions out of which the space $\mathbb{R}^3 \wedge \mathbb{R}^3$ is built. Below (Sec. 2.3.2) we will show formally that the set of these expressions is a basis in the space $\mathbb{R}^3 \wedge \mathbb{R}^3$.

Let us also compute the 3-vector $\mathbf{a} \wedge \mathbf{b} \wedge \mathbf{c}$,

$$\begin{aligned}
\mathbf{a} \wedge \mathbf{b} \wedge \mathbf{c} &= (\mathbf{a} \wedge \mathbf{b}) \wedge \mathbf{c} \\
&= (-\mathbf{e}_1 \wedge \mathbf{e}_2 + \mathbf{e}_1 \wedge \mathbf{e}_3 - \mathbf{e}_2 \wedge \mathbf{e}_3) \wedge (-2\mathbf{e}_1 + 5\mathbf{e}_2 - 3\mathbf{e}_3).
\end{aligned}$$

When we expand the brackets here, terms such as $\mathbf{e}_1 \wedge \mathbf{e}_2 \wedge \mathbf{e}_1$ will vanish because

$$\mathbf{e}_1 \wedge \mathbf{e}_2 \wedge \mathbf{e}_1 = -\mathbf{e}_2 \wedge \mathbf{e}_1 \wedge \mathbf{e}_1 = 0,$$

so only terms containing all different vectors need to be kept, and we find

$$\begin{aligned}
\mathbf{a} \wedge \mathbf{b} \wedge \mathbf{c} &= 3\mathbf{e}_1 \wedge \mathbf{e}_2 \wedge \mathbf{e}_3 + 5\mathbf{e}_1 \wedge \mathbf{e}_3 \wedge \mathbf{e}_2 + 2\mathbf{e}_2 \wedge \mathbf{e}_3 \wedge \mathbf{e}_1 \\
&= (3 - 5 + 2)\, \mathbf{e}_1 \wedge \mathbf{e}_2 \wedge \mathbf{e}_3 = 0.
\end{aligned}$$

We note that all the terms are proportional to the 3-vector $\mathbf{e}_1 \wedge \mathbf{e}_2 \wedge \mathbf{e}_3$, so only the coefficient in front of $\mathbf{e}_1 \wedge \mathbf{e}_2 \wedge \mathbf{e}_3$ was needed; then, by coincidence, that coefficient turned out to be zero. So the result is the zero 3-vector. ■

Question: Our original goal was to introduce a bilinear, antisymmetric product of vectors in order to obtain a geometric representation of oriented areas. Instead, $\mathbf{a} \wedge \mathbf{b}$ was defined algebraically, through tensor products. It is clear that $\mathbf{a} \wedge \mathbf{b}$ is antisymmetric and bilinear, but why does it represent an oriented area?

Answer: Indeed, it may not be immediately clear why oriented areas should be elements of $V \wedge V$. We have seen that the oriented area $A(\mathbf{x}, \mathbf{y})$ is an antisymmetric and bilinear function of the two vectors $\mathbf{x}$ and $\mathbf{y}$. Right now we have constructed the space $V \wedge V$ simply as the *space of antisymmetric products*. By constructing that space merely out of the axioms of the antisymmetric product, we already covered *every possible* bilinear antisymmetric product. This means that *any* antisymmetric and bilinear function of the two vectors $\mathbf{x}$ and $\mathbf{y}$ is proportional to $\mathbf{x} \wedge \mathbf{y}$ or, more generally, is a *linear function* of $\mathbf{x} \wedge \mathbf{y}$ (perhaps with values in a different space). Therefore, the space of oriented areas (that is, the space of linear combinations of $A(\mathbf{x}, \mathbf{y})$ for various $\mathbf{x}$ and $\mathbf{y}$) is in any case mapped to a subspace of $V \wedge V$. We have also seen that oriented areas in N dimensions can be represented through $\binom{N}{2}$ projections, which indicates that they are vectors in some $\binom{N}{2}$-dimensional space. We will see below that the space $V \wedge V$ has exactly this dimension (Theorem 2 in Sec. 2.3.2). Therefore, we can expect that the space of oriented areas coincides with $V \wedge V$. Below we will be working in a space V with a scalar product, where the notions of area and volume are well defined. Then we will see (Sec. 5.5.2) that tensors from $V \wedge V$ and the higher exterior powers of V indeed correspond in a natural way to oriented areas, or more generally to oriented volumes of a certain dimension.

Remark: Origin of the name "exterior." The construction of the exterior product is a modern formulation of the ideas dating back to H. Grassmann (1844). A 2-vector $\mathbf{a} \wedge \mathbf{b}$ is interpreted geometrically as the oriented area of the parallelogram spanned by the vectors $\mathbf{a}$ and $\mathbf{b}$. Similarly, a 3-vector $\mathbf{a} \wedge \mathbf{b} \wedge \mathbf{c}$ represents the oriented 3-volume of a parallelepiped spanned by $\{\mathbf{a}, \mathbf{b}, \mathbf{c}\}$. Due to the antisymmetry of the exterior product, we have $(\mathbf{a} \wedge \mathbf{b}) \wedge (\mathbf{a} \wedge \mathbf{c}) = 0$, $(\mathbf{a} \wedge \mathbf{b} \wedge \mathbf{c}) \wedge (\mathbf{b} \wedge \mathbf{d}) = 0$, etc. We can interpret this geometrically by saying that the "product" of two volumes is zero if these volumes have a vector in common. This motivated Grassmann to call his antisymmetric product "exterior." In his reasoning, the product of two "extensive quantities" (such as lines, areas, or volumes) is nonzero only when each of the two quantities is geometrically "to the exterior" (outside) of the other.

Exercise 2: Show that in a *two*-dimensional space V, any 3-vector such as $\mathbf{a} \wedge \mathbf{b} \wedge \mathbf{c}$ can be simplified to the zero 3-vector. Prove the same for n-vectors in N-dimensional spaces when $n > N$. ■

One can also consider the exterior powers of the *dual* space V^*. Tensors from $\wedge^n V^*$ are usually (for historical reasons) called n-**forms** (rather than "n-covectors").

Question: Where is the star here, really? Is the space $\wedge^n (V^*)$ different from $(\wedge^n V)^*$?

Answer: Good that you asked. These spaces are canonically isomorphic, but there is a subtle technical issue worth mentioning. Consider an example: $\mathbf{a}^* \wedge \mathbf{b}^* \in \wedge^2(V^*)$ can act upon $\mathbf{u} \wedge \mathbf{v} \in \wedge^2 V$ by the standard tensor product rule, namely $\mathbf{a}^* \otimes \mathbf{b}^*$ acts on $\mathbf{u} \otimes \mathbf{v}$ as

$$(\mathbf{a}^* \otimes \mathbf{b}^*)(\mathbf{u} \otimes \mathbf{v}) = \mathbf{a}^*(\mathbf{u})\, \mathbf{b}^*(\mathbf{v}),$$

so by using the definition of $\mathbf{a}^* \wedge \mathbf{b}^*$ and $\mathbf{u} \wedge \mathbf{v}$ through the tensor product, we find

$$\begin{aligned}(\mathbf{a}^* \wedge \mathbf{b}^*)(\mathbf{u} \wedge \mathbf{v}) &= (\mathbf{a}^* \otimes \mathbf{b}^* - \mathbf{b}^* \otimes \mathbf{a}^*)(\mathbf{u} \otimes \mathbf{v} - \mathbf{v} \otimes \mathbf{u}) \\ &= 2\mathbf{a}^*(\mathbf{u})\, \mathbf{b}^*(\mathbf{v}) - 2\mathbf{b}^*(\mathbf{u})\, \mathbf{a}^*(\mathbf{v}).\end{aligned}$$

We got a **combinatorial factor** 2, that is, a factor that arises because we have *two* permutations of the set $(\mathbf{a}, \mathbf{b})$. With $\wedge^n(V^*)$ and $(\wedge^n V)^*$ we get a factor $n!$. It is not always convenient to have this combinatorial factor. For example, in a finite number field the number $n!$ might be *equal to zero* for large enough n. In these cases we could *redefine* the action of $\mathbf{a}^* \wedge \mathbf{b}^*$ on $\mathbf{u} \wedge \mathbf{v}$ as

$$(\mathbf{a}^* \wedge \mathbf{b}^*)(\mathbf{u} \wedge \mathbf{v}) \equiv \mathbf{a}^*(\mathbf{u})\, \mathbf{b}^*(\mathbf{v}) - \mathbf{b}^*(\mathbf{u})\, \mathbf{a}^*(\mathbf{v}).$$

If we are not working in a finite number field, we are able to divide by any integer, so we may keep combinatorial factors in the denominators of expressions where such factors appear. For example, if $\{\mathbf{e}_j\}$ is a basis in V and $\omega = \mathbf{e}_1 \wedge ... \wedge \mathbf{e}_N$ is the corresponding basis tensor in the one-dimensional space $\wedge^N V$, the dual basis tensor in $(\wedge^N V)^*$ could be defined by

$$\omega^* = \frac{1}{N!} \mathbf{e}_1^* \wedge ... \wedge \mathbf{e}_N^*, \quad \text{so that } \omega^*(\omega) = 1.$$

The need for such combinatorial factors is a minor technical inconvenience that does not arise too often. We may give the following definition that avoids dividing by combinatorial factors (but now we use permutations; see Appendix B).

Definition 3: The action of a k-form $\mathbf{f}_1^* \wedge ... \wedge \mathbf{f}_k^*$ on a k-vector $\mathbf{v}_1 \wedge ... \wedge \mathbf{v}_k$ is defined by

$$\sum_\sigma (-1)^{|\sigma|} \mathbf{f}_1^*(\mathbf{v}_{\sigma(1)}) ... \mathbf{f}_k^*(\mathbf{v}_{\sigma(k)}),$$

where the summation is performed over all permutations σ of the ordered set $(1, ..., k)$.

Example 2: With $k = 3$ we have

$$\begin{aligned}&(\mathbf{p}^* \wedge \mathbf{q}^* \wedge \mathbf{r}^*)(\mathbf{a} \wedge \mathbf{b} \wedge \mathbf{c}) \\ &= \mathbf{p}^*(\mathbf{a})\mathbf{q}^*(\mathbf{b})\mathbf{r}^*(\mathbf{c}) - \mathbf{p}^*(\mathbf{b})\mathbf{q}^*(\mathbf{a})\mathbf{r}^*(\mathbf{c}) \\ &+ \mathbf{p}^*(\mathbf{b})\mathbf{q}^*(\mathbf{c})\mathbf{r}^*(\mathbf{a}) - \mathbf{p}^*(\mathbf{c})\mathbf{q}^*(\mathbf{b})\mathbf{r}^*(\mathbf{a}) \\ &+ \mathbf{p}^*(\mathbf{c})\mathbf{q}^*(\mathbf{a})\mathbf{r}^*(\mathbf{b}) - \mathbf{p}^*(\mathbf{c})\mathbf{q}^*(\mathbf{b})\mathbf{r}^*(\mathbf{a}).\end{aligned}$$

Exercise 3: a) Show that $\mathbf{a} \wedge \mathbf{b} \wedge \omega = \omega \wedge \mathbf{a} \wedge \mathbf{b}$ where ω is any antisymmetric tensor (e.g. $\omega = \mathbf{x} \wedge \mathbf{y} \wedge \mathbf{z}$).

b) Show that

$$\omega_1 \wedge \mathbf{a} \wedge \omega_2 \wedge \mathbf{b} \wedge \omega_3 = -\omega_1 \wedge \mathbf{b} \wedge \omega_2 \wedge \mathbf{a} \wedge \omega_3,$$

where $\omega_1, \omega_2, \omega_3$ are arbitrary antisymmetric tensors and $\mathbf{a}, \mathbf{b}$ are vectors.

c) Due to antisymmetry, $\mathbf{a} \wedge \mathbf{a} = 0$ for any vector $\mathbf{a} \in V$. Is it also true that $\omega \wedge \omega = 0$ for any bivector $\omega \in \wedge^2 V$?

2.2.2 * Symmetric tensor product

Question: At this point it is still unclear why the antisymmetric definition is at all useful. Perhaps we could define something else, say the symmetric product, instead of the exterior product? We could try to define a product, say $\mathbf{a} \odot \mathbf{b}$, with some other property, such as

$$\mathbf{a} \odot \mathbf{b} = 2\mathbf{b} \odot \mathbf{a}.$$

Answer: This does not work because, for example, we would have

$$\mathbf{b} \odot \mathbf{a} = 2\mathbf{a} \odot \mathbf{b} = 4\mathbf{b} \odot \mathbf{a},$$

so all the "$\odot$" products would have to vanish.

We can define the *symmetric* tensor product, $\otimes_S$, with the property

$$\mathbf{a} \otimes_S \mathbf{b} = \mathbf{b} \otimes_S \mathbf{a},$$

but it is impossible to define anything else in a similar fashion.[2]

The antisymmetric tensor product is the eigenspace (within $V \otimes V$) of the exchange operator $\hat{T}$ with eigenvalue -1. That operator has only eigenvectors with eigenvalues ± 1, so the only other possibility is to consider the eigenspace with eigenvalue $+1$. This eigenspace is spanned by symmetric tensors of the form $\mathbf{u} \otimes \mathbf{v} + \mathbf{v} \otimes \mathbf{u}$, and can be considered as the space of symmetric tensor products. We could write

$$\mathbf{a} \otimes_S \mathbf{b} \equiv \mathbf{a} \otimes \mathbf{b} + \mathbf{b} \otimes \mathbf{a}$$

and develop the properties of this product. However, it turns out that the symmetric tensor product is much less useful for the purposes of linear algebra than the antisymmetric subspace. This book derives most of the results of linear algebra using the antisymmetric product as the main tool!

2.3 Properties of spaces $\wedge^k V$

As we have seen, tensors from the space $V \otimes V$ are representable by linear combinations of the form $\mathbf{a} \otimes \mathbf{b} + \mathbf{c} \otimes \mathbf{d} + ...$, but not *uniquely* representable

[2] This is a theorem due to Grassmann (1862).

because one can transform one such linear combination into another by using the axioms of the tensor product. Similarly, n-vectors are not uniquely representable by linear combinations of exterior products. For example,

$$\mathbf{a} \wedge \mathbf{b} + \mathbf{a} \wedge \mathbf{c} + \mathbf{b} \wedge \mathbf{c} = (\mathbf{a} + \mathbf{b}) \wedge (\mathbf{b} + \mathbf{c})$$

since $\mathbf{b} \wedge \mathbf{b} = 0$. In other words, the 2-vector $\omega \equiv \mathbf{a} \wedge \mathbf{b} + \mathbf{a} \wedge \mathbf{c} + \mathbf{b} \wedge \mathbf{c}$ has an alternative representation containing only a single-term exterior product, $\omega = \mathbf{r} \wedge \mathbf{s}$ where $\mathbf{r} = \mathbf{a} + \mathbf{b}$ and $\mathbf{s} = \mathbf{b} + \mathbf{c}$.

Exercise: Show that any 2-vector in a *three*-dimensional space is representable by a single-term exterior product, i.e. to a 2-vector of the form $\mathbf{a} \wedge \mathbf{b}$.

Hint: Choose a basis $\{\mathbf{e}_1, \mathbf{e}_2, \mathbf{e}_3\}$ and show that $\alpha \mathbf{e}_1 \wedge \mathbf{e}_2 + \beta \mathbf{e}_1 \wedge \mathbf{e}_3 + \gamma \mathbf{e}_2 \wedge \mathbf{e}_3$ is equal to a single-term product. ■

What about higher-dimensional spaces? We will show (see the Exercise at the end of Sec. 2.3.2) that n-vectors cannot be in general reduced to a single-term product. This is, however, always possible for $(N-1)$-vectors in an N-dimensional space. (You showed this for $N = 3$ in the exercise above.)

Statement: Any $(N-1)$-vector in an N-dimensional space can be written as a single-term exterior product of the form $\mathbf{a}_1 \wedge ... \wedge \mathbf{a}_{N-1}$.

Proof: We prove this by using induction in N. The basis of induction is $N = 2$, where there is nothing to prove. The induction step: Suppose that the statement is proved for $(N-1)$-vectors in N-dimensional spaces, we need to prove it for N-vectors in $(N+1)$-dimensional spaces. Choose a basis $\{\mathbf{e}_1, ..., \mathbf{e}_{N+1}\}$ in the space. Any N-vector ω can be written as a linear combination of exterior product terms,

$$\begin{aligned}\omega = \; & \alpha_1 \mathbf{e}_2 \wedge ... \wedge \mathbf{e}_{N+1} + \alpha_2 \mathbf{e}_1 \wedge \mathbf{e}_3 \wedge ... \wedge \mathbf{e}_{N+1} + ... \\ & + \alpha_N \mathbf{e}_1 \wedge ... \wedge \mathbf{e}_{N-1} \wedge \mathbf{e}_{N+1} + \alpha_{N+1} \mathbf{e}_1 \wedge ... \wedge \mathbf{e}_N,\end{aligned}$$

where $\{\alpha_i\}$ are some constants.

Note that any tensor $\omega \in \wedge^{N-1} V$ can be written in this way simply by expressing every vector through the basis and by expanding the exterior products. The result will be a linear combination of the form shown above, containing at most $N+1$ single-term exterior products of the form $\mathbf{e}_1 \wedge ... \wedge \mathbf{e}_N$, $\mathbf{e}_2 \wedge ... \wedge \mathbf{e}_{N+1}$, and so on. We do not yet know whether these single-term exterior products constitute a linearly independent set; this will be established in Sec. 2.3.2. Presently, we will not need this property.

Now we would like to transform the expression above to a single term. We move $\mathbf{e}_{N+1}$ outside brackets in the first N terms:

$$\begin{aligned}\omega = & \left(\alpha_1 \mathbf{e}_2 \wedge ... \wedge \mathbf{e}_N + ... + \alpha_N \mathbf{e}_1 \wedge ... \wedge \mathbf{e}_{N-1}\right) \wedge \mathbf{e}_{N+1} \\ & + \alpha_{N+1} \mathbf{e}_1 \wedge ... \wedge \mathbf{e}_N \\ \equiv \; & \psi \wedge \mathbf{e}_{N+1} + \alpha_{N+1} \mathbf{e}_1 \wedge ... \wedge \mathbf{e}_N,\end{aligned}$$

where in the last line we have introduced an auxiliary $(N-1)$-vector ψ. If it happens that $\psi = 0$, there is nothing left to prove. Otherwise, at least one of the α_i must be nonzero; without loss of generality, suppose that $\alpha_N \neq 0$ and rewrite ω as

$$\omega = \psi \wedge \mathbf{e}_{N+1} + \alpha_{N+1}\mathbf{e}_1 \wedge ... \wedge \mathbf{e}_N = \psi \wedge \left(\mathbf{e}_{N+1} + \frac{\alpha_{N+1}}{\alpha_N}\mathbf{e}_N\right).$$

Now we note that ψ belongs to the space of $(N-1)$-vectors over the N-dimensional subspace spanned by $\{\mathbf{e}_1, ..., \mathbf{e}_N\}$. By the inductive assumption, ψ can be written as a single-term exterior product, $\psi = \mathbf{a}_1 \wedge ... \wedge \mathbf{a}_{N-1}$, of some vectors $\{\mathbf{a}_i\}$. Denoting

$$\mathbf{a}_N \equiv \mathbf{e}_{N+1} + \frac{\alpha_{N+1}}{\alpha_N}\mathbf{e}_N,$$

we obtain

$$\omega = \mathbf{a}_1 \wedge ... \wedge \mathbf{a}_{N-1} \wedge \mathbf{a}_N,$$

i.e. ω can be represented as a single-term exterior product. ■

2.3.1 Linear maps between spaces $\wedge^k V$

Since the spaces $\wedge^k V$ are vector spaces, we may consider linear maps between them.

A simplest example is a map

$$L_{\mathbf{a}} : \omega \mapsto \mathbf{a} \wedge \omega,$$

mapping $\wedge^k V \to \wedge^{k+1} V$; here the vector $\mathbf{a}$ is *fixed*. It is important to check that $L_{\mathbf{a}}$ is a *linear* map between these spaces. How do we check this? We need to check that $L_{\mathbf{a}}$ maps a linear combination of tensors into linear combinations; this is easy to see,

$$\begin{aligned} L_{\mathbf{a}}(\omega + \lambda\omega') &= \mathbf{a} \wedge (\omega + \lambda\omega') \\ &= \mathbf{a} \wedge \omega + \lambda\mathbf{a} \wedge \omega' = L_{\mathbf{a}}\omega + \lambda L_{\mathbf{a}}\omega'. \end{aligned}$$

Let us now fix a covector $\mathbf{a}^*$. A covector is a map $V \to \mathbb{K}$. In Lemma 2 of Sec. 1.7.3 we have used covectors to define linear maps $\mathbf{a}^* : V \otimes W \to W$ according to Eq. (1.21), mapping $\mathbf{v} \otimes \mathbf{w} \mapsto \mathbf{a}^*(\mathbf{v})\,\mathbf{w}$. Now we will apply the analogous construction to exterior powers and construct a map $V \wedge V \to V$. Let us denote this map by $\iota_{\mathbf{a}^*}$.

It would be incorrect to define the map $\iota_{\mathbf{a}^*}$ by the formula $\iota_{\mathbf{a}^*}(\mathbf{v} \wedge \mathbf{w}) = \mathbf{a}^*(\mathbf{v})\,\mathbf{w}$ because such a definition does not respect the antisymmetry of the wedge product and thus violates the linearity condition,

$$\iota_{\mathbf{a}^*}(\mathbf{w} \wedge \mathbf{v}) \overset{!}{=} \iota_{\mathbf{a}^*}((-1)\,\mathbf{v} \wedge \mathbf{w}) = -\iota_{\mathbf{a}^*}(\mathbf{v} \wedge \mathbf{w}) \neq \mathbf{a}^*(\mathbf{v})\mathbf{w}.$$

So we need to act with $\mathbf{a}^*$ on *each* of the vectors in a wedge product and make sure that the correct minus sign comes out. An acceptable formula for the map $\iota_{\mathbf{a}^*} : \wedge^2 V \to V$ is

$$\iota_{\mathbf{a}^*}(\mathbf{v} \wedge \mathbf{w}) \equiv \mathbf{a}^*(\mathbf{v})\,\mathbf{w} - \mathbf{a}^*(\mathbf{w})\,\mathbf{v}.$$

(Please check that the linearity condition now holds!) This is how we will define the map $\iota_{\mathbf{a}^*}$ on $\wedge^2 V$.

Let us now extend $\iota_{\mathbf{a}^*} : \wedge^2 V \to V$ to a map

$$\iota_{\mathbf{a}^*} : \wedge^k V \to \wedge^{k-1} V,$$

defined as follows:

$$\iota_{\mathbf{a}^*} \mathbf{v} \equiv \mathbf{a}^*(\mathbf{v}),$$
$$\iota_{\mathbf{a}^*}(\mathbf{v} \wedge \omega) \equiv \mathbf{a}^*(\mathbf{v})\omega - \mathbf{v} \wedge (\iota_{\mathbf{a}^*}\omega). \tag{2.7}$$

This definition is *inductive*, i.e. it shows how to define $\iota_{\mathbf{a}^*}$ on $\wedge^k V$ if we know how to define it on $\wedge^{k-1} V$. The action of $\iota_{\mathbf{a}^*}$ on a sum of terms is defined by requiring linearity,

$$\iota_{\mathbf{a}^*}(A + \lambda B) \equiv \iota_{\mathbf{a}^*}(A) + \lambda \iota_{\mathbf{a}^*}(B), \quad A, B \in \wedge^k V.$$

We can convert this inductive definition into a more explicit formula: if $\omega = \mathbf{v}_1 \wedge ... \wedge \mathbf{v}_k \in \wedge^k V$ then

$$\iota_{\mathbf{a}^*}(\mathbf{v}_1 \wedge ... \wedge \mathbf{v}_k) \equiv \mathbf{a}^*(\mathbf{v}_1)\mathbf{v}_2 \wedge ... \wedge \mathbf{v}_k - \mathbf{a}^*(\mathbf{v}_2)\mathbf{v}_1 \wedge \mathbf{v}_3 \wedge ... \wedge \mathbf{v}_k$$
$$+ ... + (-1)^{k-1} \mathbf{a}^*(\mathbf{v}_k)\mathbf{v}_1 \wedge ... \wedge \mathbf{v}_{k-1}.$$

This map is called the **interior product** or the **insertion** map. This is a useful operation in linear algebra. The insertion map $\iota_{\mathbf{a}^*}\psi$ "inserts" the covector $\mathbf{a}^*$ into the tensor $\psi \in \wedge^k V$ by acting with $\mathbf{a}^*$ on each of the vectors in the exterior product that makes up ψ.

Let us check formally that the insertion map is linear.

Statement: The map $\iota_{\mathbf{a}^*} : \wedge^k V \to \wedge^{k-1} V$ for $1 \leq k \leq N$ is a well-defined linear map, according to the inductive definition.

Proof: First, we need to check that it maps linear combinations into linear combinations; this is quite easy to see by induction, using the fact that $\mathbf{a}^* : V \to \mathbb{K}$ is linear. However, this type of linearity is not sufficient; we also need to check that the *result* of the map, i.e. the tensor $\iota_{\mathbf{a}^*}(\omega)$, is defined *independently of the representation* of ω through vectors such as $\mathbf{v}_i$. The problem is, there are many such representations, for example some tensor $\omega \in \wedge^3 V$ might be written using different vectors as

$$\omega = \mathbf{v}_1 \wedge \mathbf{v}_2 \wedge \mathbf{v}_3 = \mathbf{v}_2 \wedge (\mathbf{v}_3 - \mathbf{v}_1) \wedge (\mathbf{v}_3 + \mathbf{v}_2) \equiv \tilde{\mathbf{v}}_1 \wedge \tilde{\mathbf{v}}_2 \wedge \tilde{\mathbf{v}}_3.$$

We need to verify that any such equivalent representation yields the same resulting tensor $\iota_{\mathbf{a}^*}(\omega)$, despite the fact that the definition of $\iota_{\mathbf{a}^*}$ *appears* to depend on the choice of the vectors $\mathbf{v}_i$. Only then will it be proved that $\iota_{\mathbf{a}^*}$ is a linear map $\wedge^k V \to \wedge^{k-1} V$.

An equivalent representation of a tensor ω can be obtained only by using the properties of the exterior product, namely linearity and antisymmetry. Therefore, we need to verify that $\iota_{\mathbf{a}^*}(\omega)$ does not change when we change the representation of ω in these two ways: 1) expanding a linear combination,

$$(\mathbf{x} + \lambda \mathbf{y}) \wedge ... \mapsto \mathbf{x} \wedge ... + \lambda \mathbf{y} \wedge ...; \tag{2.8}$$

2) interchanging the order of two vectors in the exterior product and change the sign,

$$\mathbf{x} \wedge \mathbf{y} \wedge ... \mapsto -\mathbf{y} \wedge \mathbf{x} \wedge ... \tag{2.9}$$

It is clear that $\mathbf{a}^*(\mathbf{x} + \lambda\mathbf{y}) = \mathbf{a}^*(\mathbf{x}) + \lambda\mathbf{a}^*(\mathbf{y})$; it follows by induction that $\iota_{\mathbf{a}^*}\omega$ does not change under a change of representation of the type (2.8). Now we consider the change of representation of the type (2.9). We have, by definition of $\iota_{\mathbf{a}^*}$,

$$\iota_{\mathbf{a}^*}(\mathbf{v}_1 \wedge \mathbf{v}_2 \wedge \chi) = \mathbf{a}^*(\mathbf{v}_1)\mathbf{v}_2 \wedge \chi - \mathbf{a}^*(\mathbf{v}_2)\mathbf{v}_1 \wedge \chi + \mathbf{v}_1 \wedge \mathbf{v}_2 \wedge \iota_{\mathbf{a}^*}(\chi),$$

where we have denoted by χ the rest of the exterior product. It is clear from the above expression that

$$\iota_{\mathbf{a}^*}(\mathbf{v}_1 \wedge \mathbf{v}_2 \wedge \chi) = -\iota_{\mathbf{a}^*}(\mathbf{v}_2 \wedge \mathbf{v}_1 \wedge \chi) = \iota_{\mathbf{a}^*}(-\mathbf{v}_2 \wedge \mathbf{v}_1 \wedge \chi).$$

This proves that $\iota_{\mathbf{a}^*}(\omega)$ does not change under a change of representation of ω of the type (2.9). This concludes the proof. ■

Remark: It is apparent from the proof that the *minus sign* in the inductive definition (2.7) is crucial for the linearity of the map $\iota_{\mathbf{a}^*}$. Indeed, if we attempt to define a map by a formula such as

$$\mathbf{v}_1 \wedge \mathbf{v}_2 \mapsto \mathbf{a}^*(\mathbf{v}_1)\mathbf{v}_2 + \mathbf{a}^*(\mathbf{v}_2)\mathbf{v}_1,$$

the result will *not* be a linear map $\wedge^2 V \to V$ despite the appearance of linearity. The correct formula must take into account the fact that $\mathbf{v}_1 \wedge \mathbf{v}_2 = -\mathbf{v}_2 \wedge \mathbf{v}_1$.

Exercise: Show by induction in k that

$$L_{\mathbf{x}}\iota_{\mathbf{a}^*}\omega + \iota_{\mathbf{a}^*}L_{\mathbf{x}}\omega = \mathbf{a}^*(\mathbf{x})\omega, \quad \forall \omega \in \wedge^k V.$$

In other words, the linear operator $L_{\mathbf{x}}\iota_{\mathbf{a}^*} + \iota_{\mathbf{a}^*}L_{\mathbf{x}} : \wedge^k V \to \wedge^k V$ is simply the multiplication by the number $\mathbf{a}^*(\mathbf{x})$.

2.3.2 Exterior product and linear dependence

The exterior product is useful in many ways. One powerful property of the exterior product is its close relation to linear independence of sets of vectors. For example, if $\mathbf{u} = \lambda\mathbf{v}$ then $\mathbf{u} \wedge \mathbf{v} = 0$. More generally:

Theorem 1: A set $\{\mathbf{v}_1, ..., \mathbf{v}_k\}$ of vectors from V is linearly independent if and only if $(\mathbf{v}_1 \wedge \mathbf{v}_2 \wedge ... \wedge \mathbf{v}_k) \neq 0$, i.e. it is a nonzero tensor from $\wedge^k V$.

Proof: If $\{\mathbf{v}_j\}$ is linearly dependent then without loss of generality we may assume that $\mathbf{v}_1$ is a linear combination of other vectors, $\mathbf{v}_1 = \sum_{j=2}^k \lambda_j \mathbf{v}_j$. Then

$$\begin{aligned}\mathbf{v}_1 \wedge \mathbf{v}_2 \wedge ... \wedge \mathbf{v}_k &= \sum_{j=2}^{k} \lambda_j \mathbf{v}_j \wedge \mathbf{v}_2 \wedge ... \wedge \mathbf{v}_j \wedge ... \wedge \mathbf{v}_k \\ &= \sum_{j=2}^{k} (-1)^{j-1}\, \mathbf{v}_2 \wedge ... \mathbf{v}_j \wedge \mathbf{v}_j \wedge ... \wedge \mathbf{v}_k = 0.\end{aligned}$$

Conversely, we need to prove that the tensor $\mathbf{v}_1 \wedge ... \wedge \mathbf{v}_k \neq 0$ if $\{\mathbf{v}_j\}$ is linearly *in*dependent. The proof is by induction in k. The basis of induction is $k = 1$: if $\{\mathbf{v}_1\}$ is linearly independent then clearly $\mathbf{v}_1 \neq 0$. The induction step: Assume that the statement is proved for $k-1$ and that $\{\mathbf{v}_1, ..., \mathbf{v}_k\}$ is a linearly independent set. By Exercise 1 in Sec. 1.6 there exists a covector $\mathbf{f}^* \in V^*$ such that $\mathbf{f}^*(\mathbf{v}_1) = 1$ and $\mathbf{f}^*(\mathbf{v}_i) = 0$ for $2 \leq i \leq k$. Now we apply the interior product map $\iota_{\mathbf{f}^*} : \wedge^k V \to \wedge^{k-1} V$ constructed in Sec. 2.3.1 to the tensor $\mathbf{v}_1 \wedge ... \wedge \mathbf{v}_k$ and find

$$\iota_{\mathbf{f}^*}(\mathbf{v}_1 \wedge ... \wedge \mathbf{v}_k) = \mathbf{v}_2 \wedge ... \wedge \mathbf{v}_k.$$

By the induction step, the linear independence of $k-1$ vectors $\{\mathbf{v}_2, ..., \mathbf{v}_k\}$ entails $\mathbf{v}_2 \wedge ... \wedge \mathbf{v}_k \neq 0$. The map $\iota_{\mathbf{f}^*}$ is linear and cannot map a zero tensor into a nonzero tensor, therefore $\mathbf{v}_1 \wedge ... \wedge \mathbf{v}_k \neq 0$. ■

It is also important to know that any tensor from the highest exterior power $\wedge^N V$ can be represented as just a *single-term* exterior product of N vectors. (Note that the same property for $\wedge^{N-1} V$ was already established in Sec. 2.3.)

Lemma 1: For any tensor $\omega \in \wedge^N V$ there exist vectors $\{\mathbf{v}_1, ..., \mathbf{v}_N\}$ such that $\omega = \mathbf{v}_1 \wedge ... \wedge \mathbf{v}_N$.

Proof: If $\omega = 0$ then there is nothing to prove, so we assume $\omega \neq 0$. By definition, the tensor ω has a representation as a sum of *several* exterior products, say

$$\omega = \mathbf{v}_1 \wedge ... \wedge \mathbf{v}_N + \mathbf{v}'_1 \wedge ... \wedge \mathbf{v}'_N + ...$$

Let us simplify this expression to just one exterior product. First, let us omit any zero terms in this expression (for instance, $\mathbf{a} \wedge \mathbf{a} \wedge \mathbf{b} \wedge ... = 0$). Then by Theorem 1 the set $\{\mathbf{v}_1, ..., \mathbf{v}_N\}$ is linearly independent (or else the term $\mathbf{v}_1 \wedge ... \wedge \mathbf{v}_N$ would be zero). Hence, $\{\mathbf{v}_1, ..., \mathbf{v}_N\}$ is a basis in V. All other vectors such as $\mathbf{v}'_i$ can be decomposed as linear combinations of vectors in that basis. Let us denote $\psi \equiv \mathbf{v}_1 \wedge ... \wedge \mathbf{v}_N$. By expanding the brackets in exterior products such as $\mathbf{v}'_1 \wedge ... \wedge \mathbf{v}'_N$, we will obtain every time the tensor ψ with different coefficients. Therefore, the final result of simplification will be that ω equals ψ multiplied with some coefficient. This is sufficient to prove Lemma 1. ■

Now we would like to build a basis in the space $\wedge^m V$. For this we need to determine which sets of tensors from $\wedge^m V$ are linearly independent within that space.

Lemma 2: If $\{\mathbf{e}_1, ..., \mathbf{e}_N\}$ is a basis in V then any tensor $A \in \wedge^m V$ can be decomposed as a linear combination of the tensors $\mathbf{e}_{k_1} \wedge \mathbf{e}_{k_2} \wedge ... \wedge \mathbf{e}_{k_m}$ with some indices k_j, $1 \leq j \leq m$.

Proof: The tensor A is a linear combination of expressions of the form $\mathbf{v}_1 \wedge ... \wedge \mathbf{v}_m$, and each vector $\mathbf{v}_i \in V$ can be decomposed in the basis $\{\mathbf{e}_j\}$. Expanding the brackets around the wedges using the rules (2.2)–(2.4), we obtain a decomposition of an arbitrary tensor through the basis tensors. For example,

$$\begin{aligned}(\mathbf{e}_1 + 2\mathbf{e}_2) \wedge (\mathbf{e}_1 - \mathbf{e}_2 + \mathbf{e}_3) - 2(\mathbf{e}_2 - \mathbf{e}_3) \wedge (\mathbf{e}_1 - \mathbf{e}_3) \\ = -\mathbf{e}_1 \wedge \mathbf{e}_2 - \mathbf{e}_1 \wedge \mathbf{e}_3 + 4\mathbf{e}_2 \wedge \mathbf{e}_3\end{aligned}$$

(please verify this yourself!). ■

By Theorem 1, all tensors $\mathbf{e}_{k_1} \wedge \mathbf{e}_{k_2} \wedge ... \wedge \mathbf{e}_{k_m}$ constructed out of subsets of vectors from the basis $\{\mathbf{e}_1, ..., \mathbf{e}_k\}$ are nonzero, and by Lemma 2 any tensor can be decomposed into a linear combination of these tensors. But are these tensors a basis in the space $\wedge^m V$? Yes:

Lemma 3: If $\{\mathbf{v}_1, ..., \mathbf{v}_n\}$ is a linearly independent set of vectors (not necessarily a basis in V since $n \leq N$), then:

(1) The set of $\binom{n}{2}$ tensors

$$\{\mathbf{v}_j \wedge \mathbf{v}_k,\ 1 \leq j < k \leq n\} \equiv \{\mathbf{v}_1 \wedge \mathbf{v}_2, \mathbf{v}_1 \wedge \mathbf{v}_3, ..., \mathbf{v}_{n-1} \wedge \mathbf{v}_n\}$$

is linearly independent in the space $\wedge^2 V$.

(2) The set of $\binom{n}{m}$ tensors

$$\{\mathbf{v}_{k_1} \wedge \mathbf{v}_{k_2} \wedge ... \wedge \mathbf{v}_{k_m},\ 1 \leq k_1 < k_2 < ... < k_m \leq n\}$$

is linearly independent in the space $\wedge^m V$ for $2 \leq m \leq n$.

Proof: **(1)** The proof is similar to that of Lemma 3 in Sec. 1.7.3. Suppose the set $\{\mathbf{v}_j\}$ is linearly independent but the set $\{\mathbf{v}_j \wedge \mathbf{v}_k\}$ is linearly *dependent*, so that there exists a linear combination

$$\sum_{1 \leq j < k \leq n} \lambda_{jk} \mathbf{v}_j \wedge \mathbf{v}_k = 0$$

with at least some $\lambda_{jk} \neq 0$. Without loss of generality, $\lambda_{12} \neq 0$ (or else we can renumber the vectors $\mathbf{v}_j$). There exists a covector $\mathbf{f}^* \in V^*$ such that $\mathbf{f}^*(\mathbf{v}_1) = 1$ and $\mathbf{f}^*(\mathbf{v}_i) = 0$ for $2 \leq i \leq n$. Apply the interior product with this covector to the above tensor,

$$0 = \iota_{\mathbf{f}^*} \left[\sum_{1 \leq j < k \leq n} \lambda_{jk} \mathbf{v}_j \wedge \mathbf{v}_k \right] = \sum_{k=2}^{n} \lambda_{1k} \mathbf{v}_k,$$

therefore by linear independence of $\{\mathbf{v}_k\}$ all $\lambda_{1k} = 0$, contradicting the assumption $\lambda_{12} \neq 0$.

(2) The proof of part (1) is straightforwardly generalized to the space $\wedge^m V$, using induction in m. We have just proved the basis of induction, $m = 2$. Now the induction step: assume that the statement is proved for $m - 1$ and consider a set $\{\mathbf{v}_{k_1} \wedge ... \wedge \mathbf{v}_{k_m}\}$, of tensors of rank m, where $\{\mathbf{v}_j\}$ is a basis. Suppose that this set is linearly dependent; then there is a linear combination

$$\omega \equiv \sum_{k_1, ..., k_m} \lambda_{k_1 ... k_m} \mathbf{v}_{k_1} \wedge ... \wedge \mathbf{v}_{k_m} = 0$$

with some nonzero coefficients, e.g. $\lambda_{12...m} \neq 0$. There exists a covector $\mathbf{f}^*$ such that $\mathbf{f}^*(\mathbf{v}_1) = 1$ and $\mathbf{f}^*(\mathbf{v}_i) = 0$ for $2 \leq i \leq n$. Apply this covector to the tensor ω and obtain $\iota_{\mathbf{f}^*} \omega = 0$, which yields a vanishing linear combination of tensors $\mathbf{v}_{k_1} \wedge ... \wedge \mathbf{v}_{k_{m-1}}$ of rank $m - 1$ with *some* nonzero coefficients. But this contradicts the induction assumption, which says that any set of tensors $\mathbf{v}_{k_1} \wedge ... \wedge \mathbf{v}_{k_{m-1}}$ of rank $m - 1$ is linearly independent. ■

Now we are ready to compute the dimension of $\wedge^m V$.

Theorem 2: The dimension of the space $\wedge^m V$ is

$$\dim \wedge^m V = \binom{N}{m} = \frac{N!}{m!\,(N-m)!},$$

where $N \equiv \dim V$. For $m > N$ we have $\dim \wedge^m V = 0$, i.e. the spaces $\wedge^m V$ for $m > N$ consist solely of the zero tensor.

Proof: We will explicitly construct a basis in the space $\wedge^m V$. First choose a basis $\{\mathbf{e}_1, ..., \mathbf{e}_N\}$ in V. By Lemma 3, the set of $\binom{N}{m}$ tensors

$$\{\mathbf{e}_{k_1} \wedge \mathbf{e}_{k_2} \wedge ... \wedge \mathbf{e}_{k_m},\ 1 \leq k_1 < k_2 < ... < k_m \leq N\}$$

is linearly independent, and by Lemma 2 any tensor $A \in \wedge^m V$ is a linear combination of these tensors. Therefore the set $\{\mathbf{e}_{k_1} \wedge \mathbf{e}_{k_2} \wedge ... \wedge \mathbf{e}_{k_m}\}$ is a basis in $\wedge^m V$. By Theorem 1.1.5, the dimension of space is equal to the number of vectors in any basis, therefore $\dim \wedge^m N = \binom{N}{m}$.

For $m > N$, the existence of a nonzero tensor $\mathbf{v}_1 \wedge ... \wedge \mathbf{v}_m$ contradicts Theorem 1: The set $\{\mathbf{v}_1, ..., \mathbf{v}_m\}$ cannot be linearly independent since it has more vectors than the dimension of the space. Therefore all such tensors are equal to zero (more pedantically, to the *zero tensor*), which is thus the only element of $\wedge^m V$ for every $m > N$. ∎

Exercise 1: It is given that the set of four vectors $\{\mathbf{a}, \mathbf{b}, \mathbf{c}, \mathbf{d}\}$ is linearly independent. Show that the tensor $\omega \equiv \mathbf{a} \wedge \mathbf{b} + \mathbf{c} \wedge \mathbf{d} \in \wedge^2 V$ *cannot* be equal to a single-term exterior product of the form $\mathbf{x} \wedge \mathbf{y}$.

Outline of solution:

1. Constructive solution. There exists $\mathbf{f}^* \in V^*$ such that $\mathbf{f}^*(\mathbf{a}) = 1$ and $\mathbf{f}^*(\mathbf{b}) = 0$, $\mathbf{f}^*(\mathbf{c}) = 0$, $\mathbf{f}^*(\mathbf{d}) = 0$. Compute $\iota_{\mathbf{f}^*}\omega = \mathbf{b}$. If $\omega = \mathbf{x} \wedge \mathbf{y}$, it will follow that a linear combination of $\mathbf{x}$ and $\mathbf{y}$ is equal to $\mathbf{b}$, i.e. $\mathbf{b}$ belongs to the two-dimensional space $\text{Span}\,\{\mathbf{x}, \mathbf{y}\}$. Repeat this argument for the remaining three vectors ($\mathbf{a}$, $\mathbf{c}$, $\mathbf{d}$) and obtain a contradiction.

2. Non-constructive solution. Compute $\omega \wedge \omega = 2\mathbf{a} \wedge \mathbf{b} \wedge \mathbf{c} \wedge \mathbf{d} \neq 0$ by linear independence of $\{\mathbf{a}, \mathbf{b}, \mathbf{c}, \mathbf{d}\}$. If we could express $\omega = \mathbf{x} \wedge \mathbf{y}$ then we would have $\omega \wedge \omega = 0$. ∎

Remark: While $\mathbf{a} \wedge \mathbf{b}$ is interpreted geometrically as the oriented area of a parallelogram spanned by $\mathbf{a}$ and $\mathbf{b}$, a general linear combination such as $\mathbf{a} \wedge \mathbf{b} + \mathbf{c} \wedge \mathbf{d} + \mathbf{e} \wedge \mathbf{f}$ does not have this interpretation (unless it can be reduced to a single-term product $\mathbf{x} \wedge \mathbf{y}$). If not reducible to a single-term product, $\mathbf{a} \wedge \mathbf{b} + \mathbf{c} \wedge \mathbf{d}$ can be interpreted only as a *formal* linear combination of two areas.

Exercise 2: Suppose that $\psi \in \wedge^k V$ and $\mathbf{x} \in V$ are such that $\mathbf{x} \wedge \psi = 0$ while $\mathbf{x} \neq 0$. Show that there exists $\chi \in \wedge^{k-1} V$ such that $\psi = \mathbf{x} \wedge \chi$. Give an example where ψ and χ are *not* representable as a single-term exterior product.

Outline of solution: There exists $\mathbf{f}^* \in V^*$ such that $\mathbf{f}^*(\mathbf{x}) = 1$. Apply $\iota_{\mathbf{f}^*}$ to the given equality $\mathbf{x} \wedge \psi = 0$:

$$0 \stackrel{!}{=} \iota_{\mathbf{f}^*}(\mathbf{x} \wedge \psi) = \psi - \mathbf{x} \wedge \iota_{\mathbf{f}^*}\psi,$$

which means that $\psi = \mathbf{x} \wedge \chi$ with $\chi \equiv \iota_{\mathbf{f}^*}\psi$. An example can be found with $\chi = \mathbf{a} \wedge \mathbf{b} + \mathbf{c} \wedge \mathbf{d}$ as in Exercise 1, and $\mathbf{x}$ such that the set $\{\mathbf{a}, \mathbf{b}, \mathbf{c}, \mathbf{d}, \mathbf{x}\}$ is linearly independent; then $\psi \equiv \mathbf{x} \wedge \psi$ is also not reducible to a single-term product.

2.3.3 Computing the dual basis

The exterior product allows us to compute explicitly the dual basis for a given basis.

We begin with some motivation. Suppose $\{\mathbf{v}_1, ..., \mathbf{v}_N\}$ is a given basis; we would like to compute its dual basis. For instance, the covector $\mathbf{v}_1^*$ of the dual basis is the linear function such that $\mathbf{v}_1^*(\mathbf{x})$ is equal to the coefficient at $\mathbf{v}_1$ in the decomposition of $\mathbf{x}$ in the basis $\{\mathbf{v}_j\}$,

$$\mathbf{x} = \sum_{i=1}^{N} x_i \mathbf{v}_i; \quad \mathbf{v}_1^*(\mathbf{x}) = x_1.$$

We start from the observation that the tensor $\omega \equiv \mathbf{v}_1 \wedge ... \wedge \mathbf{v}_N$ is nonzero since $\{\mathbf{v}_j\}$ is a basis. The exterior product $\mathbf{x} \wedge \mathbf{v}_2 \wedge ... \wedge \mathbf{v}_N$ is equal to zero if $\mathbf{x}$ is a linear combination only of $\mathbf{v}_2$, ..., $\mathbf{v}_N$, with a zero coefficient x_1. This suggests that the exterior product of $\mathbf{x}$ with the $(N-1)$-vector $\mathbf{v}_2 \wedge ... \wedge \mathbf{v}_N$ is quite similar to the covector $\mathbf{v}_1^*$ we are looking for. Indeed, let us compute

$$\mathbf{x} \wedge \mathbf{v}_2 \wedge ... \wedge \mathbf{v}_N = x_1 \mathbf{v}_1 \wedge \mathbf{v}_2 \wedge ... \wedge \mathbf{v}_N = x_1 \omega.$$

Therefore, exterior multiplication with $\mathbf{v}_2 \wedge ... \wedge \mathbf{v}_N$ acts quite similarly to $\mathbf{v}_1^*$. To make the notation more concise, let us introduce a special **complement** operation[3] denoted by a star:

$$*(\mathbf{v}_1) \equiv \mathbf{v}_2 \wedge ... \wedge \mathbf{v}_N.$$

Then we can write $\mathbf{v}_1^*(\mathbf{x})\omega = \mathbf{x} \wedge *(\mathbf{v}_1)$. This equation can be used for computing $\mathbf{v}_1^*$: namely, for any $\mathbf{x} \in V$ the number $\mathbf{v}_1^*(\mathbf{x})$ is equal to the constant λ in the equation $\mathbf{x} \wedge *(\mathbf{v}_1) = \lambda\omega$. To make this kind of equation more convenient, let us write

$$\lambda \equiv \mathbf{v}_1^*(\mathbf{x}) = \frac{\mathbf{x} \wedge \mathbf{v}_2 \wedge ... \wedge \mathbf{v}_N}{\mathbf{v}_1 \wedge \mathbf{v}_2 \wedge ... \wedge \mathbf{v}_N} = \frac{\mathbf{x} \wedge *(\mathbf{v}_1)}{\omega},$$

where the "division" of one tensor by another is to be understood as follows: We first compute the tensor $\mathbf{x} \wedge *(\mathbf{v}_1)$; this tensor is proportional to the tensor ω since both belong to the one-dimensional space $\wedge^N V$, so we can determine the number λ such that $\mathbf{x} \wedge *(\mathbf{v}_1) = \lambda\omega$; the proportionality coefficient λ is then the result of the division of $\mathbf{x} \wedge *(\mathbf{v}_1)$ by ω.

For $\mathbf{v}_2$ we have

$$\mathbf{v}_1 \wedge \mathbf{x} \wedge \mathbf{v}_3 \wedge ... \wedge \mathbf{v}_N = x_2 \omega = \mathbf{v}_2^*(\mathbf{x})\omega.$$

[3]The complement operation was introduced by H. Grassmann (1844).

If we would like to have $x_2\omega = \mathbf{x}\wedge*(\mathbf{v}_2)$, we need to add an extra minus sign and define

$$*\,(\mathbf{v}_2) \equiv -\mathbf{v}_1 \wedge \mathbf{v}_3 \wedge ... \wedge \mathbf{v}_N.$$

Then we indeed obtain $\mathbf{v}_2^*(\mathbf{x})\omega = \mathbf{x}\wedge*(\mathbf{v}_2)$.

It is then clear that we can define the tensors $*(\mathbf{v}_i)$ for $i = 1, ..., N$ in this way. The tensor $*(\mathbf{v}_i)$ is obtained from ω by removing the vector $\mathbf{v}_i$ and by adding a sign that corresponds to shifting the vector $\mathbf{v}_i$ to the left position in the exterior product. The "complement" map, $* : V \to \wedge^{N-1}V$, satisfies $\mathbf{v}_j \wedge *(\mathbf{v}_j) = \omega$ for each *basis* vector $\mathbf{v}_j$. (Once defined on the basis vectors, the complement map can be then extended to all vectors from V by requiring linearity. However, we will apply the complement operation only to basis vectors right now.)

With these definitions, we may express the dual basis as

$$\mathbf{v}_i^*(\mathbf{x})\omega = \mathbf{x}\wedge*(\mathbf{v}_i), \quad \mathbf{x}\in V,\ i = 1, ..., N.$$

Remark: The notation $*(\mathbf{v}_i)$ suggests that e.g. $*(\mathbf{v}_1)$ is some operation applied to $\mathbf{v}_1$ and is a function only of the vector $\mathbf{v}_1$, but this is not so: The "complement" of a vector depends on the entire basis and not merely on the single vector! Also, the property $\mathbf{v}_1 \wedge *(\mathbf{v}_1) = \omega$ is not sufficient to define the tensor $*\mathbf{v}_1$. The proper definition of $*(\mathbf{v}_i)$ is the tensor obtained from ω by removing $\mathbf{v}_i$ as just explained.

Example: In the space $\mathbb{R}^2$, let us compute the dual basis to the basis $\{\mathbf{v}_1, \mathbf{v}_2\}$ where $\mathbf{v}_1 = \binom{2}{1}$ and $\mathbf{v}_2 = \binom{-1}{1}$.

Denote by $\mathbf{e}_1$ and $\mathbf{e}_2$ the standard basis vectors $\binom{1}{0}$ and $\binom{0}{1}$. We first compute the 2-vector

$$\omega = \mathbf{v}_1 \wedge \mathbf{v}_2 = (2\mathbf{e}_1 + \mathbf{e}_2) \wedge (-\mathbf{e}_1 + \mathbf{e}_2) = 3\mathbf{e}_1 \wedge \mathbf{e}_2.$$

The "complement" operation for the basis $\{\mathbf{v}_1, \mathbf{v}_2\}$ gives $*(\mathbf{v}_1) = \mathbf{v}_2$ and $*(\mathbf{v}_2) = -\mathbf{v}_1$. We now define the covectors $\mathbf{v}_{1,2}^*$ by their action on arbitrary vector $\mathbf{x} \equiv x_1\mathbf{e}_1 + x_2\mathbf{e}_2$,

$$\begin{aligned}
\mathbf{v}_1^*(\mathbf{x})\omega &= \mathbf{x}\wedge\mathbf{v}_2 = (x_1\mathbf{e}_1 + x_2\mathbf{e}_2)\wedge(-\mathbf{e}_1+\mathbf{e}_2)\\
&= (x_1 + x_2)\,\mathbf{e}_1\wedge\mathbf{e}_2 = \frac{x_1+x_2}{3}\omega,\\
\mathbf{v}_2^*(\mathbf{x})\omega &= -\mathbf{x}\wedge\mathbf{v}_1 = -\,(x_1\mathbf{e}_1 + x_2\mathbf{e}_2)\wedge(2\mathbf{e}_1+\mathbf{e}_2)\\
&= (-x_1 + 2x_2)\,\mathbf{e}_1\wedge\mathbf{e}_2 = \frac{-x_1+2x_2}{3}\omega.
\end{aligned}$$

Therefore, $\mathbf{v}_1^* = \frac{1}{3}\mathbf{e}_1^* + \frac{1}{3}\mathbf{e}_2^*$ and $\mathbf{v}_2^* = -\frac{1}{3}\mathbf{e}_1^* + \frac{2}{3}\mathbf{e}_2^*$.

Question: Can we define the complement operation for all $\mathbf{x}\in V$ by the equation $\mathbf{x}\wedge*(\mathbf{x}) = \omega$ where $\omega\in\wedge^N V$ is a fixed tensor? Does the complement really depend on the entire basis? Or perhaps a choice of ω is sufficient?

Answer: No, yes, no. Firstly, $*(\mathbf{x})$ is not uniquely specified by that equation alone, since $\mathbf{x} \wedge A = \omega$ defines A only up to tensors of the form $\mathbf{x} \wedge ...$; secondly, the equation $\mathbf{x} \wedge *(\mathbf{x}) = \omega$ indicates that $*(\lambda\mathbf{x}) = \frac{1}{\lambda} *(\mathbf{x})$, so the complement map would not be linear if defined like that. It is important to keep in mind that the complement map requires an entire basis for its definition and depends not only on the choice of a tensor ω, but also on the choice of all the basis vectors. For example, in two dimensions we have $*(\mathbf{e}_1) = \mathbf{e}_2$; it is clear that $*(\mathbf{e}_1)$ depends on the choice of $\mathbf{e}_2$!

Remark: The situation is different when the vector space is equipped with a scalar product (see Sec. 5.4.2 below). In that case, one usually chooses an *orthonormal* basis to define the complement map; then the complement map is called the **Hodge star**. It turns out that the Hodge star is independent of the choice of the basis as long as the basis is orthonormal with respect to the given scalar product, and as long as the orientation of the basis is unchanged (i.e. as long as the tensor ω does not change sign). In other words, the Hodge star operation is invariant under orthogonal and orientation-preserving transformations of the basis; these transformations preserve the tensor ω. So the Hodge star operation depends not quite on the detailed choice of the basis, but rather on the choice of the scalar product and on the orientation of the basis (the sign of ω). However, right now we are working with a general space without a scalar product. In this case, the complement map depends on the entire basis.

2.3.4 Gaussian elimination

Question: How much computational effort is actually needed to compute the exterior product of n vectors? It looks easy in two or three dimensions, but in N dimensions the product of n vectors $\{\mathbf{x}_1, ..., \mathbf{x}_n\}$ gives expressions such as

$$\bigwedge_{i=1}^{n} \mathbf{x}_n = (x_{11}\mathbf{e}_1 + ... + x_{1N}\mathbf{e}_N) \wedge ... \wedge (x_{n1}\mathbf{e}_1 + ... + x_{nN}\mathbf{e}_N),$$

which will be reduced to an exponentially large number (of order N^n) of elementary tensor products when we expand all brackets.

Answer: Of course, expanding all brackets is not the best way to compute long exterior products. We can instead use a procedure similar to the Gaussian elimination for computing determinants. The key observation is that

$$\mathbf{x}_1 \wedge \mathbf{x}_2 \wedge ... = \mathbf{x}_1 \wedge (\mathbf{x}_2 - \lambda\mathbf{x}_1) \wedge ...$$

for any number λ, and that it is easy to compute an exterior product of the form

$$(\alpha_1\mathbf{e}_1 + \alpha_2\mathbf{e}_2 + \alpha_3\mathbf{e}_3) \wedge (\beta_2\mathbf{e}_2 + \beta_3\mathbf{e}_3) \wedge \mathbf{e}_3 = \alpha_1\beta_2\mathbf{e}_1 \wedge \mathbf{e}_2 \wedge \mathbf{e}_3.$$

It is easy to compute this exterior product because the second vector $(\beta_2\mathbf{e}_2 + \beta_3\mathbf{e}_3)$ does not contain the basis vector $\mathbf{e}_1$ and the third vector does not contain $\mathbf{e}_1$ or $\mathbf{e}_2$. So we can simplify the computation of a long exterior product if we rewrite

$$\begin{aligned}\bigwedge_{i=1}^{n} \mathbf{x}_n &= \mathbf{x}_1 \wedge \tilde{\mathbf{x}}_2 \wedge ... \wedge \tilde{\mathbf{x}}_n \\ &\equiv \mathbf{x}_1 \wedge (\mathbf{x}_2 - \lambda_{11}\mathbf{x}_1) \wedge ... \wedge (\mathbf{x}_n - \lambda_{n1}\mathbf{x}_1 - ... - \lambda_{n-1,n-1}\mathbf{x}_{n-1}),\end{aligned}$$

where the coefficients $\{\lambda_{ij} \,|\, 1 \leq i \leq n-1,\ 1 \leq j \leq i\}$ are chosen appropriately such that the vector $\tilde{\mathbf{x}}_2 \equiv \mathbf{x}_2 - \lambda_{11}\mathbf{x}_1$ does not contain the basis vector $\mathbf{e}_1$, and generally the vector

$$\tilde{\mathbf{x}}_k \equiv \mathbf{x}_k - \lambda_{k1}\mathbf{x}_1 - ... - \lambda_{k-1,k-1}\mathbf{x}_{k-1}$$

does not contain the basis vectors $\mathbf{e}_1$,..., $\mathbf{e}_{k-1}$. (That is, these basis vectors have been "eliminated" from the vector $\mathbf{x}_k$, hence the name of the method.) Eliminating $\mathbf{e}_1$ from $\mathbf{x}_2$ can be done with $\lambda_{11} = \frac{x_{21}}{x_{11}}$, which is possible provided that $x_{11} \neq 0$; if $x_{11} = 0$, we need to renumber the vectors $\{\mathbf{x}_j\}$. If none of them contains $\mathbf{e}_1$, we skip $\mathbf{e}_1$ and proceed with $\mathbf{e}_2$ instead. Elimination of other basis vectors proceeds similarly. After performing this algorithm, we will either find that some vector $\tilde{\mathbf{x}}_k$ is itself zero, which means that the entire exterior product vanishes, or we will find the product of vectors of the form

$$\tilde{\mathbf{x}}_1 \wedge ... \wedge \tilde{\mathbf{x}}_n,$$

where the vectors $\tilde{\mathbf{x}}_i$ are linear combinations of $\mathbf{e}_i$, ..., $\mathbf{e}_N$ (not containing $\mathbf{e}_1$, ..., $\mathbf{e}_i$).

If $n = N$, the product can be evaluated immediately since the last vector, $\tilde{\mathbf{x}}_N$, is proportional to $\mathbf{e}_N$, so

$$\begin{aligned}\tilde{\mathbf{x}}_1 \wedge ... \wedge \tilde{\mathbf{x}}_n &= (c_{11}\mathbf{e}_1 + ...) \wedge ... \wedge (c_{nn}\mathbf{e}_N) \\ &= c_{11}c_{22}...c_{nn}\mathbf{e}_1 \wedge ... \wedge \mathbf{e}_N.\end{aligned}$$

The computation is somewhat longer if $n < N$, so that

$$\tilde{\mathbf{x}}_n = c_{nn}\mathbf{e}_n + ... + c_{nN}\mathbf{e}_N.$$

In that case, we may eliminate, say, $\mathbf{e}_n$ from $\tilde{\mathbf{x}}_1$, ..., $\tilde{\mathbf{x}}_{n-1}$ by subtracting a multiple of $\tilde{\mathbf{x}}_n$ from them, but we cannot simplify the product any more; at that point we need to expand the last bracket (containing $\mathbf{x}_n$) and write out the terms.

Example 1: We will calculate the exterior product

$$\begin{aligned}&\mathbf{a} \wedge \mathbf{b} \wedge \mathbf{c} \\ &\equiv (7\mathbf{e}_1 - 8\mathbf{e}_2 + \mathbf{e}_3) \wedge (\mathbf{e}_1 - 2\mathbf{e}_2 - 15\mathbf{e}_3) \wedge (2\mathbf{e}_1 - 5\mathbf{e}_2 - \mathbf{e}_3).\end{aligned}$$

We will eliminate $\mathbf{e}_1$ from $\mathbf{a}$ and $\mathbf{c}$ (just to keep the coefficients simpler):

$$\begin{aligned}\mathbf{a} \wedge \mathbf{b} \wedge \mathbf{c} &= (\mathbf{a} - 7\mathbf{b}) \wedge \mathbf{b} \wedge (\mathbf{c} - 2\mathbf{b}) \\ &= (6\mathbf{e}_2 + 106\mathbf{e}_3) \wedge \mathbf{b} \wedge (-\mathbf{e}_2 + 9\mathbf{e}_3) \\ &\equiv \mathbf{a}_1 \wedge \mathbf{b} \wedge \mathbf{c}_1.\end{aligned}$$

Now we eliminate $\mathbf{e}_2$ from $\mathbf{a}_1$, and then the product can be evaluated quickly:

$$\begin{aligned}\mathbf{a}\wedge\mathbf{b}\wedge\mathbf{c}&=\mathbf{a}_1\wedge\mathbf{b}\wedge\mathbf{c}_1=(\mathbf{a}_1+6\mathbf{c}_1)\wedge\mathbf{b}\wedge\mathbf{c}_1\\&=(160\mathbf{e}_3)\wedge(\mathbf{e}_1-2\mathbf{e}_2-5\mathbf{e}_3)\wedge(-\mathbf{e}_2+9\mathbf{e}_3)\\&=160\mathbf{e}_3\wedge\mathbf{e}_1\wedge(-\mathbf{e}_2)=-160\mathbf{e}_1\wedge\mathbf{e}_2\wedge\mathbf{e}_3.\end{aligned}$$

Example 2: Consider

$$\begin{aligned}\mathbf{a}\wedge\mathbf{b}\wedge\mathbf{c}\equiv&(\mathbf{e}_1+2\mathbf{e}_2-\mathbf{e}_3+\mathbf{e}_4)\\&\wedge(2\mathbf{e}_1+\mathbf{e}_2-\mathbf{e}_3+3\mathbf{e}_4)\wedge(-\mathbf{e}_1-\mathbf{e}_2+\mathbf{e}_4).\end{aligned}$$

We eliminate $\mathbf{e}_1$ and $\mathbf{e}_2$:

$$\begin{aligned}\mathbf{a}\wedge\mathbf{b}\wedge\mathbf{c}&=\mathbf{a}\wedge(\mathbf{b}-2\mathbf{a})\wedge(\mathbf{c}+\mathbf{a})\\&=\mathbf{a}\wedge(-3\mathbf{e}_2+\mathbf{e}_3+\mathbf{e}_4)\wedge(\mathbf{e}_2-\mathbf{e}_3+2\mathbf{e}_4)\\&\equiv\mathbf{a}\wedge\mathbf{b}_1\wedge\mathbf{c}_1=\mathbf{a}\wedge\mathbf{b}_1\wedge(\mathbf{c}_1+3\mathbf{b}_1)\\&=\mathbf{a}\wedge\mathbf{b}_1\wedge(2\mathbf{e}_3+5\mathbf{e}_4)\equiv\mathbf{a}\wedge\mathbf{b}_1\wedge\mathbf{c}_2.\end{aligned}$$

We can now eliminate $\mathbf{e}_3$ from $\mathbf{a}$ and $\mathbf{b}_1$:

$$\begin{aligned}\mathbf{a}\wedge\mathbf{b}_1\wedge\mathbf{c}_2&=(\mathbf{a}+\frac{1}{2}\mathbf{c}_2)\wedge(\mathbf{b}_1-\frac{1}{2}\mathbf{c}_2)\wedge\mathbf{c}_2\equiv\mathbf{a}_2\wedge\mathbf{b}_2\wedge\mathbf{c}_2\\&=(\mathbf{e}_1+2\mathbf{e}_2+\frac{7}{2}\mathbf{e}_4)\wedge(-3\mathbf{e}_2-\frac{3}{2}\mathbf{e}_4)\wedge(2\mathbf{e}_3+5\mathbf{e}_4).\end{aligned}$$

Now we cannot eliminate any more vectors, so we expand the last bracket and simplify the result by omitting the products of equal vectors:

$$\begin{aligned}\mathbf{a}_2\wedge\mathbf{b}_2\wedge\mathbf{c}_2&=\mathbf{a}_2\wedge\mathbf{b}_2\wedge 2\mathbf{e}_3+\mathbf{a}_2\wedge\mathbf{b}_2\wedge 5\mathbf{e}_4\\&=(\mathbf{e}_1+2\mathbf{e}_2)\wedge(-\frac{3}{2}\mathbf{e}_4)\wedge 2\mathbf{e}_3+\mathbf{e}_1\wedge(-3\mathbf{e}_2)\wedge 2\mathbf{e}_3\\&\quad+\mathbf{e}_1\wedge(-3\mathbf{e}_2)\wedge 5\mathbf{e}_4\\&=3\mathbf{e}_1\wedge\mathbf{e}_3\wedge\mathbf{e}_4+6\mathbf{e}_2\wedge\mathbf{e}_3\wedge\mathbf{e}_4-6\mathbf{e}_1\wedge\mathbf{e}_2\wedge\mathbf{e}_3-15\mathbf{e}_1\wedge\mathbf{e}_2\wedge\mathbf{e}_4.\end{aligned}$$

2.3.5 Rank of a set of vectors

We have defined the rank of a map (Sec. 1.8.4) as the dimension of the image of the map, and we have seen that the rank is equal to the minimum number of tensor product terms needed to represent the map as a tensor. An analogous concept can be introduced for sets of vectors.

Definition: If $S=\{\mathbf{v}_1,...,\mathbf{v}_n\}$ is a set of vectors (where n is not necessarily smaller than the dimension N of space), the **rank** of the set S is the dimension of the subspace spanned by the vectors $\{\mathbf{v}_1,...,\mathbf{v}_n\}$. Written as a formula,

$$\text{rank}\,(S)=\dim\text{Span}\,S.$$

The rank of a set S is equal to the maximum number of vectors in any linearly independent subset of S. For example, consider the set $\{0,\mathbf{v},2\mathbf{v},3\mathbf{v}\}$

where $\mathbf{v} \neq 0$. The rank of this set is 1 since these four vectors span a one-dimensional subspace,

$$\text{Span}\{0, \mathbf{v}, 2\mathbf{v}, 3\mathbf{v}\} = \text{Span}\{\mathbf{v}\}.$$

Any subset of S having two or more vectors is linearly dependent.

We will now show how to use the exterior product for computing the rank of a given (finite) set $S = \{\mathbf{v}_1, ..., \mathbf{v}_n\}$.

According to Theorem 1 in Sec. 2.3.2, the set S is linearly independent if and only if $\mathbf{v}_1 \wedge ... \wedge \mathbf{v}_n \neq 0$. So we first compute the tensor $\mathbf{v}_1 \wedge ... \wedge \mathbf{v}_n$. If this tensor is nonzero then the set S is linearly independent, and the rank of S is equal to n. If, on the other hand, $\mathbf{v}_1 \wedge ... \wedge \mathbf{v}_n = 0$, the rank is less than n. We can determine the rank of S by the following procedure. First, we assume that all $\mathbf{v}_j \neq 0$ (any zero vectors can be omitted without changing the rank of S). Then we compute $\mathbf{v}_1 \wedge \mathbf{v}_2$; if the result is zero, we may omit $\mathbf{v}_2$ since $\mathbf{v}_2$ is proportional to $\mathbf{v}_1$ and try $\mathbf{v}_1 \wedge \mathbf{v}_3$. If $\mathbf{v}_1 \wedge \mathbf{v}_2 \neq 0$, we try $\mathbf{v}_1 \wedge \mathbf{v}_2 \wedge \mathbf{v}_3$, and so on. The procedure can be formulated using induction in the obvious way. Eventually we will arrive at a subset $\{\mathbf{v}_{i_1}, ..., \mathbf{v}_{i_k}\} \subset S$ such that $\mathbf{v}_{i_1} \wedge ... \wedge ...\mathbf{v}_{i_k} \neq 0$ but $\mathbf{v}_{i_1} \wedge ... \wedge ...\mathbf{v}_{i_k} \wedge \mathbf{v}_j = 0$ for any other $\mathbf{v}_j$. Thus, there are no linearly independent subsets of S having $k+1$ or more vectors. Then the rank of S is equal to k.

The subset $\{\mathbf{v}_{i_1}, ..., \mathbf{v}_{i_k}\}$ is built by a procedure that depends on the order in which the vectors $\mathbf{v}_j$ are selected. However, the next statement says that the resulting subspace spanned by $\{\mathbf{v}_{i_1}, ..., \mathbf{v}_{i_k}\}$ is the same regardless of the order of vectors $\mathbf{v}_j$. Hence, the subset $\{\mathbf{v}_{i_1}, ..., \mathbf{v}_{i_k}\}$ yields a basis in Span S.

Statement: Suppose a set S of vectors has rank k and contains *two* different linearly independent subsets, say $S_1 = \{\mathbf{v}_1, ..., \mathbf{v}_k\}$ and $S_2 = \{\mathbf{u}_1, ..., \mathbf{u}_k\}$, both having k vectors (but no linearly independent subsets having $k+1$ or more vectors). Then the tensors $\mathbf{v}_1 \wedge ... \wedge \mathbf{v}_k$ and $\mathbf{u}_1 \wedge ... \wedge \mathbf{u}_k$ are proportional to each other (as tensors from $\wedge^k V$).

Proof: The tensors $\mathbf{v}_1 \wedge ... \wedge \mathbf{v}_k$ and $\mathbf{u}_1 \wedge ... \wedge \mathbf{u}_k$ are both nonzero by Theorem 1 in Sec. 2.3.2. We will now show that it is possible to replace $\mathbf{v}_1$ by one of the vectors from the set S_2, say $\mathbf{u}_l$, such that the new tensor $\mathbf{u}_l \wedge \mathbf{v}_2 \wedge ... \wedge \mathbf{v}_k$ is nonzero and proportional to the original tensor $\mathbf{v}_1 \wedge ... \wedge \mathbf{v}_k$. It will follow that this procedure can be repeated for every other vector $\mathbf{v}_i$, until we replace all $\mathbf{v}_i$'s by some $\mathbf{u}_i$'s and thus prove that the tensors $\mathbf{v}_1 \wedge ... \wedge \mathbf{v}_k$ and $\mathbf{u}_1 \wedge ... \wedge \mathbf{u}_k$ are proportional to each other.

It remains to prove that the vector $\mathbf{v}_1$ can be replaced. We need to find a suitable vector $\mathbf{u}_l$. Let $\mathbf{u}_l$ be one of the vectors from S_2, and let us check whether $\mathbf{v}_1$ could be replaced by $\mathbf{u}_l$. We first note that $\mathbf{v}_1 \wedge ... \wedge \mathbf{v}_k \wedge \mathbf{u}_l = 0$ since there are no linearly independent subsets of S having $k+1$ vectors. Hence the set $\{\mathbf{v}_1, ..., \mathbf{v}_k, \mathbf{u}_l\}$ is linearly *dependent*. It follows (since the set $\{\mathbf{v}_i \,|\, i = 1, ..., k\}$ was linearly independent before we added $\mathbf{u}_l$ to it) that $\mathbf{u}_l$ can be expressed as a linear combination of the $\mathbf{v}_i$'s with some coefficients α_i:

$$\mathbf{u}_l = \alpha_1 \mathbf{v}_1 + ... + \alpha_k \mathbf{v}_k.$$

If $\alpha_1 \neq 0$ then we will have

$$\mathbf{u}_l \wedge \mathbf{v}_2 \wedge ... \wedge \mathbf{v}_k = \alpha_1 \mathbf{v}_1 \wedge \mathbf{v}_2 \wedge ... \wedge \mathbf{v}_k.$$

The new tensor is nonzero and proportional to the old tensor, so we can replace $\mathbf{v}_1$ by $\mathbf{u}_l$.

However, it could also happen that $\alpha_1 = 0$. In that case we need to choose a different vector $\mathbf{u}_{l'} \in S_2$ such that the corresponding coefficient α_1 is nonzero. It remains to prove that such a choice is possible. If this were impossible then all $\mathbf{u}_i$'s would have been expressible as linear combinations of $\mathbf{v}_i$'s with zero coefficients at the vector $\mathbf{v}_1$. In that case, the exterior product $\mathbf{u}_1 \wedge ... \wedge \mathbf{u}_k$ would be equal to a linear combination of exterior products of vectors $\mathbf{v}_i$ with $i = 2, ..., k$. These exterior products contain k vectors among which only $(k-1)$ vectors are different. Such exterior products are all equal to zero. However, this contradicts the assumption $\mathbf{u}_1 \wedge ... \wedge \mathbf{u}_k \neq 0$. Therefore, at least one vector $\mathbf{u}_l$ exists such that $\alpha_1 \neq 0$, and the required replacement is always possible. ■

Remark: It follows from the above Statement that the subspace spanned by S can be uniquely characterized by a nonzero tensor such as $\mathbf{v}_1 \wedge ... \wedge \mathbf{v}_k$ in which the constituents — the vectors $\mathbf{v}_1, ..., \mathbf{v}_k$ — form a basis in the subspace Span S. It does not matter which linearly independent subset we choose for this purpose. We also have a computational procedure for determining the subspace Span S together with its dimension. Thus, we find that a k-dimensional subspace is adequately specified by selecting a nonzero tensor $\omega \in \wedge^k V$ of the form $\omega = \mathbf{v}_1 \wedge ... \wedge \mathbf{v}_k$. For a given subspace, this tensor ω is unique up to a nonzero constant factor. Of course, the decomposition of ω into an exterior product of vectors $\{\mathbf{v}_i \,|\, i = 1, ..., k\}$ is not unique, but any such decomposition yields a set $\{\mathbf{v}_i \,|\, i = 1, ..., k\}$ spanning the same subspace.

Exercise 1: Let $\{\mathbf{v}_1, ..., \mathbf{v}_n\}$ be a linearly independent set of vectors, $\omega \equiv \mathbf{v}_1 \wedge ... \wedge \mathbf{v}_n \neq 0$, and $\mathbf{x}$ be a given vector such that $\omega \wedge \mathbf{x} = 0$. Show that $\mathbf{x}$ belongs to the subspace Span $\{\mathbf{v}_1, ..., \mathbf{v}_n\}$.

Exercise 2: Given a nonzero covector $\mathbf{f}^*$ and a vector $\mathbf{n}$ such that $\mathbf{f}^*(\mathbf{n}) \neq 0$, show that the operator $\hat{P}$ defined by

$$\hat{P}\mathbf{x} = \mathbf{x} - \mathbf{n}\frac{\mathbf{f}^*(\mathbf{x})}{\mathbf{f}^*(\mathbf{n})}$$

is a projector onto the subspace $\mathbf{f}^{*\perp}$, i.e. that $\mathbf{f}^*(\hat{P}\mathbf{x}) = 0$ for all $\mathbf{x} \in V$. Show that

$$(\hat{P}\mathbf{x}) \wedge \mathbf{n} = \mathbf{x} \wedge \mathbf{n}, \quad \forall \mathbf{x} \in V.$$

2.3.6 Exterior product in index notation

Here I show how to perform calculations with the exterior product using the index notation (see Sec. 1.9), although I will not use this later because the index-free notation is more suitable for the purposes of this book.

Let us choose a basis $\{\mathbf{e}_j\}$ in V; then the dual basis $\{\mathbf{e}_j^*\}$ in V and the basis $\{\mathbf{e}_{k_1} \wedge ... \wedge \mathbf{e}_{k_m}\}$ in $\wedge^m V$ are fixed. By definition, the exterior product of two vectors $\mathbf{u}$ and $\mathbf{v}$ is

$$A \equiv \mathbf{u} \wedge \mathbf{v} = \mathbf{u} \otimes \mathbf{v} - \mathbf{v} \otimes \mathbf{u},$$

therefore it is written in the index notation as $A^{ij} = u^i v^j - u^j v^i$. Note that the matrix A^{ij} is antisymmetric: $A^{ij} = -A^{ji}$.

Another example: The 3-vector $\mathbf{u} \wedge \mathbf{v} \wedge \mathbf{w}$ can be expanded in the basis as

$$\mathbf{u} \wedge \mathbf{v} \wedge \mathbf{w} = \sum_{i,j,k=1}^{N} B^{ijk} \mathbf{e}_i \wedge \mathbf{e}_j \wedge \mathbf{e}_k.$$

What is the relation between the components u^i, v^i, w^i of the vectors and the components B^{ijk}? A direct calculation yields

$$B^{ijk} = u^i v^j w^k - u^i v^k w^j + u^k v^i w^j - u^k w^j v^i + u^j w^k v^i - u^j w^i w^k. \tag{2.10}$$

In other words, every permutation of the set (i, j, k) of indices enters with the sign corresponding to the parity of that permutation.

Remark: Readers familiar with the standard definition of the matrix determinant will recognize a formula quite similar to the determinant of a 3×3 matrix. The connection between determinants and exterior products will be fully elucidated in Chapter 3.

Remark: The "three-dimensional array" B^{ijk} is antisymmetric with respect to *any* pair of indices:

$$B^{ijk} = -B^{jik} = -B^{ikj} = ...$$

Such arrays are called **totally antisymmetric**. ■

The formula (2.10) for the components B^{ijk} of $\mathbf{u} \wedge \mathbf{v} \wedge \mathbf{w}$ is not particularly convenient and cannot be easily generalized. We will now rewrite Eq. (2.10) in a different form that will be more suitable for expressing exterior products of arbitrary tensors.

Let us first consider the exterior product of three vectors as a map $\hat{E} : V \otimes V \otimes V \rightarrow \wedge^3 V$. This map is linear and can be represented, in the index notation, in the following way:

$$u^i v^j w^k \mapsto (\mathbf{u} \wedge \mathbf{v} \wedge \mathbf{w})^{ijk} = \sum_{l,m,n} E^{ijk}_{lmn} u^l v^m w^n,$$

where the array E^{ijk}_{lmn} is the component representation of the map E. Comparing with the formula (2.10), we find that E^{ijk}_{lmn} can be expressed through the Kronecker δ-symbol as

$$E^{ijk}_{lmn} = \delta^i_l \delta^j_m \delta^k_n - \delta^i_l \delta^k_m \delta^j_n + \delta^k_l \delta^i_m \delta^j_n - \delta^k_l \delta^j_m \delta^i_n + \delta^j_l \delta^k_m \delta^i_n - \delta^j_l \delta^i_m \delta^k_n.$$

It is now clear that the exterior product of two vectors can be also written as

$$(\mathbf{u} \wedge \mathbf{v})^{ij} = \sum_{l,m} E^{ij}_{lm} u^l v^m,$$

where

$$E^{ij}_{lm} = \delta^i_l \delta^j_m - \delta^j_l \delta^i_m .$$

By analogy, the map $\hat{E} : V \otimes ... \otimes V \to \wedge^n V$ (for $2 \leq n \leq N$) can be represented in the index notation by the array of components $E^{i_1 ... i_n}_{j_1 ... j_n}$. This array is totally antisymmetric with respect to all the indices $\{i_s\}$ and separately with respect to all $\{j_s\}$. Using this array, the exterior product of two general antisymmetric tensors, say $\phi \in \wedge^m V$ and $\psi \in \wedge^n V$, such that $m + n \leq N$, can be represented in the index notation by

$$(\phi \wedge \psi)^{i_1 ... i_{m+n}} = \frac{1}{m!n!} \sum_{(j_s, k_s)} E^{i_1 ... i_{m+n}}_{j_1 ... j_m k_1 ... k_n} \phi^{j_1 ... j_m} \psi^{k_1 ... k_n} .$$

The combinatorial factor $m!n!$ is needed to compensate for the $m!$ equal terms arising from the summation over $(j_1, ..., j_m)$ due to the fact that $\phi^{j_1 ... j_m}$ is totally antisymmetric, and similarly for the $n!$ equal terms arising from the summation over $(k_1, ..., k_m)$.

It is useful to have a general formula for the array $E^{i_1 ... i_n}_{j_1 ... j_n}$. One way to define it is

$$E^{i_1 ... i_n}_{j_1 ... j_n} = \begin{cases} (-1)^{|\sigma|} & \text{if } (i_1, ..., i_n) \text{ is a permutation } \sigma \text{ of } (j_1, ..., j_n) \, ; \\ 0 & \text{otherwise.} \end{cases}$$

We will now show how one can express $E^{i_1 ... i_n}_{j_1 ... j_n}$ through the Levi-Civita symbol ε.

The **Levi-Civita symbol** is defined as a totally antisymmetric array with N indices, whose values are 0 or ± 1 according to the formula

$$\varepsilon^{i_1 ... i_N} = \begin{cases} (-1)^{|\sigma|} & \text{if } (i_1, ..., i_N) \text{ is a permutation } \sigma \text{ of } (1, ..., N) \, ; \\ 0 & \text{otherwise.} \end{cases}$$

Comparing this with the definition of $E^{i_1 ... i_n}_{j_1 ... j_n}$, we notice that

$$\varepsilon^{i_1 ... i_N} = E^{i_1 ... i_N}_{1 ... N} .$$

Depending on convenience, we may write ε with upper or lower indices since ε is just an array of numbers in this calculation.

In order to express $E^{i_1 ... i_n}_{j_1 ... j_n}$ through $\varepsilon^{i_1 ... i_N}$, we obviously need to use at least two copies of ε — one with upper and one with lower indices. Let us therefore consider the expression

$$\tilde{E}^{i_1 ... i_n}_{j_1 ... j_n} \equiv \sum_{k_1, ..., k_{N-n}} \varepsilon^{i_1 ... i_n k_1 ... k_{N-n}} \varepsilon_{j_1 ... j_n k_1 ... k_{N-n}} , \tag{2.11}$$

where the summation is performed *only* over the $N - n$ indices $\{k_s\}$. This expression has $2n$ free indices i_1, ..., i_n and j_1, ..., j_n, and is totally antisymmetric in these free indices (since ε is totally antisymmetric in all indices).

Statement: The exterior product operator $E^{i_1 \ldots i_n}_{j_1 \ldots j_n}$ is expressed through the Levi-Civita symbol as

$$E^{i_1 \ldots i_n}_{j_1 \ldots j_n} = \frac{1}{(N-n)!} \tilde{E}^{i_1 \ldots i_n}_{j_1 \ldots j_n}, \tag{2.12}$$

where $\tilde{E}$ is defined by Eq. (2.11).

Proof: Let us compare the values of $E^{i_1 \ldots i_n}_{j_1 \ldots j_n}$ and $\tilde{E}^{i_1 \ldots i_n}_{j_1 \ldots j_n}$, where the indices $\{i_s\}$ and $\{j_s\}$ have some fixed values. There are two cases: either the set $(i_1, ..., i_n)$ is a permutation of the set $(j_1, ..., j_n)$; in that case we may denote this permutation by σ; or $(i_1, ..., i_n)$ is not a permutation of $(j_1, ..., j_n)$.

Considering the case when a permutation σ brings $(j_1, ..., j_n)$ into $(i_1, ..., i_n)$, we find that the symbols ε in Eq. (2.11) will be nonzero only if the indices $(k_1, ..., k_{N-n})$ are a permutation of the complement of the set $(i_1, ..., i_n)$. There are $(N-n)!$ such permutations, each contributing the same value to the sum in Eq. (2.11). Hence, we may write[4] the sum as

$$\tilde{E}^{i_1 \ldots i_n}_{j_1 \ldots j_n} = (N-n)!\, \varepsilon^{i_1 \ldots i_n k_1 \ldots k_{N-n}} \varepsilon_{j_1 \ldots j_n k_1 \ldots k_{N-n}} \text{ (no sums!)},$$

where the indices $\{k_s\}$ are chosen such that the values of ε are nonzero. Since

$$\sigma\left(j_1, ..., j_n\right) = \left(i_1, ..., i_n\right),$$

we may permute the first n indices in $\varepsilon_{j_1 \ldots j_n k_1 \ldots k_{N-n}}$

$$\begin{aligned} \tilde{E}^{i_1 \ldots i_n}_{j_1 \ldots j_n} &= (N-n)!(-1)^{|\sigma|} \varepsilon^{i_1 \ldots i_n k_1 \ldots k_{N-n}} \varepsilon_{i_1 \ldots i_n k_1 \ldots k_{N-n}} \text{ (no sums!)} \\ &= (N-n)!(-1)^{|\sigma|}. \end{aligned}$$

(In the last line, we replaced the squared ε by 1.) Thus, the required formula for $\tilde{E}$ is valid in the first case.

In the case when σ does not exist, we note that

$$\tilde{E}^{i_1 \ldots i_n}_{j_1 \ldots j_n} = 0,$$

because in that case one of the ε's in Eq. (2.11) will have at least some indices equal and thus will be zero. Therefore $\tilde{E}$ and E are equal to zero for the same sets of indices. ■

Note that the formula for the top exterior power ($n = N$) is simple and involves no summations and no combinatorial factors:

$$E^{i_1 \ldots i_N}_{j_1 \ldots j_N} = \varepsilon^{i_1 \ldots i_N} \varepsilon_{j_1 \ldots j_N}.$$

Exercise: The operator $\hat{E} : V \otimes V \otimes V \to \wedge^3 V$ can be considered within the subspace $\wedge^3 V \subset V \otimes V \otimes V$, which yields an operator $\hat{E} : \wedge^3 V \to \wedge^3 V$. Show that in this subspace,

$$\hat{E} = 3!\, \hat{1}_{\wedge^3 V}.$$

Generalize to $\wedge^n V$ in the natural way.

Hint: Act with $\hat{E}$ on $\mathbf{a} \wedge \mathbf{b} \wedge \mathbf{c}$.

[4] In the equation below, I have put the warning "no sums" for clarity: A summation over all repeated indices is often *implicitly* assumed in the index notation.

Remark: As a rule, a summation of the Levi-Civita symbol ε with any antisymmetric tensor (e.g. another ε) gives rise to a combinatorial factor $n!$ when the summation goes over n indices.

2.3.7 * Exterior algebra (Grassmann algebra)

The formalism of exterior algebra is used e.g. in physical theories of quantum fermionic fields and supersymmetry.
Definition: An **algebra** is a vector space with a distributive multiplication. In other words, $\mathcal{A}$ is an algebra if it is a vector space over a field $\mathbb{K}$ and if for any $a, b \in \mathcal{A}$ their product $ab \in \mathcal{A}$ is defined, such that $a(b+c) = ab + ac$ and $(a+b)c = ac + bc$ and $\lambda(ab) = (\lambda a)b = a(\lambda b)$ for $\lambda \in \mathbb{K}$. An algebra is called **commutative** if $ab = ba$ for all a, b.

The properties of the multiplication in an algebra can be summarized by saying that for any fixed element $a \in \mathcal{A}$, the transformations $x \mapsto ax$ and $x \mapsto xa$ are linear maps of the algebra into itself.
Examples of algebras:

1. All $N \times N$ matrices with coefficients from $\mathbb{K}$ are a N^2-dimensional algebra. The multiplication is defined by the usual matrix multiplication formula. This algebra is not commutative because not all matrices commute.

2. The field $\mathbb{K}$ is a one-dimensional algebra over itself. (Not a very exciting example.) This algebra is commutative.

Statement: If $\omega \in \wedge^m V$ then we can define the map $L_\omega : \wedge^k V \to \wedge^{k+m} V$ by the formula

$$L_\omega (\mathbf{v}_1 \wedge ... \wedge \mathbf{v}_k) \equiv \omega \wedge \mathbf{v}_1 \wedge ... \wedge \mathbf{v}_k.$$

For elements of $\wedge^0 V \equiv \mathbb{K}$, we define $L_\lambda \omega \equiv \lambda \omega$ and also $L_\omega \lambda \equiv \lambda \omega$ for any $\omega \in \wedge^k V$, $\lambda \in \mathbb{K}$. Then the map L_ω is linear for any $\omega \in \wedge^m V$, $0 \leq m \leq N$.

Proof: Left as exercise. ■

Definition: The **exterior algebra** (also called the **Grassmann algebra**) based on a vector space V is the space $\wedge V$ defined as the direct sum,

$$\wedge V \equiv \mathbb{K} \oplus V \oplus \wedge^2 V \oplus ... \oplus \wedge^N V,$$

with the multiplication defined by the map L, which is extended to the whole of $\wedge V$ by linearity.

For example, if $\mathbf{u}, \mathbf{v} \in V$ then $1 + \mathbf{u} \in \wedge V$,

$$A \equiv 3 - \mathbf{v} + \mathbf{u} - 2\mathbf{v} \wedge \mathbf{u} \in \wedge V,$$

and

$$L_{1+\mathbf{u}} A = (1 + \mathbf{u}) \wedge (3 - \mathbf{v} + \mathbf{u} - 2\mathbf{v} \wedge \mathbf{u}) = 3 - \mathbf{v} + 4\mathbf{u} - \mathbf{v} \wedge \mathbf{u}.$$

Note that we still write the symbol $\wedge$ to denote multiplication in $\wedge V$ although now it is not necessarily anticommutative; for instance, $1 \wedge x = x \wedge 1 = x$ for any x in this algebra.

Remark: The summation in expressions such as $1+\mathbf{u}$ above is *formal* in the usual sense: $1+\mathbf{u}$ is not a new vector or a new tensor, but an element of a *new space*. The exterior algebra is thus the space of formal linear combinations of numbers, vectors, 2-vectors, etc., all the way to N-vectors. ■

Since $\wedge V$ is a direct sum of $\wedge^0 V$, $\wedge^1 V$, etc., the elements of $\wedge V$ are sums of scalars, vectors, bivectors, etc., i.e. of objects having a definite "grade" — scalars being "of grade" 0, vectors of grade 1, and generally k-vectors being of grade k. It is easy to see that k-vectors and l-vectors either commute or anticommute, for instance

$$(\mathbf{a}\wedge\mathbf{b})\wedge\mathbf{c}=\mathbf{c}\wedge(\mathbf{a}\wedge\mathbf{b}),$$
$$(\mathbf{a}\wedge\mathbf{b}\wedge\mathbf{c})\wedge 1=1\wedge(\mathbf{a}\wedge\mathbf{b}\wedge\mathbf{c}),$$
$$(\mathbf{a}\wedge\mathbf{b}\wedge\mathbf{c})\wedge\mathbf{d}=-\mathbf{d}\wedge(\mathbf{a}\wedge\mathbf{b}\wedge\mathbf{c}).$$

The general law of commutation and anticommutation can be written as

$$\omega_k\wedge\omega_l=(-1)^{kl}\,\omega_l\wedge\omega_k,$$

where $\omega_k\in\wedge^k V$ and $\omega_l\in\wedge^l V$. However, it is important to note that sums of elements having different grades, such as $1+\mathbf{a}$, are elements of $\wedge V$ that do *not* have a definite grade, because they do not belong to any single subspace $\wedge^k V\subset\wedge V$. Elements that do not have a definite grade can of course still be multiplied within $\wedge V$, but they *neither* commute *nor* anticommute, for example:

$$(1+\mathbf{a})\wedge(1+\mathbf{b})=1+\mathbf{a}+\mathbf{b}+\mathbf{a}\wedge\mathbf{b},$$
$$(1+\mathbf{b})\wedge(1+\mathbf{a})=1+\mathbf{a}+\mathbf{b}-\mathbf{a}\wedge\mathbf{b}.$$

So $\wedge V$ is a *noncommutative* (but associative) algebra. Nevertheless, the fact that elements of $\wedge V$ having a pure grade either commute or anticommute is important, so this kind of algebra is called a **graded algebra**.

Exercise 1: Compute the dimension of the algebra $\wedge V$ as a vector space, if $\dim V=N$.

Answer: $\dim(\wedge V)=\sum_{i=0}^{N}\binom{N}{i}=2^N$.

Exercise 2: Suppose that an element $x\in\wedge V$ is a sum of elements of *pure even* grade, e.g. $x=1+\mathbf{a}\wedge\mathbf{b}$. Show that x commutes with any other element of $\wedge V$.

Exercise 3: Compute $\exp(\mathbf{a})$ and $\exp(\mathbf{a}\wedge\mathbf{b}+\mathbf{c}\wedge\mathbf{d})$ by writing the Taylor series using the multiplication within the algebra $\wedge V$.

Hint: Simplify the expression $\exp(x)=1+x+\frac{1}{2}x\wedge x+...$ for the particular x as given.

Answer: $\exp(\mathbf{a})=1+\mathbf{a}$;

$$\exp(\mathbf{a}\wedge\mathbf{b}+\mathbf{c}\wedge\mathbf{d})=1+\mathbf{a}\wedge\mathbf{b}+\mathbf{c}\wedge\mathbf{d}+\mathbf{a}\wedge\mathbf{b}\wedge\mathbf{c}\wedge\mathbf{d}.$$

3 Basic applications

In this section we will consider finite-dimensional vector spaces V without a scalar product. We will denote by N the dimensionality of V, i.e. $N = \dim V$.

3.1 Determinants through permutations: the hard way

In textbooks on linear algebra, the following definition is found.
Definition D0: The **determinant** of a square $N \times N$ matrix A_{ij} is the number

$$\det(A_{ij}) \equiv \sum_{\sigma} (-1)^{|\sigma|} A_{\sigma(1)1}...A_{\sigma(N)N}, \tag{3.1}$$

where the summation goes over all permutations $\sigma : (1, ..., N) \mapsto (k_1, ..., k_N)$ of the ordered set $(1, ..., N)$, and the parity function $|\sigma|$ is equal to 0 if the permutation σ is even and to 1 if it is odd. (An **even** permutation is reducible to an even number of elementary exchanges of adjacent numbers; for instance, the permutation $(1, 3, 2)$ is odd while $(3, 1, 2)$ is even. See Appendix B if you need to refresh your knowledge of permutations.)

Let us illustrate Eq. (3.1) with 2×2 and 3×3 matrices. Since there are only two permutations of the set $(1, 2)$, namely

$$(1, 2) \mapsto (1, 2) \text{ and } (1, 2) \mapsto (2, 1),$$

and six permutations of the set $(1, 2, 3)$, namely

$$(1, 2, 3), (1, 3, 2), (2, 1, 3), (2, 3, 1), (3, 1, 2), (3, 2, 1),$$

we can write explicit formulas for these determinants:

$$\det \begin{pmatrix} a_{11} & a_{12} \\ a_{21} & a_{22} \end{pmatrix} = a_{11}a_{22} - a_{21}a_{12};$$

$$\det \begin{pmatrix} a_{11} & a_{12} & a_{13} \\ a_{21} & a_{22} & a_{23} \\ a_{31} & a_{32} & a_{33} \end{pmatrix} = a_{11}a_{22}a_{33} - a_{11}a_{32}a_{23} - a_{21}a_{12}a_{33} \\ + a_{21}a_{32}a_{13} + a_{31}a_{12}a_{23} - a_{31}a_{22}a_{13}.$$

We note that the determinant of an $N \times N$ matrix has $N!$ terms in this type of formula, because there are $N!$ different permutations of the set $(1, ..., N)$. A numerical evaluation of the determinant of a large matrix using this formula is prohibitively long.

Using the definition D0 and the properties of permutations, one can directly prove various properties of determinants, for instance their antisymmetry with respect to exchanges of matrix rows or columns, and finally the relevance of $\det(A_{ij})$ to linear equations $\sum_j A_{ij}x_j = a_i$, as well as the important property

$$\det(AB) = (\det A)(\det B).$$

Deriving these properties in this way will require long calculations.

Question: To me, definition D0 seems unmotivated and strange. It is not clear why this complicated combination of matrix elements has any useful properties at all. Even if so then maybe there exists another complicated combination of matrix elements that is even more useful?

Answer: Yes, indeed: There exist other complicated combinations that are also useful. All this is best understood if we do not begin by studying the definition (3.1). Instead, we will proceed in a coordinate-free manner and build upon geometric intuition.

We will interpret the matrix A_{jk} not as a "table of numbers" but as a coordinate representation of a linear transformation $\hat{A}$ in some vector space V with respect to some given basis. We will define an action of the operator $\hat{A}$ on the exterior product space $\wedge^N V$ in a certain way. That action will allow us to understand the properties and the uses of determinants without long calculations.

Another useful interpretation of the matrix A_{jk} is to regard it as a table of components of a *set* of N vectors $\mathbf{v}_1, ..., \mathbf{v}_N$ in a given basis $\{\mathbf{e}_j\}$, that is,

$$\mathbf{v}_j = \sum_{k=1}^{N} A_{jk}\mathbf{e}_k, \quad j = 1, ..., N.$$

The determinant of the matrix A_{jk} is then naturally related to the exterior product $\mathbf{v}_1 \wedge ... \wedge \mathbf{v}_N$. This construction is especially useful for solving linear equations.

These constructions and related results occupy the present chapter. Most of the derivations are straightforward and short but require some facility with calculations involving the exterior product. I recommend that you repeat all the calculations yourself.

Exercise: If $\{\mathbf{v}_1, ..., \mathbf{v}_N\}$ are N vectors and σ is a permutation of the ordered set $(1, ..., N)$, show that

$$\mathbf{v}_1 \wedge ... \wedge \mathbf{v}_N = (-1)^{|\sigma|}\, \mathbf{v}_{\sigma(1)} \wedge ... \wedge \mathbf{v}_{\sigma(N)}.$$

3.2 The space $\wedge^N V$ and oriented volume

Of all the exterior power spaces $\wedge^k V$ ($k = 1, 2, ...$), the last nontrivial space is $\wedge^N V$ where $N \equiv \dim V$, for it is impossible to have a nonzero exterior product of $(N+1)$ or more vectors. In other words, the spaces $\wedge^{N+1} V$, $\wedge^{N+2} V$ etc. are all zero-dimensional and thus do not contain any nonzero tensors.

By Theorem 2 from Sec. 2.3.2, the space $\wedge^N V$ is one-dimensional. Therefore, all nonzero tensors from $\wedge^N V$ are proportional to each other. Hence, any nonzero tensor $\omega_1 \in \wedge^N V$ can serve as a basis tensor in $\wedge^N V$.

The space $\wedge^N V$ is extremely useful because it is so simple and yet is directly related to determinants and volumes; this idea will be developed now. We begin by considering an example.

Example: In a two-dimensional space V, let us choose a basis $\{\mathbf{e}_1, \mathbf{e}_2\}$ and consider two arbitrary vectors $\mathbf{v}_1$ and $\mathbf{v}_2$. These vectors can be decomposed in the basis as

$$\mathbf{v}_1 = a_{11}\mathbf{e}_1 + a_{12}\mathbf{e}_2, \quad \mathbf{v}_2 = a_{21}\mathbf{e}_1 + a_{22}\mathbf{e}_2,$$

where $\{a_{ij}\}$ are some coefficients. Let us now compute the 2-vector $\mathbf{v}_1 \wedge \mathbf{v}_2 \in \wedge^2 V$:

$$\begin{aligned}\mathbf{v}_1 \wedge \mathbf{v}_2 &= (a_{11}\mathbf{e}_1 + a_{12}\mathbf{e}_2) \wedge (a_{21}\mathbf{e}_1 + a_{22}\mathbf{e}_2)\\ &= a_{11}a_{22}\mathbf{e}_1 \wedge \mathbf{e}_2 + a_{12}a_{21}\mathbf{e}_2 \wedge \mathbf{e}_1\\ &= (a_{11}a_{22} - a_{12}a_{21})\,\mathbf{e}_1 \wedge \mathbf{e}_2.\end{aligned}$$

We may observe that firstly, the 2-vector $\mathbf{v}_1 \wedge \mathbf{v}_2$ is proportional to $\mathbf{e}_1 \wedge \mathbf{e}_2$, and secondly, the proportionality coefficient is equal to the determinant of the matrix a_{ij}.

If we compute the exterior product $\mathbf{v}_1 \wedge \mathbf{v}_2 \wedge \mathbf{v}_3$ of three vectors in a 3-dimensional space, we will similarly notice that the result is proportional to $\mathbf{e}_1 \wedge \mathbf{e}_2 \wedge \mathbf{e}_3$, and the proportionality coefficient is again equal to the determinant of the matrix a_{ij}. ■

Let us return to considering a general, N-dimensional space V. The examples just given motivate us to study N-vectors (i.e. tensors from the top exterior power space $\wedge^N V$) and their relationships of the form $\mathbf{v}_1 \wedge ... \wedge \mathbf{v}_N = \lambda \mathbf{e}_1 \wedge ... \wedge \mathbf{e}_N$.

By Lemma 1 from Sec. 2.3.2, every nonzero element of $\wedge^N V$ must be of the form $\mathbf{v}_1 \wedge ... \wedge \mathbf{v}_N$, where the set $\{\mathbf{v}_1, ..., \mathbf{v}_N\}$ is linearly independent and thus a basis in V. Conversely, each basis $\{\mathbf{v}_j\}$ in V yields a nonzero tensor $\mathbf{v}_1 \wedge ... \wedge \mathbf{v}_N \in \wedge^N V$. This tensor has a useful geometric interpretation because, in some sense, it represents the *volume* of the N-dimensional parallelepiped spanned by the vectors $\{\mathbf{v}_j\}$. I will now explain this idea.

A rigorous definition of "volume" in N-dimensional space requires much background work in geometry and measure theory; I am not prepared to explain all this here. However, we can motivate the interpretation of the tensor $\mathbf{v}_1 \wedge ... \wedge \mathbf{v}_N$ as the volume by appealing to the visual notion of the volume of a parallelepiped.[1]

[1] In this text, we do not actually need a mathematically rigorous notion of "volume" — it is used purely to develop geometrical intuition. All formulations and proofs in this text are completely algebraic.

Statement: Consider an N-dimensional space V where the (N-dimensional) volume of solid bodies can be computed through some reasonable[2] geometric procedure. Then:

(1) Two parallelepipeds spanned by the sets of vectors $\{\mathbf{u}_1, \mathbf{u}_2, ..., \mathbf{u}_N\}$ and $\{\mathbf{v}_1, \mathbf{v}_2, ..., \mathbf{v}_N\}$ have equal volumes if and only if the corresponding tensors from $\wedge^N V$ are equal up to a sign,

$$\mathbf{u}_1 \wedge ... \wedge \mathbf{u}_N = \pm \mathbf{v}_1 \wedge ... \wedge \mathbf{v}_N. \tag{3.2}$$

Here "two bodies have equal volumes" means (in the style of ancient Greek geometry) that the bodies can be cut into suitable pieces, such that the volumes are found to be identical by inspection after a rearrangement of the pieces.

(2) If $\mathbf{u}_1 \wedge ... \wedge \mathbf{u}_N = \lambda \mathbf{v}_1 \wedge ... \wedge \mathbf{v}_N$, where $\lambda \in \mathbb{K}$ is a number, $\lambda \neq 0$, then the volumes of the two parallelepipeds differ by a factor of $|\lambda|$.

To prove these statements, we will use the following lemma.

Lemma: In an N-dimensional space:

(1) The volume of a parallelepiped spanned by $\{\lambda \mathbf{v}_1, \mathbf{v}_2 ..., \mathbf{v}_N\}$ is λ times greater than that of $\{\mathbf{v}_1, \mathbf{v}_2, ..., \mathbf{v}_N\}$.

(2) Two parallelepipeds spanned by the sets of vectors $\{\mathbf{v}_1, \mathbf{v}_2, ..., \mathbf{v}_N\}$ and $\{\mathbf{v}_1 + \lambda \mathbf{v}_2, \mathbf{v}_2, ..., \mathbf{v}_N\}$ have equal volume.

Proof of Lemma: **(1)** This is clear from geometric considerations: When a parallelepiped is stretched λ times in one direction, its volume must increase by the factor λ. **(2)** First, we ignore the vectors $\mathbf{v}_3$,...,$\mathbf{v}_N$ and consider the two-dimensional plane containing $\mathbf{v}_1$ and $\mathbf{v}_2$. In Fig. 3.1 one can see that the parallelograms spanned by $\{\mathbf{v}_1, \mathbf{v}_2\}$ and by $\{\mathbf{v}_1 + \lambda \mathbf{v}_2, \mathbf{v}_2\}$ can be cut into appropriate pieces to demonstrate the equality of their area. Now, we consider the N-dimensional volume (a three-dimensional example is shown in Fig. 3.2). Similarly to the two-dimensional case, we find that the N-dimensional parallelepipeds spanned by $\{\mathbf{v}_1, \mathbf{v}_2, ..., \mathbf{v}_N\}$ and by $\{\mathbf{v}_1 + \lambda \mathbf{v}_2, \mathbf{v}_2, ..., \mathbf{v}_N\}$ have equal N-dimensional volume. ■

Proof of Statement: **(1)** To prove that the volumes are equal when the tensors are equal, we will transform the first basis $\{\mathbf{u}_1, \mathbf{u}_2, ..., \mathbf{u}_N\}$ into the second basis $\{\mathbf{v}_1, \mathbf{v}_2, ..., \mathbf{v}_N\}$ by a sequence of transformations of two types: either we will multiply one of the vectors $\mathbf{v}_j$ by a number λ, or add $\lambda \mathbf{v}_j$ to another vector $\mathbf{v}_k$. We first need to demonstrate that any basis can be transformed into any other basis by this procedure. To demonstrate this, recall the proof of Theorem 1.1.5 in which vectors from the first basis were systematically replaced by vectors of the second one. Each replacement can be implemented by a certain sequence of replacements of the kind $\mathbf{u}_j \to \lambda \mathbf{u}_j$ or $\mathbf{u}_j \to \mathbf{u}_j + \lambda \mathbf{u}_i$. Note that the tensor $\mathbf{u}_1 \wedge ... \wedge \mathbf{u}_N$ changes in the same way as the volume under these replacements: The tensor $\mathbf{u}_1 \wedge ... \wedge \mathbf{u}_N$ gets multiplied by λ after $\mathbf{u}_j \to \lambda \mathbf{u}_j$ and remains unchanged after $\mathbf{u}_j \to \mathbf{u}_j + \lambda \mathbf{u}_i$. At the end of the

[2]Here by "reasonable" I mean that the volume has the usual properties: for instance, the volume of a body consisting of two parts equals the sum of the volumes of the parts. An example of such procedure would be the N-fold integral $\int dx_1 ... \int dx_N$, where x_j are coordinates of points in an orthonormal basis.

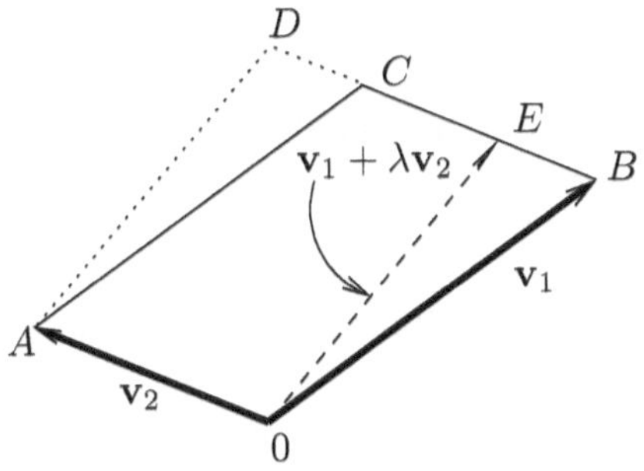

Figure 3.1: The area of the parallelogram $0ACB$ spanned by $\{\mathbf{v}_1, \mathbf{v}_2\}$ is equal to the area of the parallelogram $0ADE$ spanned by $\{\mathbf{v}_1 + \lambda \mathbf{v}_2, \mathbf{v}_2\}$.

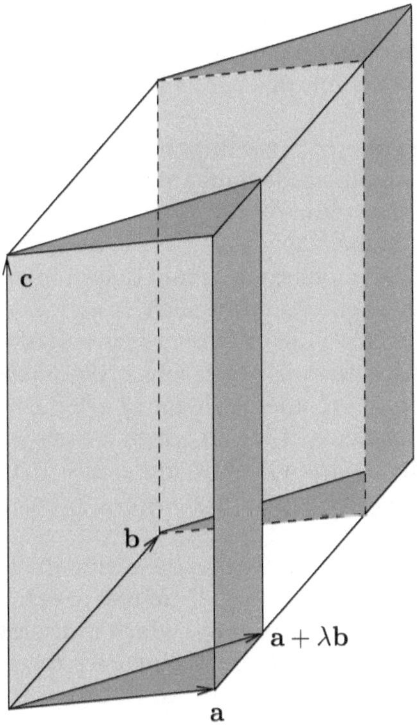

Figure 3.2: Parallelepipeds spanned by $\{\mathbf{a}, \mathbf{b}, \mathbf{c}\}$ and by $\{\mathbf{a} + \lambda \mathbf{b}, \mathbf{b}, \mathbf{c}\}$ have equal volume since the volumes of the shaded regions are equal.

replacement procedure, the basis $\{\mathbf{u}_j\}$ becomes the basis $\{\mathbf{v}_j\}$ (up to the ordering of vectors), while the volume is multiplied by the same factor as the tensor $\mathbf{u}_1 \wedge ... \wedge \mathbf{u}_N$. The ordering of the vectors in the set $\{\mathbf{v}_j\}$ can be changed with possibly a sign change in the tensor $\mathbf{u}_1 \wedge ... \wedge \mathbf{u}_N$. Therefore the statement (3.2) is equivalent to the assumption that the volumes of $\{\mathbf{v}_j\}$ and $\{\mathbf{u}_j\}$ are equal. **(2)** A transformation $\mathbf{v}_1 \to \lambda \mathbf{v}_1$ increases the volume by a factor of $|\lambda|$ and makes the two tensors equal, therefore the volumes differ by a factor of $|\lambda|$. ■

Let us now consider the interpretation of the above Statement. Suppose we somehow know that the parallelepiped spanned by the vectors $\{\mathbf{u}_1, ..., \mathbf{u}_N\}$ has unit volume. Given this knowledge, the volume of any other parallelepiped spanned by some other vectors $\{\mathbf{v}_1, ..., \mathbf{v}_N\}$ is easy to compute. Indeed, we can compute the tensors $\mathbf{u}_1 \wedge ... \wedge \mathbf{u}_N$ and $\mathbf{v}_1 \wedge ... \wedge \mathbf{v}_N$. Since the space $\wedge^N V$ is one-dimensional, these two tensors must be proportional to each other. By expanding the vectors $\mathbf{v}_j$ in the basis $\{\mathbf{u}_j\}$, it is straightforward to compute the coefficient λ in the relationship

$$\mathbf{v}_1 \wedge ... \wedge \mathbf{v}_N = \lambda \mathbf{u}_1 \wedge ... \wedge \mathbf{u}_N.$$

The Statement now says that the volume of a parallelepiped spanned by the vectors $\{\mathbf{v}_1, ..., \mathbf{v}_N\}$ is equal to $|\lambda|$.

Exercise 1: The volume of a parallelepiped spanned by vectors **a**, **b**, **c** is equal to 19. Compute the volume of a parallelepiped spanned by the vectors $2\mathbf{a} - \mathbf{b}$, $\mathbf{c} + 3\mathbf{a}$, $\mathbf{b}$.

Solution: Since $(2\mathbf{a} - \mathbf{b}) \wedge (\mathbf{c} + 3\mathbf{a}) \wedge \mathbf{b} = 2\mathbf{a} \wedge \mathbf{c} \wedge \mathbf{b} = -2\mathbf{a} \wedge \mathbf{b} \wedge \mathbf{c}$, the volume is 38 (twice 19; we ignored the minus sign since we are interested only in the absolute value of the volume). ■

It is also clear that the tensor $\mathbf{v}_1 \wedge ... \wedge \mathbf{v}_N$ allows us only to *compare* the volumes of two parallelepipeds; we cannot determine the volume of one parallelepiped taken by itself. A tensor such as $\mathbf{v}_1 \wedge ... \wedge \mathbf{v}_N$ can be used to determine the numerical value of the volume only if we can compare it with another given tensor, $\mathbf{u}_1 \wedge ... \wedge \mathbf{u}_N$, which (*by assumption*) corresponds to a parallelepiped of unit volume. A choice of a "reference" tensor $\mathbf{u}_1 \wedge ... \wedge \mathbf{u}_N$ can be made, for instance, if we are given a basis in V; without this choice, there is no natural map from $\wedge^N V$ to numbers ($\mathbb{K}$). In other words, the space $\wedge^N V$ is *not canonically isomorphic* to the space $\mathbb{K}$ (even though both $\wedge^N V$ and $\mathbb{K}$ are one-dimensional vector spaces). Indeed, a canonical isomorphism between $\wedge^N V$ and $\mathbb{K}$ would imply that the element $1 \in \mathbb{K}$ has a corresponding canonically defined tensor $\omega_1 \in \wedge^N V$. In that case there would be some basis $\{\mathbf{e}_j\}$ in V such that $\mathbf{e}_1 \wedge ... \wedge \mathbf{e}_N = \omega_1$, which indicates that the basis $\{\mathbf{e}_j\}$ is in some sense "preferred" or "natural." However, there is no "natural" or "preferred" choice of basis in a vector space V, unless some additional structure is given (such as a scalar product). Hence, no canonical choice of $\omega_1 \in \wedge^N V$ is possible.

Remark: When a scalar product is defined in V, there is a preferred choice of basis, namely an orthonormal basis $\{\mathbf{e}_j\}$ such that $\langle \mathbf{e}_i, \mathbf{e}_j \rangle = \delta_{ij}$ (see Sec. 5.1). Since the length of each of the basis vectors is 1, and the basis vectors are orthogonal to each other, the volume of the parallelepiped spanned by $\{\mathbf{e}_j\}$ is equal to 1. (This is the usual Euclidean definition of volume.) Then the tensor $\omega_1 \equiv \bigwedge_{j=1}^N \mathbf{e}_j$ can be computed using this basis and used as a unit volume tensor. We will see below (Sec. 5.5.2) that this tensor does not depend on the choice of the orthonormal basis, up to the orientation. The isomorphism between $\wedge^N V$ and $\mathbb{K}$ is then fixed (up to the sign), thanks to the scalar product. ■

In the absence of a scalar product, one can say that the *value of the volume* in an abstract vector space is not a number but a tensor from the space $\wedge^N V$. It is sufficient to regard the element $\mathbf{v}_1 \wedge ... \wedge \mathbf{v}_N \in \wedge^N V$ as the *definition* of the "$\wedge^N V$-valued volume" of the parallelepiped spanned by $\{\mathbf{v}_j\}$. The space $\wedge^N V$ is one-dimensional, so the "tensor-valued volume" has the familiar properties we expect (it is "almost a number"). One thing is unusual about this "volume": It is **oriented**, that is, it changes sign if we exchange the order of two vectors from the set $\{\mathbf{v}_j\}$.

Exercise 2: Suppose $\{\mathbf{u}_1, ..., \mathbf{u}_N\}$ is a basis in V. Let $\mathbf{x}$ be some vector whose components in the basis $\{\mathbf{u}_j\}$ are given, $\mathbf{x} = \sum_j \alpha_j \mathbf{u}_j$. Compute the (tensor-valued) volume of the parallelepiped spanned by $\{\mathbf{u}_1 + \mathbf{x}, ..., \mathbf{u}_N + \mathbf{x}\}$.

Hints: Use the linearity property, $(\mathbf{a} + \mathbf{x}) \wedge ... = \mathbf{a} \wedge ... + \mathbf{x} \wedge ...$, and notice the simplification

$$\mathbf{x} \wedge (\mathbf{a} + \mathbf{x}) \wedge (\mathbf{b} + \mathbf{x}) \wedge ... \wedge (\mathbf{c} + \mathbf{x}) = \mathbf{x} \wedge \mathbf{a} \wedge \mathbf{b} \wedge ... \wedge \mathbf{c}.$$

Answer: The volume tensor is

$$(\mathbf{u}_1 + \mathbf{x}) \wedge ... \wedge (\mathbf{u}_N + \mathbf{x}) = (1 + \alpha_1 + ... + \alpha_N)\, \mathbf{u}_1 \wedge ... \wedge \mathbf{u}_N.$$

Remark: tensor-valued area. The idea that the volume is "oriented" can be understood perhaps more intuitively by considering the area of the parallelogram spanned by two vectors $\mathbf{a}, \mathbf{b}$ in the familiar 3-dimensional space. It is customary to draw the vector product $\mathbf{a} \times \mathbf{b}$ as the representation of this area, since the length $|\mathbf{a} \times \mathbf{b}|$ is equal to the area, and the direction of $\mathbf{a} \times \mathbf{b}$ is normal to the area. Thus, the vector $\mathbf{a} \times \mathbf{b}$ can be understood as the "oriented area" of the parallelogram. However, note that the direction of the vector $\mathbf{a} \times \mathbf{b}$ depends not only on the angular orientation of the parallelogram in space, but also on the order of the vectors $\mathbf{a}, \mathbf{b}$. The 2-vector $\mathbf{a} \wedge \mathbf{b}$ is the natural analogue of the vector product $\mathbf{a} \times \mathbf{b}$ in higher-dimensional spaces. Hence, it is algebraically natural to regard the tensor $\mathbf{a} \wedge \mathbf{b} \in \wedge^2 V$ as the "tensor-valued" representation of the area of the parallelogram spanned by $\{\mathbf{a}, \mathbf{b}\}$.

Consider now a parallelogram spanned by $\mathbf{a}, \mathbf{b}$ in a *two*-dimensional plane. We can still represent the oriented area of this parallelogram by the vector product $\mathbf{a} \times \mathbf{b}$, where we imagine that the plane is embedded in a three-dimensional space. The area of the parallelogram does not have a nontrivial angular orientation any more since the vector product $\mathbf{a} \times \mathbf{b}$ is always orthogonal to the plane; the only feature left from the orientation is the positive or negative sign of $\mathbf{a} \times \mathbf{b}$ relative to an arbitrarily chosen vector $\mathbf{n}$ normal to the plane. Hence, we may say that the sign of the oriented volume of a parallelepiped is the only remnant of the angular orientation of the parallelepiped in space when the dimension of the parallelepiped is equal to the dimension of space.

(See Sec. 2.1 for more explanations about the geometrical interpretation of volume in terms of exterior product.) ■

3.3 Determinants of operators

Let $\hat{A} \in \text{End } V$ be a linear operator. Consider its action on tensors from the space $\wedge^N V$ defined in the following way, $\mathbf{v}_1 \wedge ... \wedge ... \mathbf{v}_N \mapsto \hat{A}\mathbf{v}_1 \wedge ... \wedge \hat{A}\mathbf{v}_N$. I denote this operation by $\wedge^N \hat{A}^N$, so

$$\wedge^N \hat{A}^N (\mathbf{v}_1 \wedge ... \wedge \mathbf{v}_N) \equiv (\hat{A}\mathbf{v}_1) \wedge ... \wedge (\hat{A}\mathbf{v}_N).$$

The notation $\wedge^N \hat{A}^N$ underscores the fact that there are N copies of $\hat{A}$ acting simultaneously.

We have just defined $\wedge^N \hat{A}^N$ on single-term products $\mathbf{v}_1 \wedge ... \wedge \mathbf{v}_N$; the action of $\wedge^N \hat{A}^N$ on linear combinations of such products is obtained by requiring linearity.

Let us verify that $\wedge^N \hat{A}^N$ is a linear map; it is sufficient to check that it is compatible with the exterior product axioms:

$$\begin{aligned} \hat{A}(\mathbf{v} + \lambda\mathbf{u}) \wedge \hat{A}\mathbf{v}_2 \wedge ... \wedge \hat{A}\mathbf{v}_N &= \hat{A}\mathbf{v} \wedge \hat{A}\mathbf{v}_2 \wedge ... \wedge \hat{A}\mathbf{v}_N \\ &\quad + \lambda\hat{A}\mathbf{u} \wedge \hat{A}\mathbf{v}_2 \wedge ... \wedge \hat{A}\mathbf{v}_N ; \\ \hat{A}\mathbf{v}_1 \wedge \hat{A}\mathbf{v}_2 \wedge ... \wedge \hat{A}\mathbf{v}_N &= -\hat{A}\mathbf{v}_2 \wedge \hat{A}\mathbf{v}_1 \wedge ... \wedge \hat{A}\mathbf{v}_N . \end{aligned}$$

Therefore, $\wedge^N \hat{A}^N$ is now defined as a linear operator $\wedge^N V \to \wedge^N V$.

By Theorem 2 in Sec. 2.3.2, the space $\wedge^N V$ is one-dimensional. So $\wedge^N \hat{A}^N$, being a linear operator in a one-dimensional space, must act simply as multiplication by a number. (*Every* linear operator in a one-dimensional space must act as multiplication by a number!) Thus we can write

$$\wedge^N \hat{A}^N = \alpha \hat{1}_{\wedge^N V},$$

where $\alpha \in \mathbb{K}$ is a number which is somehow associated with the operator $\hat{A}$. What is the significance of this number α? This number is actually equal to the *determinant* of the operator $\hat{A}$ as given by Definition D0. But let us pretend that we do not know anything about determinants; it is very convenient to use this construction to *define* the determinant and to derive its properties.

Definition D1: The **determinant** $\det \hat{A}$ of an operator $\hat{A} \in \text{End } V$ is the number by which any nonzero tensor $\omega \in \wedge^N V$ is multiplied when $\wedge^N \hat{A}^N$ acts on it:

$$(\wedge^N \hat{A}^N)\omega = (\det \hat{A})\omega. \tag{3.3}$$

In other words, $\wedge^N A^N = (\det \hat{A})\hat{1}_{\wedge^N V}$.

We can immediately put this definition to use; here are the first results.

Statement 1: The determinant of a product is the product of determinants: $\det(\hat{A}\hat{B}) = (\det \hat{A})(\det \hat{B})$.

Proof: Act with $\wedge^N \hat{A}^N$ and then with $\wedge^N \hat{B}^N$ on a nonzero tensor $\omega \in \wedge^N V$. Since these operators act as multiplication by a number, the result is the multiplication by the product of these numbers. We thus have

$$(\wedge^N \hat{A}^N)(\wedge^N \hat{B}^N)\omega = (\wedge^N \hat{A}^N)(\det \hat{B})\omega = (\det \hat{A})(\det \hat{B})\omega.$$

On the other hand, for $\omega = \mathbf{v}_1 \wedge ... \wedge \mathbf{v}_N$ we have

$$\begin{aligned}(\wedge^N \hat{A}^N)(\wedge^N \hat{B}^N)\omega &= (\wedge^N \hat{A}^N)\hat{B}\mathbf{v}_1 \wedge ... \wedge \hat{B}\mathbf{v}_N \\ &= \hat{A}\hat{B}\mathbf{v}_1 \wedge ... \wedge \hat{A}\hat{B}\mathbf{v}_N = \wedge^N (\hat{A}\hat{B})^N \omega \\ &= (\det(\hat{A}\hat{B}))\omega.\end{aligned}$$

Therefore, $\det(\hat{A}\hat{B}) = (\det \hat{A})(\det \hat{B})$. ■

Exercise 1: Prove that $\det(\lambda\hat{A}) = \lambda^N \det \hat{A}$ for any $\lambda \in \mathbb{K}$ and $\hat{A} \in \text{End } V$.

Now let us clarify the relation between the determinant and the volume. We will prove that the determinant of a transformation $\hat{A}$ is the coefficient by which the volume of parallelepipeds will grow when we act with $\hat{A}$ on the vector space. After proving this, I will *derive* the relation (3.1) for the determinant through the matrix coefficients of $\hat{A}$ in some basis; it will follow that the formula (3.1) gives the same results in any basis.

Statement 2: When a parallelepiped spanned by the vectors $\{\mathbf{v}_1, ..., \mathbf{v}_N\}$ is transformed by a linear operator $\hat{A}$, so that $\mathbf{v}_j \mapsto \hat{A}\mathbf{v}_j$, the volume of the parallelepiped grows by the factor $|\det \hat{A}|$.

Proof: Suppose the volume of the parallelepiped spanned by the vectors $\{\mathbf{v}_1, ..., \mathbf{v}_N\}$ is v. The transformed parallelepiped is spanned by vectors $\{\hat{A}\mathbf{v}_1, ..., \hat{A}\mathbf{v}_N\}$. According to the definition of the determinant, $\det \hat{A}$ is a number such that

$$\hat{A}\mathbf{v}_1 \wedge ... \wedge \hat{A}\mathbf{v}_N = (\det \hat{A})\mathbf{v}_1 \wedge ... \wedge \mathbf{v}_N.$$

By Statement 3.2, the volume of the transformed parallelepiped is $|\det \hat{A}|$ times the volume of the original parallelepiped. ■

If we consider the oriented (i.e. tensor-valued) volume, we find that it grows by the factor $\det \hat{A}$ (without the absolute value). Therefore we could define the determinant also in the following way:

Definition D2: The determinant $\det \hat{A}$ of a linear transformation $\hat{A}$ is the number by which the *oriented* volume of any parallelepiped grows after the transformation. (One is then obliged to prove that this number does not depend on the choice of the initial parallelepiped! We just proved this in Statement 1 using an algebraic definition D1 of the determinant.)

With this definition of the determinant, the property

$$\det(\hat{A}\hat{B}) = (\det \hat{A})(\det \hat{B})$$

is easy to understand: The composition of the transformations $\hat{A}$ and $\hat{B}$ multiplies the volume by the product of the individual volume growth factors $\det \hat{A}$ and $\det \hat{B}$.

Finally, here is a derivation of the formula (3.1) from Definition D1.

Statement 3: If $\{\mathbf{e}_j\}$ is any basis in V, $\{\mathbf{e}_j^*\}$ is the dual basis, and a linear operator $\hat{A}$ is represented by a tensor,

$$\hat{A} = \sum_{j,k=1}^{N} A_{jk} \mathbf{e}_j \otimes \mathbf{e}_k^*, \tag{3.4}$$

then the determinant of $\hat{A}$ is given by the formula (3.1).

Proof: The operator $\hat{A}$ defined by Eq. (3.4) acts on the basis vectors $\{\mathbf{e}_j\}$ as follows,

$$\hat{A}\mathbf{e}_k = \sum_{j=1}^{N} A_{jk}\mathbf{e}_j.$$

A straightforward calculation is all that is needed to obtain the formula for the determinant. I first consider the case $N = 2$ as an illustration:

$$\begin{aligned}\wedge^2 \hat{A}^2 (\mathbf{e}_1 \wedge \mathbf{e}_2) &= \hat{A}\mathbf{e}_1 \wedge \hat{A}\mathbf{e}_2 \\ &= (A_{11}\mathbf{e}_1 + A_{21}\mathbf{e}_2) \wedge (A_{12}\mathbf{e}_1 + A_{22}\mathbf{e}_2) \\ &= A_{11}A_{22}\mathbf{e}_1 \wedge \mathbf{e}_2 + A_{21}A_{12}\mathbf{e}_2 \wedge \mathbf{e}_1 \\ &= (A_{11}A_{22} - A_{12}A_{21})\, \mathbf{e}_1 \wedge \mathbf{e}_2.\end{aligned}$$

Hence $\det \hat{A} = A_{11}A_{22} - A_{12}A_{21}$, in agreement with the usual formula.

Now I consider the general case. The action of $\wedge^N \hat{A}^N$ on the basis element $\mathbf{e}_1 \wedge ... \wedge \mathbf{e}_N \in \wedge^N V$ is

$$\begin{aligned}\wedge^N \hat{A}^N (\mathbf{e}_1 \wedge ... \wedge \mathbf{e}_N) &= \hat{A}\mathbf{e}_1 \wedge ... \wedge \hat{A}\mathbf{e}_N \\ &= \left(\sum_{j_1=1}^{N} A_{j_1 1}\mathbf{e}_{j_1}\right) \wedge ... \wedge \left(\sum_{j_N=1}^{N} A_{j_N N}\mathbf{e}_{j_N}\right) \\ &= \sum_{j_1=1}^{N} ... \sum_{j_N=1}^{N} A_{j_1 1}\mathbf{e}_{j_1} \wedge ... \wedge A_{j_N N}\mathbf{e}_{j_N} \\ &= \sum_{j_1=1}^{N} ... \sum_{j_N=1}^{N} (A_{j_1 1} ... A_{j_N N})\mathbf{e}_{j_1} \wedge ... \wedge \mathbf{e}_{j_N}.\end{aligned} \tag{3.5}$$

In the last sum, the only nonzero terms are those in which the indices $j_1, ..., j_N$ do not repeat; in other words, $(j_1, ..., j_N)$ is a *permutation* of the set $(1, ..., N)$. Let us therefore denote this permutation by σ and write $\sigma(1) \equiv j_1, ..., \sigma(N) \equiv j_N$. Using the antisymmetry of the exterior product and the definition of the parity $|\sigma|$ of the permutation σ, we can express

$$\mathbf{e}_{j_1} \wedge ... \wedge \mathbf{e}_{j_N} = \mathbf{e}_{\sigma(1)} \wedge ... \wedge \mathbf{e}_{\sigma(N)} = (-1)^{|\sigma|}\, \mathbf{e}_1 \wedge ... \wedge \mathbf{e}_N.$$

Now we can rewrite the last line in Eq. (3.5) in terms of sums over all permutations σ instead of sums over all $\{j_1, ..., j_N\}$:

$$\wedge^N \hat{A}^N (\mathbf{e}_1 \wedge ... \wedge \mathbf{e}_N) = \sum_\sigma A_{\sigma(1)1} ... A_{\sigma(N)N} \mathbf{e}_{\sigma(1)} \wedge ... \wedge \mathbf{e}_{\sigma(N)}$$

$$= \sum_\sigma A_{\sigma(1)1} ... A_{\sigma(N)N} (-1)^{|\sigma|} \mathbf{e}_1 \wedge ... \wedge \mathbf{e}_N.$$

Thus we have reproduced the formula (3.1). ■

We have seen three equivalent definitions of the determinant, each with its own advantages: first, a direct but complicated definition (3.1) in terms of matrix coefficients; second, an elegant but abstract definition (3.3) that depends on the construction of the exterior product; third, an intuitive and visual definition in terms of the volume which, however, is based on the geometric notion of "volume of an N-dimensional domain" rather than on purely algebraic constructions. All three definitions are equivalent when applied to linear operators in finite-dimensional spaces.

3.3.1 Examples: computing determinants

Question: We have been working with operators more or less in the same way as with matrices, like in Eq. (3.4). What is the advantage of the coordinate-free approach if we are again computing with the elements of matrices?

Answer: In some cases, there is no other way except to represent an operator in some basis through a matrix such as A_{ij}. However, in many cases an interesting operator can be represented *geometrically*, i.e. without choosing a basis. It is often useful to express an operator in a basis-free manner because this yields some nontrivial information that would otherwise be obscured by an unnecessary (or wrong) choice of basis. It is useful to be able to employ both the basis-free and the component-based techniques. Here are some examples where we compute determinants of operators defined without a basis.

Example 1: Operators of the form $\hat{1}_V + \mathbf{a} \otimes \mathbf{b}^*$ are useful in geometry because they can represent reflections or projections with respect to an axis or a plane if $\mathbf{a}$ and $\mathbf{b}^*$ are chosen appropriately. For instance, if $\mathbf{b}^* \neq 0$, we can define a **hyperplane** $H_{\mathbf{b}^*} \subset V$ as the subspace annihilated by the covector $\mathbf{b}^*$, i.e. the subspace consisting of vectors $\mathbf{v} \in V$ such that $\mathbf{b}^* (\mathbf{v}) = 0$. If a vector $\mathbf{a} \in V$ is such that $\mathbf{b}^* (\mathbf{a}) \neq 0$, i.e. $\mathbf{a} \notin H_{\mathbf{b}^*}$, then

$$\hat{P} \equiv \hat{1}_V - \frac{1}{\mathbf{b}^* (\mathbf{a})} \mathbf{a} \otimes \mathbf{b}^*$$

is a projector onto $H_{\mathbf{b}^*}$, while the operator

$$\hat{R} \equiv \hat{1}_V - \frac{2}{\mathbf{b}^* (\mathbf{a})} \mathbf{a} \otimes \mathbf{b}^*$$

describes a **mirror reflection** with respect to the hyperplane $H_{\mathbf{b}^*}$, in the sense that $\mathbf{v} + \hat{R}\mathbf{v} \in H_{\mathbf{b}^*}$ for any $\mathbf{v} \in V$. ■

The following statement shows how to calculate determinants of such operators. For instance, with the above definitions we would find $\det \hat{P} = 0$ and $\det \hat{R} = -1$ by a direct application of Eq. (3.6).

Statement: Let $\mathbf{a} \in V$ and $\mathbf{b}^* \in V^*$. Then

$$\det\left(\hat{1}_V + \mathbf{a} \otimes \mathbf{b}^*\right) = 1 + \mathbf{b}^*(\mathbf{a}). \tag{3.6}$$

Proof: If $\mathbf{b}^* = 0$, the formula is trivial, so we assume that $\mathbf{b}^* \neq 0$. Then we need to consider two cases: $\mathbf{b}^*(\mathbf{a}) \neq 0$ or $\mathbf{b}^*(\mathbf{a}) = 0$; however, the final formula (3.6) is the same in both cases.

Case 1. By Statement 1.6, if $\mathbf{b}^*(\mathbf{a}) \neq 0$ there exists a basis $\{\mathbf{a}, \mathbf{v}_2, ..., \mathbf{v}_N\}$ such that $\mathbf{b}^*(\mathbf{v}_i) = 0$ for $2 \leq i \leq N$, where $N = \dim V$. Then we compute the determinant by applying the operator $\wedge^N \left(\hat{1}_V + \mathbf{a} \otimes \mathbf{b}^*\right)^N$ to the tensor $\mathbf{a} \wedge \mathbf{v}_2 \wedge ... \wedge \mathbf{v}_N$: since

$$\begin{aligned} \left(\hat{1}_V + \mathbf{a} \otimes \mathbf{b}^*\right) \mathbf{a} &= \left(1 + \mathbf{b}^*(\mathbf{a})\right) \mathbf{a}, \\ \left(\hat{1}_V + \mathbf{a} \otimes \mathbf{b}^*\right) \mathbf{v}_i &= \mathbf{v}_i, \quad i = 2, ..., N, \end{aligned}$$

we get

$$\begin{aligned} &\wedge^N \left(\hat{1}_V + \mathbf{a} \otimes \mathbf{b}^*\right)^N \mathbf{a} \wedge \mathbf{v}_2 \wedge ... \wedge \mathbf{v}_N \\ &\quad = \left(1 + \mathbf{b}^*(\mathbf{a})\right) \mathbf{a} \wedge \mathbf{v}_2 \wedge ... \wedge \mathbf{v}_N. \end{aligned}$$

Therefore $\det\left(\hat{1}_V + \mathbf{a} \otimes \mathbf{b}^*\right) = 1 + \mathbf{b}^*(\mathbf{a})$, as required.

Case 2. If $\mathbf{b}^*(\mathbf{a}) = 0$, we will show that $\det\left(\hat{1}_V + \mathbf{a} \otimes \mathbf{b}^*\right) = 1$. We cannot choose the basis $\{\mathbf{a}, \mathbf{v}_2, ..., \mathbf{v}_N\}$ as in case 1, so we need to choose another basis. There exists some vector $\mathbf{w} \in V$ such that $\mathbf{b}^*(\mathbf{w}) \neq 0$ because by assumption $\mathbf{b}^* \neq 0$. It is clear that $\{\mathbf{w}, \mathbf{a}\}$ is a linearly independent set: otherwise we would have $\mathbf{b}^*(\mathbf{w}) = 0$. Therefore, we can complete this set to a basis $\{\mathbf{w}, \mathbf{a}, \mathbf{v}_3, ..., \mathbf{v}_N\}$. Further, the vectors $\mathbf{v}_3, ..., \mathbf{v}_N$ can be chosen such that $\mathbf{b}^*(\mathbf{v}_i) = 0$ for $3 \leq i \leq N$. Now we compute the determinant by acting with the operator $\wedge^N \left(\hat{1}_V + \mathbf{a} \otimes \mathbf{b}^*\right)^N$ on the tensor $\mathbf{a} \wedge \mathbf{w} \wedge \mathbf{v}_3 \wedge ... \wedge \mathbf{v}_N$: since

$$\begin{aligned} \left(\hat{1}_V + \mathbf{a} \otimes \mathbf{b}^*\right) \mathbf{a} &= \mathbf{a}, \\ \left(\hat{1}_V + \mathbf{a} \otimes \mathbf{b}^*\right) \mathbf{w} &= \mathbf{w} + \mathbf{b}^*(\mathbf{w})\, \mathbf{a}, \\ \left(\hat{1}_V + \mathbf{a} \otimes \mathbf{b}^*\right) \mathbf{v}_i &= \mathbf{v}_i, \quad i = 3, ..., N, \end{aligned}$$

we get

$$\begin{aligned} &\wedge^N \left(\hat{1}_V + \mathbf{a} \otimes \mathbf{b}^*\right)^N \mathbf{a} \wedge \mathbf{w} \wedge \mathbf{v}_3 \wedge ... \wedge \mathbf{v}_N \\ &\quad = \mathbf{a} \wedge \left(\mathbf{w} + \mathbf{b}^*(\mathbf{w})\, \mathbf{a}\right) \wedge \mathbf{v}_3 \wedge ... \wedge \mathbf{v}_N \\ &\quad = \mathbf{a} \wedge \mathbf{w} \wedge \mathbf{v}_3 \wedge ... \wedge \mathbf{v}_N. \end{aligned}$$

Therefore $\det\left(\hat{1}_V + \mathbf{a} \otimes \mathbf{b}^*\right) = 1$. ■

Exercise 1: In a similar way, prove the following statement: If $\mathbf{a}_i \in V$ and $\mathbf{b}_i^* \in V^*$ for $1 \leq i \leq n < N$ are such that $\mathbf{b}_i^*(\mathbf{a}_j) = 0$ for all $i > j$, then

$$\det\left(\hat{1}_V + \sum_{i=1}^{n} \mathbf{a}_i \otimes \mathbf{b}_i^*\right) = \prod_{i=1}^{n} \left(1 + \mathbf{b}_i^*(\mathbf{a}_i)\right).$$

Exercise 2: Consider the three-dimensional space of polynomials $p(x)$ in the variable x of degree at most 2 with real coefficients. The operators $\hat{A}$ and $\hat{B}$ are defined by

$$(\hat{A}p)(x) \equiv p(x) + x\frac{dp(x)}{dx},$$
$$(\hat{B}p)(x) \equiv x^2 p(1) + 2p(x).$$

Check that these operators are linear. Compute the determinants of $\hat{A}$ and $\hat{B}$.

Solution: The operators are linear because they are expressed as formulas containing $p(x)$ linearly. Let us use the underbar to distinguish the polynomials $\underline{1}$, $\underline{x}$ from numbers such as 1. A convenient basis tensor of the 3rd exterior power is $\underline{1} \wedge \underline{x} \wedge \underline{x}^2$, so we perform the calculation,

$$(\det \hat{A})(\underline{1} \wedge \underline{x} \wedge \underline{x}^2) = (\hat{A}\underline{1}) \wedge (\hat{A}\underline{x}) \wedge (\hat{A}\underline{x}^2)$$
$$= \underline{1} \wedge (2\underline{x}) \wedge (3\underline{x}^2) = 6(\underline{1} \wedge \underline{x} \wedge \underline{x}^2),$$

and find that $\det \hat{A} = 6$. Similarly we find $\det \hat{B} = 12$. ■

Exercise 3: Suppose the space V is decomposed into a direct sum of U and W, and an operator $\hat{A}$ is such that U and W are invariant subspaces ($\hat{A}\mathbf{x} \in U$ for all $\mathbf{x} \in U$, and the same for W). Denote by $\hat{A}_U$ the restriction of the operator $\hat{A}$ to the subspace U. Show that

$$\det \hat{A} = (\det \hat{A}_U)(\det \hat{A}_W).$$

Hint: Choose a basis in V as the union of a basis in U and a basis in W. In this basis, the operator $\hat{A}$ is represented by a **block-diagonal** matrix.

3.4 Determinants of square tables

Note that the determinant formula (3.1) applies to *any* square matrix, without referring to any transformations in any vector spaces. Sometimes it is useful to compute the determinants of matrices that do not represent linear transformations. Such matrices are really just *tables of numbers*. The properties of determinants of course remain the same whether or not the matrix represents a linear transformation in the context of the problem we are solving. The geometric construction of the determinant through the space $\wedge^N V$ is useful because it helps us understand heuristically where the properties of the determinant come from.

Given just a square table of numbers, it is often useful to *introduce* a linear transformation corresponding to the matrix in some (conveniently chosen) basis; this often helps solve problems. An example frequently used in linear algebra is a matrix consisting of the components of some vectors in a basis. Suppose $\{\mathbf{e}_j \,|\, j = 1, ..., N\}$ is a basis and $\{\mathbf{v}_j \,|\, j = 1, ..., N\}$ are some vectors. Since each of the $\mathbf{v}_j$ can be decomposed through the basis $\{\mathbf{e}_j\}$, say

$$\mathbf{v}_i = \sum_{j=1}^{N} v_{ij}\mathbf{e}_j, \quad i = 1, ..., N,$$

we may consider the coefficients v_{ij} as a square matrix. This matrix, at first glance, does not represent a linear transformation; it's just a square-shaped table of the coefficients v_{ij}. However, let us *define* a linear operator $\hat{A}$ by the condition that $\hat{A}\mathbf{e}_i = \mathbf{v}_i$ for all $i = 1, ..., N$. This condition defines $\hat{A}\mathbf{x}$ for any vector $\mathbf{x}$ if we assume the linearity of $\hat{A}$ (see Exercise 2 in Sec. 1.2.2). The operator $\hat{A}$ has the following matrix representation with respect to the basis $\{\mathbf{e}_i\}$ and the dual basis $\{\mathbf{e}_i^*\}$:

$$\hat{A} = \sum_{i=1}^{N} \mathbf{v}_i \otimes \mathbf{e}_i^* = \sum_{i=1}^{N} \sum_{j=1}^{N} v_{ij} \mathbf{e}_j \otimes \mathbf{e}_i^*.$$

So the matrix v_{ji} (the transpose of v_{ij}) is the matrix representing the transformation $\hat{A}$. Let us consider the determinant of this transformation:

$$(\det \hat{A})\mathbf{e}_1 \wedge ... \wedge \mathbf{e}_N = \hat{A}\mathbf{e}_1 \wedge ... \wedge \hat{A}\mathbf{e}_N = \mathbf{v}_1 \wedge ... \wedge \mathbf{v}_N.$$

The determinant of the matrix v_{ji} is thus equal to the determinant of the transformation $\hat{A}$. Hence, the computation of the determinant of the matrix v_{ji} is equivalent to the computation of the tensor $\mathbf{v}_1 \wedge ... \wedge \mathbf{v}_N \in \wedge^N V$ and its comparison with the basis tensor $\mathbf{e}_1 \wedge ... \wedge \mathbf{e}_N$. We have thus proved the following statement.

Statement 1: The determinant of the matrix v_{ji} made up by the components of the vectors $\{\mathbf{v}_j\}$ in a basis $\{\mathbf{e}_j\}$ $(j = 1, ..., N)$ is the number C defined as the coefficient in the tensor equality

$$\mathbf{v}_1 \wedge ... \wedge \mathbf{v}_N = C\mathbf{e}_1 \wedge ... \wedge \mathbf{e}_N.$$

Corollary: The determinant of a matrix does not change when a multiple of one row is added to another row. The determinant is linear as a function of each row. The determinant changes sign when two rows are exchanged.

Proof: We consider the matrix v_{ij} as the table of coefficients of vectors $\{\mathbf{v}_j\}$ in a basis $\{\mathbf{e}_j\}$, as explained above. Since

$$(\det v_{ji})\mathbf{e}_1 \wedge ... \wedge \mathbf{e}_N = \mathbf{v}_1 \wedge ... \wedge \mathbf{v}_N,$$

we need only to examine the properties of the tensor $\omega \equiv \mathbf{v}_1 \wedge ... \wedge \mathbf{v}_N$ under various replacements. When a multiple of row k is added to another row j, we replace $\mathbf{v}_j \mapsto \mathbf{v}_j + \lambda \mathbf{v}_k$ for fixed j, k; then the tensor ω does not change,

$$\mathbf{v}_1 \wedge ... \wedge \mathbf{v}_j \wedge ... \wedge \mathbf{v}_N = \mathbf{v}_1 \wedge ... \wedge (\mathbf{v}_j + \lambda \mathbf{v}_k) \wedge ... \wedge \mathbf{v}_N,$$

hence the determinant of v_{ij} does not change. To show that the determinant is linear as a function of each row, we consider the replacement $\mathbf{v}_j \mapsto \mathbf{u} + \lambda \mathbf{v}$ for fixed j; the tensor ω is then equal to the sum of the tensors $\mathbf{v}_1 \wedge ... \wedge \mathbf{u} \wedge ... \wedge \mathbf{v}_N$ and $\lambda \mathbf{v}_1 \wedge ... \wedge \mathbf{v} \wedge ... \wedge \mathbf{v}_N$. Finally, exchanging the rows k and l in the matrix v_{ij} corresponds to exchanging the vectors $\mathbf{v}_k$ and $\mathbf{v}_l$, and then the tensor ω changes sign. ■

It is an important property that matrix transposition leaves the determinant unchanged.

Statement 2: The determinant of the transposed operator is unchanged:

$$\det \hat{A}^T = \det \hat{A}.$$

Proof: I give two proofs, one based on Definition D0 and the properties of permutations, another entirely coordinate-free — based on Definition D1 of the determinant and definition 1.8.4 of the transposed operator.

First proof: According to Definition D0, the determinant of the transposed matrix A_{ji} is given by the formula

$$\det(A_{ji}) \equiv \sum_{\sigma} (-1)^{|\sigma|} A_{1,\sigma(1)}...A_{N,\sigma(N)}, \tag{3.7}$$

so the only difference between $\det(A_{ij})$ and $\det(A_{ji})$ is the order of indices in the products of matrix elements, namely $A_{\sigma(i),i}$ instead of $A_{i,\sigma(i)}$. We can show that the sum in Eq. (3.7) consists of exactly the same terms as the sum in Eq. (3.1), only the terms occur in a different order. This is sufficient to prove that $\det(A_{ij}) = \det(A_{ji})$.

The sum in Eq. (3.7) consists of terms of the form $A_{1,\sigma(1)}...A_{N,\sigma(N)}$, where σ is some permutation. We may reorder factors in this term,

$$A_{1,\sigma(1)}...A_{N,\sigma(N)} = A_{\sigma'(1),1}...A_{\sigma'(N),N},$$

where σ' is another permutation such that $A_{i,\sigma(i)} = A_{\sigma'(i),i}$ for $i = 1, ..., N$. This is achieved when σ' is the permutation inverse to σ, i.e. we need to use $\sigma' \equiv \sigma^{-1}$. Since there exists precisely one inverse permutation σ^{-1} for each permutation σ, we may transform the sum in Eq. (3.7) into a sum over all inverse permutations σ'; each permutation will still enter exactly once into the new sum. Since the parity of the inverse permutation σ^{-1} is the same as the parity of σ (see Statement 3 in Appendix B), the factor $(-1)^{|\sigma|}$ will remain unchanged. Therefore, the sum will remain the same.

Second proof: The transposed operator is defined as

$$(\hat{A}^T \mathbf{f}^*)(\mathbf{x}) = \mathbf{f}^*(\hat{A}\mathbf{x}), \quad \forall \mathbf{f}^* \in V^*, \ \mathbf{x} \in V.$$

In order to compare the determinants $\det \hat{A}$ and $\det(\hat{A}^T)$ according to Definition D1, we need to compare the numbers $\wedge^N \hat{A}^N$ and $\wedge^N (\hat{A}^T)^N$.

Let us choose nonzero tensors $\omega \in \wedge^N V$ and $\omega^* \in \wedge^N V^*$. By Lemma 1 in Sec. 2.3.2, these tensors have representations of the form $\omega = \mathbf{v}_1 \wedge ... \wedge \mathbf{v}_N$ and $\omega^* = \mathbf{f}_1^* \wedge ... \wedge \mathbf{f}_N^*$. We have

$$(\det \hat{A})\mathbf{v}_1 \wedge ... \wedge \mathbf{v}_N = \hat{A}\mathbf{v}_1 \wedge ... \wedge \hat{A}\mathbf{v}_N.$$

Now we would like to relate this expression with the analogous expression for $\hat{A}^T$. In order to use the definition of $\hat{A}^T$, we need to act on the vectors $\hat{A}\mathbf{v}_i$ by the covectors $\mathbf{f}_j^*$. Therefore, we act with the N-form $\omega^* \in \wedge^N V^* \cong (\wedge^N V)^*$ on the N-vector $\wedge^N \hat{A}^N \omega \in \wedge^N V$ (this canonical action was defined by Definition 3 in Sec. 2.2). Since this action is linear, we find

$$\omega^*(\wedge^N \hat{A}^N \omega) = (\det \hat{A})\omega^*(\omega).$$

(Note that $\omega^*(\omega) \neq 0$ since by assumption the tensors ω and ω^* are nonzero.) On the other hand,

$$\begin{aligned}\omega^*(\wedge^N \hat{A}^N \omega) &= \sum_\sigma (-1)^{|\sigma|} \mathbf{f}_1^*(\hat{A}\mathbf{v}_{\sigma(1)})...\mathbf{f}_N^*(\hat{A}\mathbf{v}_{\sigma(N)}) \\ &= \sum_\sigma (-1)^{|\sigma|} (\hat{A}^T\mathbf{f}_1^*)(\mathbf{v}_{\sigma(1)})...(\hat{A}^T\mathbf{f}_N^*)(\mathbf{v}_{\sigma(N)}) \\ &= (\wedge^N(\hat{A}^T)^N\omega^*)(\omega) = (\det \hat{A}^T)\omega^*(\omega).\end{aligned}$$

Hence $\det \hat{A}^T = \det \hat{A}$. ■

Exercise* (Laplace expansion): As shown in the Corollary above, the determinant of the matrix v_{ij} is a linear function of each of the vectors $\{\mathbf{v}_i\}$. Consider $\det(v_{ij})$ as a linear function of the first vector, $\mathbf{v}_1$; this function is a *covector* that we may temporarily denote by $\mathbf{f}_1^*$. Show that $\mathbf{f}_1^*$ can be represented in the dual basis $\{\mathbf{e}_j^*\}$ as

$$\mathbf{f}_1^* = \sum_{i=1}^N (-1)^{i-1} B_{1i}\mathbf{e}_i^*,$$

where the coefficients B_{1i} are **minors** of the matrix v_{ij}, that is, determinants of the matrix v_{ij} from which row 1 and column i have been deleted.

Solution: Consider one of the coefficients, for example $B_{11} \equiv \mathbf{f}_1^*(\mathbf{e}_1)$. This coefficient can be determined from the tensor equality

$$\mathbf{e}_1 \wedge \mathbf{v}_2 \wedge ... \wedge \mathbf{v}_N = B_{11}\mathbf{e}_1 \wedge ... \wedge \mathbf{e}_N. \quad (3.8)$$

We could reduce B_{11} to a determinant of an $(N-1)\times(N-1)$ matrix if we could cancel $\mathbf{e}_1$ on both sides of Eq. (3.8). We would be able to cancel $\mathbf{e}_1$ if we had a tensor equality of the form

$$\mathbf{e}_1 \wedge \psi = B_{11}\mathbf{e}_1 \wedge \mathbf{e}_2 \wedge ... \wedge \mathbf{e}_N,$$

where the $(N-1)$-vector ψ were proportional to $\mathbf{e}_2 \wedge ... \wedge \mathbf{e}_N$. However, $\mathbf{v}_2 \wedge ... \wedge \mathbf{v}_N$ in Eq. (3.8) is not necessarily proportional to $\mathbf{e}_2 \wedge ... \wedge \mathbf{e}_N$; so we need to transform Eq. (3.8) to a suitable form. In order to do this, we transform the vectors $\mathbf{v}_i$ into vectors that belong to the subspace spanned by $\{\mathbf{e}_2, ..., \mathbf{e}_N\}$. We subtract from each $\mathbf{v}_i$ $(i = 2, ..., N)$ a suitable multiple of $\mathbf{e}_1$ and define the vectors $\tilde{\mathbf{v}}_i$ $(i = 2, ..., N)$ such that $\mathbf{e}_1^*(\tilde{\mathbf{v}}_i) = 0$:

$$\tilde{\mathbf{v}}_i \equiv \mathbf{v}_i - \mathbf{e}_1^*(\mathbf{v}_i)\mathbf{e}_1, \quad i = 2, ..., N.$$

Then $\tilde{\mathbf{v}}_i \in \text{Span}\{\mathbf{e}_2, ..., \mathbf{e}_N\}$ and also

$$\mathbf{e}_1 \wedge \mathbf{v}_2 \wedge ... \wedge \mathbf{v}_N = \mathbf{e}_1 \wedge \tilde{\mathbf{v}}_2 \wedge ... \wedge \tilde{\mathbf{v}}_N.$$

Now Eq. (3.8) is rewritten as

$$\mathbf{e}_1 \wedge \tilde{\mathbf{v}}_2 \wedge ... \wedge \tilde{\mathbf{v}}_N = B_{11}\mathbf{e}_1 \wedge \mathbf{e}_2 \wedge ... \wedge \mathbf{e}_N.$$

Since $\tilde{\mathbf{v}}_i \in \text{Span}\{\mathbf{e}_2, ..., \mathbf{e}_N\}$, the tensors $\tilde{\mathbf{v}}_2 \wedge ... \wedge \tilde{\mathbf{v}}_N$ and $\mathbf{e}_2 \wedge ... \wedge \mathbf{e}_N$ are proportional to each other. Now we are allowed to cancel $\mathbf{e}_1$ and obtain

$$\tilde{\mathbf{v}}_2 \wedge ... \wedge \tilde{\mathbf{v}}_N = B_{11}\mathbf{e}_2 \wedge ... \wedge \mathbf{e}_N.$$

Note that the vectors $\tilde{\mathbf{v}}_i$ have the first components equal to zero. In other words, B_{11} is equal to the determinant of the matrix v_{ij} from which row 1 (i.e. the vector $\mathbf{v}_1$) and column 1 (the coefficients at $\mathbf{e}_1$) have been deleted. The coefficients B_{1j} for $j = 2, ..., N$ are calculated similarly. ■

3.4.1 * Index notation for $\wedge^N V$ and determinants

Let us see how determinants are written in the index notation.

In order to use the index notation, we need to fix a basis $\{\mathbf{e}_j\}$ and represent each vector and each tensor by their components in that basis. Determinants are related to the space $\wedge^N V$. Let us consider a set of vectors $\{\mathbf{v}_1, ..., \mathbf{v}_N\}$ and the tensor

$$\psi \equiv \mathbf{v}_1 \wedge ... \wedge \mathbf{v}_N \in \wedge^N V.$$

Since the space $\wedge^N V$ is one-dimensional and its basis consists of the single tensor $\mathbf{e}_1 \wedge ... \wedge \mathbf{e}_N$, the index representation of ψ consists, in principle, of the single number C in a formula such as

$$\psi = C\mathbf{e}_1 \wedge ... \wedge \mathbf{e}_N.$$

However, it is more convenient to use a totally antisymmetric array of numbers having N indices, $\psi^{i_1...i_N}$, so that

$$\psi = \frac{1}{N!} \sum_{i_1,...,i_N=1}^{N} \psi^{i_1...i_N} \mathbf{e}_{i_1} \wedge ... \wedge \mathbf{e}_{i_N}.$$

Then the coefficient C is $C \equiv \psi^{12...N}$. In the formula above, the combinatorial factor $N!$ compensates the fact that we are summing an antisymmetric product of vectors with a totally antisymmetric array of coefficients.

To write such arrays more conveniently, one can use Levi-Civita symbol $\varepsilon^{i_1...i_N}$ (see Sec. 2.3.6). It is clear that any other totally antisymmetric array of numbers with N indices, such as $\psi^{i_1...i_N}$, is proportional to $\varepsilon^{i_1...i_N}$: For indices $\{i_1, ..., i_N\}$ that correspond to a permutation σ we have

$$\psi^{i_1...i_N} = \psi^{12...N}(-1)^{|\sigma|},$$

and hence

$$\psi^{i_1...i_N} = (\psi^{12...N})\varepsilon^{i_1...i_N}.$$

How to compute the index representation of ψ given the array v_j^k of the components of the vectors $\{\mathbf{v}_j\}$? We need to represent the tensor

$$\psi \equiv \sum_{\sigma} (-1)^{|\sigma|} \mathbf{v}_{\sigma(1)} \otimes \mathbf{v}_{\sigma(2)} \otimes ... \otimes \mathbf{v}_{\sigma(N)}.$$

Hence, we can use the Levi-Civita symbol and write

$$\psi^{12...N} = \sum_{\sigma} (-1)^{|\sigma|} v^1_{\sigma(1)} \otimes v^2_{\sigma(2)} \otimes ... \otimes v^N_{\sigma(N)}$$
$$= \sum_{i_1,...,i_N=1}^{N} \varepsilon^{i_1...i_N} v^1_{i_1} ... v^N_{i_N}.$$

The component $\psi^{12...N}$ is the only number we need to represent ψ in the basis $\{\mathbf{e}_j\}$.

The Levi-Civita symbol itself can be seen as the index representation of the tensor

$$\omega \equiv \mathbf{e}_1 \wedge ... \wedge \mathbf{e}_N$$

in the basis $\{\mathbf{e}_j\}$. (The components of ω in a different basis will, of course, differ from $\varepsilon^{i_1...i_N}$ by a constant factor.)

Now let us construct the index representation of the determinant of an operator $\hat{A}$. The operator is given by its matrix A^i_j and acts on a vector $\mathbf{v}$ with components v^i yielding a vector $\mathbf{u} \equiv \hat{A}\mathbf{v}$ with components

$$u^k = \sum_{i=1}^{N} A^k_i v^i.$$

Hence, the operator $\wedge^N \hat{A}^N$ acting on ψ yields an antisymmetric tensor whose component with the indices $k_1...k_N$ is

$$\left[(\wedge^N \hat{A}^N)\psi\right]^{k_1...k_N} = \left[\hat{A}\mathbf{v}_1 \wedge ... \wedge \hat{A}\mathbf{v}_N\right]^{k_1...k_N}$$
$$= \sum_{i_s,j_s} \varepsilon^{i_1...i_N} A^{k_1}_{j_1} v^{j_1}_{i_1} ... A^{k_N}_{j_N} v^{j_N}_{i_N}.$$

Since the tensor $\wedge^N \hat{A}^N \psi$ is proportional to ψ with the coefficient $\det \hat{A}$, the same proportionality holds for the components of these tensors:

$$\sum_{i_s,j_s} \varepsilon^{i_1...i_N} A^{k_1}_{j_1} v^{j_1}_{i_1} ... A^{k_N}_{j_N} v^{j_N}_{i_N} = (\det \hat{A}) \psi^{k_1...k_N}$$
$$= (\det \hat{A}) \sum_{i_s} \varepsilon^{i_1...i_N} v^{k_1}_{i_1} ... v^{k_N}_{i_N}.$$

The relation above must hold for arbitrary vectors $\{\mathbf{v}_j\}$. This is sufficient to derive a formula for $\det \hat{A}$. Since $\{\mathbf{v}_j\}$ are arbitrary, we may select $\{\mathbf{v}_j\}$ as the basis vectors $\{\mathbf{e}_j\}$, so that $v^k_i = \delta^k_i$. Substituting this into the equation above, we find

$$\sum_{i_s,j_s} \varepsilon^{i_1...i_N} A^{k_1}_{i_1} ... A^{k_N}_{i_N} = (\det \hat{A}) \varepsilon^{k_1...k_N}.$$

We can now solve for $\det \hat{A}$ by multiplying with another Levi-Civita symbol $\varepsilon_{k_1...k_N}$, written this time with lower indices to comply with the summation convention, and summing over all k_s. By elementary combinatorics (there are $N!$ possibilities to choose the indices $k_1, ..., k_N$ such that they are all different), we have

$$\sum_{k_1,...,k_N} \varepsilon_{k_1...k_N} \varepsilon^{k_1...k_N} = N!,$$

and therefore

$$\det(\hat{A}) = \frac{1}{N!} \sum_{i_s,k_s} \varepsilon_{k_1...k_N} \varepsilon^{i_1...i_N} A_{i_1}^{k_1} ... A_{i_N}^{k_N}.$$

This formula can be seen as the index representation of

$$\det \hat{A} = \omega^* (\wedge^N \hat{A}^N \omega),$$

where $\omega^* \in (\wedge^N V)^*$ is the tensor dual to ω and such that $\omega^*(\omega) = 1$. The components of ω^* are

$$\frac{1}{N!} \varepsilon_{k_1...k_N}.$$

We have shown how the index notation can express calculations with determinants and tensors in the space $\wedge^N V$. Such calculations in the index notation are almost always more cumbersome than in the index-free notation.

3.5 Solving linear equations

Determinants allow us to "determine" whether a system of linear equations has solutions. I will now explain this using exterior products. I will also show how to use exterior products for actually finding the solutions of linear equations when they exist.

A system of N linear equations for N unknowns $x_1, ..., x_N$ can be written in the matrix form,

$$\sum_{j=1}^{N} A_{ij} x_j = b_i, \quad i = 1, ..., N. \tag{3.9}$$

Here A_{ij} is a given matrix of coefficients, and the N numbers b_i are also given.

The first step in studying Eq. (3.9) is to interpret it in a geometric way, so that A_{ij} is not merely a "table of numbers" but a geometric object. We introduce an N-dimensional vector space $V = \mathbb{R}^N$, in which a basis $\{\mathbf{e}_i\}$ is fixed. There are two options (both will turn out to be useful). The first option is to interpret A_{ij}, b_j, and x_j as the coefficients representing some linear operator $\hat{A}$ and some vectors $\mathbf{b}, \mathbf{x}$ in the basis $\{\mathbf{e}_j\}$:

$$\hat{A} \equiv \sum_{i,j=1}^{N} A_{ij} \mathbf{e}_i \otimes \mathbf{e}_j^*, \quad \mathbf{b} \equiv \sum_{j=1}^{N} b_j \mathbf{e}_j, \quad \mathbf{x} \equiv \sum_{j=1}^{N} x_j \mathbf{e}_j.$$

Then we reformulate Eq. (3.9) as the vector equation

$$\hat{A}\mathbf{x} = \mathbf{b}, \tag{3.10}$$

from which we would like to find the unknown vector $\mathbf{x}$.

The second option is to interpret A_{ij} as the components of a *set* of N vectors $\{\mathbf{a}_1, ..., \mathbf{a}_N\}$ with respect to the basis,

$$\mathbf{a}_j \equiv \sum_{i=1}^{N} A_{ij}\mathbf{e}_i, \quad j = 1, ..., N,$$

to define $\mathbf{b}$ as before,

$$\mathbf{b} \equiv \sum_{j=1}^{N} b_j\mathbf{e}_j,$$

and to rewrite Eq. (3.9) as an equation expressing $\mathbf{b}$ as a linear combination of $\{\mathbf{a}_j\}$ with unknown coefficients $\{x_j\}$,

$$\sum_{j=1}^{N} x_j\mathbf{a}_j = \mathbf{b}. \tag{3.11}$$

In this interpretation, $\{x_j\}$ is just a set of N unknown numbers. These numbers could be interpreted the set of components of the vector $\mathbf{b}$ in the basis $\{\mathbf{a}_j\}$ if $\{\mathbf{a}_j\}$ were actually a basis, which is not necessarily the case.

3.5.1 Existence of solutions

Let us begin with the first interpretation, Eq. (3.10). When does Eq. (3.10) have solutions? The solution certainly exists when the operator $\hat{A}$ is **invertible**, i.e. the **inverse operator** $\hat{A}^{-1}$ exists such that $\hat{A}\hat{A}^{-1} = \hat{A}^{-1}\hat{A} = \hat{1}_V$; then the solution is found as $\mathbf{x} = \hat{A}^{-1}\mathbf{b}$. The condition for the existence of $\hat{A}^{-1}$ is that the determinant of $\hat{A}$ is nonzero. When the determinant of $\hat{A}$ is zero, the solution may or may not exist, and the solution is more complicated. I will give a proof of these statements based on the new definition D1 of the determinant.

Theorem 1: If $\det \hat{A} \neq 0$, the equation $\hat{A}\mathbf{x} = \mathbf{b}$ has a unique solution $\mathbf{x}$ for any $\mathbf{b} \in V$. There exists a linear operator $\hat{A}^{-1}$ such that the solution $\mathbf{x}$ is expressed as $\mathbf{x} = \hat{A}^{-1}\mathbf{b}$.

Proof: Suppose $\{\mathbf{e}_i \,|\, i = 1, ..., N\}$ is a basis in V. It follows from $\det \hat{A} \neq 0$ that

$$\wedge^N \hat{A}^N (\mathbf{e}_1 \wedge ... \wedge \mathbf{e}_N) = (\hat{A}\mathbf{e}_1) \wedge ... \wedge (\hat{A}\mathbf{e}_N) \neq 0.$$

By Theorem 1 of Sec. 2.3.2, the set of vectors $\{\hat{A}\mathbf{e}_1, ..., \hat{A}\mathbf{e}_N\}$ is linearly independent and therefore is a basis in V. Thus there exists a unique set of coefficients $\{c_i\}$ such that

$$\mathbf{b} = \sum_{i=1}^{N} c_i(\hat{A}\mathbf{e}_i).$$

Then due to linearity of $\hat{A}$ we have

$$\mathbf{b} = \hat{A} \sum_{i=1}^{N} c_i \mathbf{e}_i;$$

in other words, the solution of the equation $\hat{A}\mathbf{x} = \mathbf{b}$ is $\mathbf{x} \equiv \sum_{i=1}^{N} c_i \mathbf{e}_i$. Since the coefficients $\{c_i\}$ are determined uniquely, the solution $\mathbf{x}$ is unique.

The solution $\mathbf{x}$ can be expressed as a function of $\mathbf{b}$ as follows. Since $\{\hat{A}\mathbf{e}_i\}$ is a basis, there exists the corresponding dual basis, which we may denote by $\{\mathbf{v}_j^*\}$. Then the coefficients c_i can be expressed as $c_i = \mathbf{v}_i^*(\mathbf{b})$, and the vector $\mathbf{x}$ as

$$\mathbf{x} = \sum_{i=1}^{N} c_i \mathbf{e}_i = \sum_{i=1}^{N} \mathbf{e}_i \mathbf{v}_i^*(\mathbf{b}) = \big(\sum_{i=1}^{N} \mathbf{e}_i \otimes \mathbf{v}_i^*\big)\mathbf{b} \equiv \hat{A}^{-1}\mathbf{b}.$$

This shows explicitly that the operator $\hat{A}^{-1}$ exists and is linear. ■

Corollary: If $\det \hat{A} \neq 0$, the equation $\hat{A}\mathbf{v} = 0$ has only the (trivial) solution $\mathbf{v} = 0$.

Proof: The zero vector $\mathbf{v} = 0$ is a solution of $\hat{A}\mathbf{v} = 0$. By the above theorem the solution of that equation is unique, thus there are no other solutions. ■

Theorem 2 (existence of eigenvectors): If $\det \hat{A} = 0$, there exists at least one eigenvector with eigenvalue 0, that is, at least one nonzero vector $\mathbf{v}$ such that $\hat{A}\mathbf{v} = 0$.

Proof: Choose a basis $\{\mathbf{e}_j\}$ and consider the set $\{\hat{A}\mathbf{e}_1, ..., \hat{A}\mathbf{e}_N\}$. This set must be linearly dependent since

$$\hat{A}\mathbf{e}_1 \wedge ... \wedge \hat{A}\mathbf{e}_N = (\det \hat{A})\mathbf{e}_1 \wedge ... \wedge \mathbf{e}_N = 0.$$

Hence, there must exist at least one linear combination $\sum_{i=1}^{N} \lambda_i \hat{A}\mathbf{e}_i = 0$ with λ_i not all zero. Then the vector $\mathbf{v} \equiv \sum_{i=1}^{N} \lambda_i \mathbf{e}_i$ is nonzero and satisfies $\hat{A}\mathbf{v} = 0$. ■

Remark: If $\det \hat{A} = 0$, there *may* exist more than one eigenvector $\mathbf{v}$ such that $\hat{A}\mathbf{v} = 0$; more detailed analysis is needed to fully determine the eigenspace of zero eigenvalue, but we found that at least one eigenvector $\mathbf{v}$ exists. If $\det \hat{A} = 0$ then the equation $\hat{A}\mathbf{x} = \mathbf{b}$ with $\mathbf{b} \neq 0$ may still have solutions, although not for every $\mathbf{b}$. Moreover, when a solution $\mathbf{x}$ exists it will *not* be unique because $\mathbf{x} + \lambda\mathbf{v}$ is another solution if $\mathbf{x}$ is one. The full analysis of solvability of the equation $\hat{A}\mathbf{x} = \mathbf{b}$ when $\det \hat{A} = 0$ is more complicated (see the end of Sec. 3.5.2). ■

Once the inverse operator $\hat{A}^{-1}$ is determined, it is easy to compute solutions of any number of equations $\hat{A}\mathbf{x} = \mathbf{b}_1$, $\hat{A}\mathbf{x} = \mathbf{b}_2$, etc., for any number of vectors $\mathbf{b}_1$, $\mathbf{b}_2$, etc. However, if we only need to solve *one* such equation, $\hat{A}\mathbf{x} = \mathbf{b}$, then computing the full inverse operator is too much work: We have to determine the entire dual basis $\{\mathbf{v}_j^*\}$ and construct the operator $\hat{A}^{-1} = \sum_{i=1}^{N} \mathbf{e}_i \otimes \mathbf{v}_i^*$. An easier method is then provided by Kramer's rule.

3.5.2 Kramer's rule and beyond

We will now use the second interpretation, Eq. (3.11), of a linear system. This equation claims that $\mathbf{b}$ is a linear combination of the N vectors of the set $\{\mathbf{a}_1, ..., \mathbf{a}_N\}$. Clearly, this is true for any $\mathbf{b}$ if $\{\mathbf{a}_1, ..., \mathbf{a}_N\}$ is a basis in V; in that case, the solution $\{x_j\}$ exists and is unique because the dual basis, $\{\mathbf{a}_j^*\}$, exists and allows us to write the solution as

$$x_j = \mathbf{a}_j^*(\mathbf{b}).$$

On the other hand, when $\{\mathbf{a}_1, ..., \mathbf{a}_N\}$ is not a basis in V it is not certain that some given vector $\mathbf{b}$ is a linear combination of $\mathbf{a}_j$. In that case, the solution $\{x_j\}$ may or may not exist, and when it exists it will not be unique.

We first consider the case where $\{\mathbf{a}_j\}$ is a basis in V. In this case, the solution $\{x_j\}$ exists, and we would like to determine it more explicitly. We recall that an explicit computation of the dual basis was shown in Sec. 2.3.3. Motivated by the constructions given in that section, we consider the tensor

$$\omega \equiv \mathbf{a}_1 \wedge ... \wedge \mathbf{a}_N \in \wedge^N V$$

and additionally the N tensors $\{\omega_j \,|\, j = 1, ..., N\}$, defined by

$$\omega_j \equiv \mathbf{a}_1 \wedge ... \wedge \mathbf{a}_{j-1} \wedge \mathbf{b} \wedge \mathbf{a}_{j+1} \wedge ... \wedge \mathbf{a}_N \in \wedge^N V. \tag{3.12}$$

The tensor ω_j is the exterior product of all the vectors $\mathbf{a}_1$ to $\mathbf{a}_N$ except that $\mathbf{a}_j$ is replaced by $\mathbf{b}$. Since we know that the solution x_j exists, we can substitute $\mathbf{b} = \sum_{i=1}^N x_i \mathbf{a}_i$ into Eq. (3.12) and find

$$\omega_j = \mathbf{a}_1 \wedge ... \wedge x_j \mathbf{a}_j \wedge ... \wedge \mathbf{a}_N = x_j \omega.$$

Since $\{\mathbf{a}_j\}$ is a basis, the tensor $\omega \in \wedge^N V$ is nonzero (Theorem 1 in Sec. 2.3.2). Hence x_j ($j = 1, ..., N$) can be computed as the coefficient of proportionality between ω_j and ω:

$$x_j = \frac{\omega_j}{\omega} = \frac{\mathbf{a}_1 \wedge ... \wedge \mathbf{a}_{j-1} \wedge \mathbf{b} \wedge \mathbf{a}_{j+1} \wedge ... \wedge \mathbf{a}_N}{\mathbf{a}_1 \wedge ... \wedge \mathbf{a}_N}.$$

As before, the "division" of tensors means that the nonzero tensor ω is to be factored out of the numerator and canceled with the denominator, leaving a number.

This formula represents **Kramer's rule**, which yields explicitly the coefficients x_j necessary to represent a vector $\mathbf{b}$ through vectors $\{\mathbf{a}_1, ..., \mathbf{a}_N\}$. In its matrix formulation, Kramer's rule says that x_j is equal to the determinant of the modified matrix A_{ij} where the j-th column has been replaced by the column $(b_1, ..., b_N)$, divided by the determinant of the unmodified A_{ij}.

It remains to consider the case where $\{\mathbf{a}_j\}$ is *not* a basis in V. We have seen in Statement 2.3.5 that there exists a maximal nonzero exterior product of some linearly independent subset of $\{\mathbf{a}_j\}$; this subset can be found by trying various exterior products of the $\mathbf{a}_j$'s. Let us now denote by ω this maximal exterior product. Without loss of generality, we may renumber the $\mathbf{a}_j$'s so that $\omega = \mathbf{a}_1 \wedge ... \wedge \mathbf{a}_r$, where r is the rank of the set $\{\mathbf{a}_j\}$. If the equation $\sum_{j=1}^n x_j \mathbf{a}_j = \mathbf{b}$ has a solution then $\mathbf{b}$ is expressible as a linear combination of the $\mathbf{a}_j$'s; thus we must have $\omega \wedge \mathbf{b} = 0$. We can check whether $\omega \wedge \mathbf{b} = 0$ since we have already computed ω. If we find that $\omega \wedge \mathbf{b} \neq 0$ we know that the equation $\sum_{j=1}^n x_j \mathbf{a}_j = \mathbf{b}$ has *no solutions*.

If we find that $\omega \wedge \mathbf{b} = 0$ then we can conclude that the vector $\mathbf{b}$ belongs to the subspace Span $\{\mathbf{a}_1, ..., \mathbf{a}_r\}$, and so the equation $\sum_{j=1}^n x_j \mathbf{a}_j = \mathbf{b}$ *has* solutions, — in fact infinitely many of them. To determine all solutions, we will note that the set $\{\mathbf{a}_1, ..., \mathbf{a}_r\}$ is linearly independent, so $\mathbf{b}$ is uniquely represented as a linear combination of the vectors $\mathbf{a}_1, ..., \mathbf{a}_r$. In other words, there is a unique solution of the form

$$x_i^{(1)} = (x_1^{(1)}, ..., x_r^{(1)}, 0, ..., 0)$$

that may have nonzero coefficients $x_1^{(1)}, ..., x_r^{(1)}$ only up to the component number r, after which $x_i^{(1)} = 0$ $(r + 1 \leq i \leq n)$. To obtain the coefficients $x_i^{(1)}$, we use Kramer's rule for the subspace Span $\{\mathbf{a}_1, ..., \mathbf{a}_r\}$:

$$x_i^{(1)} = \frac{\mathbf{a}_1 \wedge ... \wedge \mathbf{a}_{j-1} \wedge \mathbf{b} \wedge \mathbf{a}_{j+1} \wedge ... \wedge \mathbf{a}_r}{\mathbf{a}_1 \wedge ... \wedge \mathbf{a}_r}.$$

We can now obtain the general solution of the equation $\sum_{j=1}^n x_j \mathbf{a}_j = \mathbf{b}$ by adding to the solution $x_i^{(1)}$ an arbitrary solution $x_i^{(0)}$ of the homogeneous equation, $\sum_{j=1}^n x_j^{(0)} \mathbf{a}_j = 0$. The solutions of the homogeneous equation build a subspace that can be determined as an eigenspace of the operator $\hat{A}$ as considered in the previous subsection. We can also determine the homogeneous solutions using the method of this section, as follows.

We decompose the vectors $\mathbf{a}_{r+1}, ..., \mathbf{a}_n$ into linear combinations of $\mathbf{a}_1$, ..., $\mathbf{a}_r$ again by using Kramer's rule:

$$\mathbf{a}_k = \sum_{j=1}^r \alpha_{kj} \mathbf{a}_j, \quad k = r + 1, ..., n,$$

$$\alpha_{kj} \equiv \frac{\mathbf{a}_1 \wedge ... \wedge \mathbf{a}_{j-1} \wedge \mathbf{a}_k \wedge \mathbf{a}_{j+1} \wedge ... \wedge \mathbf{a}_r}{\mathbf{a}_1 \wedge ... \wedge \mathbf{a}_r}.$$

Having computed the coefficients α_{kj}, we determine the $(n - r)$-dimensional space of homogeneous solutions. This space is spanned by the $(n - r)$ solutions that can be chosen, for example, as follows:

$$\begin{aligned}
x_i^{(0)(r+1)} &= (\alpha_{(r+1)1}, ..., \alpha_{(r+1)r}, -1, 0, ..., 0), \\
x_i^{(0)(r+2)} &= (\alpha_{(r+2)1}, ..., \alpha_{(r+2)r}, 0, -1, ..., 0), \\
&\ldots \\
x_i^{(0)(n)} &= (\alpha_{n1}, ..., \alpha_{nr}, 0, 0, ..., -1).
\end{aligned}$$

Finally, the solution of the equation $\sum_{j=1}^{n} x_j \mathbf{a}_j = \mathbf{b}$ can be written as

$$x_i = x_i^{(1)} + \sum_{k=r+1}^{n} \beta_k x_i^{(0)(k)}, \quad i = 1, ..., n,$$

where $\{\beta_k \mid k = r+1, ...n\}$ are *arbitrary* coefficients. The formula above explicitly contains $(n-r)$ arbitrary constants and is called the general solution of $\sum_{i=1}^{n} x_i \mathbf{a}_i = \mathbf{b}$. (The **general solution** of something is a formula with arbitrary constants that describes all solutions.)

Example: Consider the linear system

$$\begin{aligned} 2x + y &= 1 \\ 2x + 2y + z &= 4 \\ y + z &= 3 \end{aligned}$$

Let us apply the procedure above to this system. We interpret this system as the vector equation $x\mathbf{a} + y\mathbf{b} + z\mathbf{c} = \mathbf{p}$ where $\mathbf{a} = (2,2,0)$, $\mathbf{b} = (1,2,1)$, $\mathbf{c} = (0,1,1)$, and $\mathbf{p} = (1,4,3)$ are given vectors. Introducing an explicit basis $\{\mathbf{e}_1, \mathbf{e}_2, \mathbf{e}_3\}$, we compute (using elimination)

$$\begin{aligned} \mathbf{a} \wedge \mathbf{b} &= (2\mathbf{e}_1 + 2\mathbf{e}_2) \wedge (\mathbf{e}_1 + 2\mathbf{e}_2 + \mathbf{e}_3) \\ &= 2\,(\mathbf{e}_1 + \mathbf{e}_2) \wedge (\mathbf{e}_1 + 2\mathbf{e}_2 + \mathbf{e}_3) \\ &= 2\,(\mathbf{e}_1 + \mathbf{e}_2) \wedge (\mathbf{e}_2 + \mathbf{e}_3) = \mathbf{a} \wedge \mathbf{c}. \end{aligned}$$

Therefore $\mathbf{a} \wedge \mathbf{b} \wedge \mathbf{c} = 0$, and the maximal nonzero exterior product can be chosen as $\omega \equiv \mathbf{a} \wedge \mathbf{b}$. Now we check whether the vector $\mathbf{p}$ belongs to the subspace Span $\{\mathbf{a}, \mathbf{b}\}$:

$$\begin{aligned} \omega \wedge \mathbf{p} &= 2\,(\mathbf{e}_1 + \mathbf{e}_2) \wedge (\mathbf{e}_2 + \mathbf{e}_3) \wedge (\mathbf{e}_1 + 4\mathbf{e}_2 + 3\mathbf{e}_3) \\ &= 2\,(\mathbf{e}_1 + \mathbf{e}_2) \wedge (\mathbf{e}_2 + \mathbf{e}_3) \wedge 3(\mathbf{e}_2 + \mathbf{e}_3) = 0. \end{aligned}$$

Therefore, $\mathbf{p}$ can be represented as a linear combination of $\mathbf{a}$ and $\mathbf{b}$. To determine the coefficients, we use Kramer's rule: $\mathbf{p} = \alpha\mathbf{a} + \beta\mathbf{b}$ where

$$\begin{aligned} \alpha &= \frac{\mathbf{p} \wedge \mathbf{b}}{\mathbf{a} \wedge \mathbf{b}} = \frac{(\mathbf{e}_1 + 4\mathbf{e}_2 + 3\mathbf{e}_3) \wedge (\mathbf{e}_1 + 2\mathbf{e}_2 + \mathbf{e}_3)}{2\,(\mathbf{e}_1 + \mathbf{e}_2) \wedge (\mathbf{e}_2 + \mathbf{e}_3)} \\ &= \frac{-2\mathbf{e}_1 \wedge \mathbf{e}_2 - 2\mathbf{e}_1 \wedge \mathbf{e}_3 - 2\mathbf{e}_2 \wedge \mathbf{e}_3}{2\,(\mathbf{e}_1 \wedge \mathbf{e}_2 + \mathbf{e}_1 \wedge \mathbf{e}_3 + \mathbf{e}_2 \wedge \mathbf{e}_3)} = -1; \\ \beta &= \frac{\mathbf{a} \wedge \mathbf{p}}{\mathbf{a} \wedge \mathbf{b}} = \frac{2\,(\mathbf{e}_1 + \mathbf{e}_2) \wedge (\mathbf{e}_1 + 4\mathbf{e}_2 + 3\mathbf{e}_3)}{2\,(\mathbf{e}_1 + \mathbf{e}_2) \wedge (\mathbf{e}_2 + \mathbf{e}_3)} \\ &= \frac{3\mathbf{e}_1 \wedge \mathbf{e}_2 + 3\mathbf{e}_1 \wedge \mathbf{e}_3 + 3\mathbf{e}_2 \wedge \mathbf{e}_3}{\mathbf{e}_1 \wedge \mathbf{e}_2 + \mathbf{e}_1 \wedge \mathbf{e}_3 + \mathbf{e}_2 \wedge \mathbf{e}_3} = 3. \end{aligned}$$

Therefore, $\mathbf{p} = -\mathbf{a} + 3\mathbf{b}$; thus the inhomogeneous solution is $\mathbf{x}^{(1)} = (-1, 3, 0)$.

To determine the space of homogeneous solutions, we decompose $\mathbf{c}$ into a linear combination of $\mathbf{a}$ and $\mathbf{b}$ by the same method; the result is $\mathbf{c} = -\frac{1}{2}\mathbf{a} + \mathbf{b}$. So the space of homogeneous solutions is spanned by the single solution

$$x_i^{(0)(1)} = \left(-\tfrac{1}{2}, 1, -1\right).$$

Finally, we write the general solution as

$$x_i = x_i^{(1)} + \beta x_i^{(0)(1)} = \left(-1 - \tfrac{1}{2}\beta, 3 + \beta, -\beta\right),$$

where β is an arbitrary constant. ■

Remark: In the calculations of the coefficients according to Kramer's rule the numerators and the denominators always contain the same tensor, such as $\mathbf{e}_1 \wedge \mathbf{e}_2 + \mathbf{e}_1 \wedge \mathbf{e}_3 + \mathbf{e}_2 \wedge \mathbf{e}_3$, multiplied by a constant factor. We have seen this in the above examples. This is guaranteed to happen in every case; it is impossible that a numerator should contain $\mathbf{e}_1 \wedge \mathbf{e}_2 + \mathbf{e}_1 \wedge \mathbf{e}_3 + 2\mathbf{e}_2 \wedge \mathbf{e}_3$ or some other tensor not proportional to ω. Therefore, in practical calculations it is sufficient to compute just one coefficient, say at $\mathbf{e}_1 \wedge \mathbf{e}_2$, in both the numerator and the denominator.

Exercise: Techniques based on Kramer's rule can be applied also to non-square systems. Consider the system

$$\begin{aligned} x + y &= 1 \\ y + z &= 1 \end{aligned}$$

This system has infinitely many solutions. Determine the general solution.

Answer: For example, the general solution can be written as

$$x_i = (1, 0, 1) + \alpha\,(1, -1, 1),$$

where α is an arbitrary number.

3.6 Vandermonde matrix

The **Vandermonde matrix** is defined by

$$\text{Vand}\,(x_1, ..., x_N) \equiv \begin{pmatrix} 1 & 1 & \cdots & 1 \\ x_1 & x_2 & & x_N \\ x_1^2 & x_2^2 & & x_N^2 \\ \vdots & \vdots & \ddots & \\ x_1^{N-1} & x_2^{N-1} & \cdots & x_N^{N-1} \end{pmatrix}.$$

It is a curious matrix that is useful in several ways. A classic result is an explicit formula for the determinant of this matrix. Let us first compute the determinant for a Vandermonde matrix of small size.

Exercise 1: Verify that the Vandermonde determinants for $N = 2$ and $N = 3$ are as follows,

$$\begin{vmatrix} 1 & 1 \\ x & y \end{vmatrix} = y - x; \quad \begin{vmatrix} 1 & 1 & 1 \\ x & y & z \\ x^2 & y^2 & z^2 \end{vmatrix} = (y - x)\,(z - x)\,(z - y)\,.$$

It now appears plausible from these examples that the determinant that we denote by $\det\left(\text{Vand}(x_1, ..., x_N)\right)$ is equal to the product of the pairwise differences between all the x_i's.

Statement 1: The determinant of the Vandermonde matrix is given by

$$\begin{aligned}
&\det(\text{Vand}(x_1, ..., x_N)) \\
&= (x_2 - x_1)(x_3 - x_1) \dots (x_N - x_{N-1}) \\
&= \prod_{1 \leq i < j \leq N} (x_j - x_i).
\end{aligned} \tag{3.13}$$

Proof: Let us represent the Vandermonde matrix as a table of the components of a set of N vectors $\{\mathbf{v}_j\}$ with respect to some basis $\{\mathbf{e}_j\}$. Looking at the Vandermonde matrix, we find that the components of the vector $\mathbf{v}_1$ are $(1, 1, ..., 1)$, so

$$\mathbf{v}_1 = \mathbf{e}_1 + ... + \mathbf{e}_N.$$

The components of the vector $\mathbf{v}_2$ are $(x_1, x_2, ..., x_N)$; the components of the vector $\mathbf{v}_3$ are $(x_1^2, x_2^2, ..., x_N^2)$. Generally, the vector $\mathbf{v}_j$ ($j = 1, ..., N$) has components $(x_1^{j-1}, ..., x_N^{j-1})$. It is convenient to introduce a linear operator $\hat{A}$ such that $\hat{A}\mathbf{e}_1 = x_1\mathbf{e}_1$, ..., $\hat{A}\mathbf{e}_N = x_N\mathbf{e}_N$; in other words, the operator $\hat{A}$ is diagonal in the basis $\{\mathbf{e}_j\}$, and $\mathbf{e}_j$ is an eigenvector of $\hat{A}$ with the eigenvalue x_j. A tensor representation of $\hat{A}$ is

$$\hat{A} = \sum_{j=1}^{N} x_j \mathbf{e}_j \otimes \mathbf{e}_j^*.$$

Then we have a short formula for $\mathbf{v}_j$:

$$\mathbf{v}_j = \hat{A}^{j-1}\mathbf{u}, \quad j = 1, ..., N; \quad \mathbf{u} \equiv \mathbf{v}_1 = \mathbf{e}_1 + ... + \mathbf{e}_N.$$

According to Statement 1 of Sec. 3.4, the determinant of the Vandermonde matrix is equal to the coefficient C in the equation

$$\mathbf{v}_1 \wedge ... \wedge \mathbf{v}_N = C\mathbf{e}_1 \wedge ... \wedge \mathbf{e}_N.$$

So our purpose now is to determine C. Let us use the formula for $\mathbf{v}_j$ to rewrite

$$\mathbf{v}_1 \wedge ... \wedge \mathbf{v}_N = \mathbf{u} \wedge \hat{A}\mathbf{u} \wedge \hat{A}^2\mathbf{u} \wedge ... \wedge \hat{A}^{N-1}\mathbf{u}. \tag{3.14}$$

Now we use the following trick: since $\mathbf{a} \wedge \mathbf{b} = \mathbf{a} \wedge (\mathbf{b} + \lambda\mathbf{a})$ for any λ, we may replace

$$\mathbf{u} \wedge \hat{A}\mathbf{u} = \mathbf{u} \wedge (\hat{A}\mathbf{u} + \lambda\mathbf{u}) = \mathbf{u} \wedge (\hat{A} + \lambda\hat{1})\mathbf{u}.$$

Similarly, we may replace the factor $\hat{A}^2\mathbf{u}$ by $(\hat{A}^2 + \lambda_1\hat{A} + \lambda_2)\mathbf{u}$, with arbitrary coefficients λ_1 and λ_2. We may pull this trick in every factor in the tensor product (3.14) starting from the second factor. In effect, we may replace $\hat{A}^k$ by an arbitrary polynomial $p_k(\hat{A})$ of degree k as long as the coefficient at $\hat{A}^k$ remains 1. (Such polynomials are called **monic polynomials**.) So we obtain

$$\begin{aligned}
&\mathbf{u} \wedge \hat{A}\mathbf{u} \wedge \hat{A}^2\mathbf{u} \wedge ... \wedge \hat{A}^{N-1}\mathbf{u} \\
&= \mathbf{u} \wedge p_1(\hat{A})\mathbf{u} \wedge p_2(\hat{A})\hat{A}\mathbf{u} \wedge ... \wedge p_{N-1}(\hat{A})\mathbf{u}.
\end{aligned}$$

Since we may choose the monic polynomials $p_j(\hat{A})$ arbitrarily, we would like to choose them such that the formula is simplified as much as possible.

Let us first choose the polynomial p_{N-1} because that polynomial has the highest degree $(N-1)$ and so affords us the most freedom. Here comes another trick: If we choose

$$p_{N-1}(x) \equiv (x - x_1)(x - x_2) \dots (x - x_{N-1}),$$

then the operator $p_{N-1}(\hat{A})$ will be much simplified:

$$p_{N-1}(\hat{A})\mathbf{e}_N = p_{N-1}(x_N)\mathbf{e}_N; \; p_{N-1}(\hat{A})\mathbf{e}_j = 0, \quad j = 1, \dots, N-1.$$

Therefore $p_{N-1}(\hat{A})\mathbf{u} = p_{N-1}(x_N)\mathbf{e}_N$. Now we repeat this trick for the polynomial p_{N-2}, choosing

$$p_{N-2}(x) \equiv (x - x_1) \dots (x - x_{N-2})$$

and finding

$$p_{N-2}(\hat{A})\mathbf{u} = p_{N-2}(x_{N-1})\mathbf{e}_{N-1} + p_{N-2}(x_N)\mathbf{e}_N.$$

We need to compute the exterior product, which simplifies:

$$\begin{aligned} & p_{N-2}(\hat{A})\mathbf{u} \wedge p_{N-1}(\hat{A})\mathbf{u} \\ & = (p_{N-2}(x_{N-1})\mathbf{e}_{N-1} + p_{N-2}(x_N)\mathbf{e}_N) \wedge p_{N-1}(x_N)\mathbf{e}_N \\ & = p_{N-2}(x_{N-1})\mathbf{e}_{N-1} \wedge p_{N-1}(x_N)\mathbf{e}_N. \end{aligned}$$

Proceeding inductively in this fashion, we find

$$\begin{aligned} & \mathbf{u} \wedge p_1(\hat{A})\mathbf{u} \wedge \dots \wedge p_{N-1}(\hat{A})\mathbf{u} \\ & = \mathbf{u} \wedge p_1(x_2)\mathbf{e}_2 \wedge \dots \wedge p_{N-1}(x_N)\mathbf{e}_N \\ & = p_1(x_2) \dots p_{N-1}(x_N)\mathbf{e}_1 \wedge \dots \wedge \mathbf{e}_N, \end{aligned}$$

where we defined each monic polynomial $p_j(x)$ as

$$p_j(x) \equiv (x - x_1) \dots (x - x_j), \quad j = 1, \dots, N-1.$$

For instance, $p_1(x) = x - x_1$. The product of the polynomials,

$$\begin{aligned} & p_1(x_2)p_2(x_3) \dots p_{N-1}(x_N) \\ & = (x_2 - x_1)(x_3 - x_1)(x_3 - x_2) \dots (x_N - x_{N-1}) \\ & = \prod_{1 \le i < j \le N} (x_j - x_i). \end{aligned}$$

yields the required formula (3.13). ■

Remark: This somewhat long argument explains the procedure of subtracting various rows of the Vandermonde matrix from each other in order to simplify the determinant. (The calculation appears long because I have motivated every step, rather than just go through the equations.) One can observe that the determinant of the Vandermonde matrix is nonzero if and only if all the values x_j are different. This property allows one to prove the Vandermonde formula in a much more elegant way.[3] Namely, one can notice that the expression $\mathbf{v}_1 \wedge ... \wedge \mathbf{v}_N$ is a polynomial in x_j of degree not more than $\frac{1}{2}N(N-1)$; that this polynomial is equal to zero unless every x_j is different; therefore this polynomial must be equal to Eq. (3.13) times a constant. To find that constant, one computes explicitly the coefficient at the term $x_2 x_3^2 ... x_N^{N-1}$, which is equal to 1, hence the constant is 1. ■

In the next two subsections we will look at two interesting applications of the Vandermonde matrix.

3.6.1 Linear independence of eigenvectors

Statement: Suppose that the vectors $\mathbf{e}_1, ..., \mathbf{e}_n$ are nonzero and are eigenvectors of an operator $\hat{A}$ with *all different* eigenvalues $\lambda_1, ..., \lambda_n$. Then the set $\{\mathbf{e}_1, ..., \mathbf{e}_n\}$ is linearly independent. (The number n may be less than the dimension N of the vector space V; the statement holds also for infinite-dimensional spaces).

Proof. Let us show that the set $\{\mathbf{e}_j \,|\, j = 1, ..., n\}$ is linearly independent. By definition of linear independence, we need to show that $\sum_{j=1}^n c_j \mathbf{e}_j = 0$ is possible only if all the coefficients c_j are equal to zero. Let us denote $\mathbf{u} = \sum_{j=1}^n c_j \mathbf{e}_j$ and assume that $\mathbf{u} = 0$. Consider the vectors $\mathbf{u}$, $\hat{A}\mathbf{u}$, ..., $\hat{A}^{n-1}\mathbf{u}$; by assumption all these vectors are equal to zero. The condition that these vectors are equal to zero is a system of vector equations that looks like this,

$$\begin{aligned} c_1 \mathbf{e}_1 + ... + c_n \mathbf{e}_n &= 0, \\ c_1 \lambda_1 \mathbf{e}_1 + ... + c_n \lambda_n \mathbf{e}_n &= 0, \\ &... \\ c_1 \lambda_1^{n-1} \mathbf{e}_1 + ... + c_n \lambda_n^{n-1} \mathbf{e}_n &= 0. \end{aligned}$$

This system of equations can be written in a matrix form with the Vandermonde matrix,

$$\begin{pmatrix} 1 & 1 & \cdots & 1 \\ \lambda_1 & \lambda_2 & & \lambda_n \\ \vdots & \vdots & \ddots & \\ \lambda_1^{n-1} & \lambda_2^{n-1} & \cdots & \lambda_n^{n-1} \end{pmatrix} \begin{bmatrix} c_1 \mathbf{e}_1 \\ c_2 \mathbf{e}_2 \\ \vdots \\ c_n \mathbf{e}_n \end{bmatrix} = \begin{bmatrix} 0 \\ 0 \\ \vdots \\ 0 \end{bmatrix}.$$

Since the eigenvalues λ_j are (by assumption) all different, the determinant of the Vandermonde matrix is nonzero. Therefore, this system of equations has

[3] I picked this up from a paper by C. Krattenthaler (see online arxiv.org/abs/math.co/9902004) where many other special determinants are evaluated using similar techniques.

only the trivial solution, $c_j \mathbf{e}_j = 0$ for all j. Since $\mathbf{e}_j \neq 0$, it is necessary that all $c_j = 0$, $j = 1, ...n$. ■

Exercise: Show that we are justified in using the matrix method for solving a system of equations with *vector-valued* unknowns $c_i \mathbf{e}_i$.

Hint: Act with an arbitrary covector $\mathbf{f}^*$ on all the equations.

3.6.2 Polynomial interpolation

The task of **polynomial interpolation** consists of finding a polynomial that passes through specified points.

Statement: If the numbers x_1, ..., x_N are all different and numbers y_1, ..., y_N are arbitrary then there exists a unique polynomial $p(x)$ of degree at most $N-1$ that has values y_j at the points x_j ($j = 1, ..., N$).

Proof. Let us try to determine the coefficients of the polynomial $p(x)$. We write a polynomial with unknown coefficients,

$$p(x) = p_0 + p_1 x + ... + p_{N-1} x^{N-1},$$

and obtain a system of N linear equations, $p(x_j) = y_j$ ($j = 1, ..., N$), for the N unknowns p_j. The crucial observation is that this system of equations has the Vandermonde matrix. For example, with $N = 3$ we have three equations,

$$\begin{aligned} p(x_1) &= p_0 + p_1 x_1 + p_2 x_1^2 = y_1, \\ p(x_2) &= p_0 + p_1 x_2 + p_2 x_2^2 = y_2, \\ p(x_3) &= p_0 + p_1 x_3 + p_2 x_3^2 = y_3, \end{aligned}$$

which can be rewritten in the matrix form as

$$\begin{pmatrix} 1 & x_1 & x_1^2 \\ 1 & x_2 & x_2^2 \\ 1 & x_3 & x_3^2 \end{pmatrix} \begin{bmatrix} p_0 \\ p_1 \\ p_2 \end{bmatrix} = \begin{bmatrix} y_1 \\ y_2 \\ y_3 \end{bmatrix}.$$

Since the determinant of the Vandermonde matrix is nonzero as long as all x_j are different, these equations always have a unique solution $\{p_j\}$. Therefore the required polynomial always exists and is unique. ■

Question: The polynomial $p(x)$ *exists*, but how can I write it explicitly?

Answer: One possibility is the **Lagrange interpolating polynomial**; let us illustrate the idea on an example with three points:

$$\begin{aligned} p(x) = y_1 & \frac{(x - x_2)(x - x_3)}{(x_1 - x_2)(x_1 - x_3)} + y_2 \frac{(x - x_1)(x - x_3)}{(x_2 - x_1)(x_2 - x_3)} \\ & + y_3 \frac{(x - x_1)(x - x_2)}{(x_3 - x_1)(x_3 - x_2)}. \end{aligned}$$

It is easy to check directly that this polynomial indeed has values $p(x_i) = y_i$ for $i = 1, 2, 3$. However, other (equivalent, but computationally more efficient) formulas are used in numerical calculations.

3.7 Multilinear actions in exterior powers

As we have seen, the action of $\hat{A}$ on the exterior power $\wedge^N V$ by

$$\mathbf{v}_1 \wedge ... \wedge \mathbf{v}_N \mapsto \hat{A}\mathbf{v}_1 \wedge ... \wedge \hat{A}\mathbf{v}_N$$

has been very useful. However, this is not the only way $\hat{A}$ can act on an N-vector. Let us explore other possibilities; we will later see that they have their uses as well.

A straightforward generalization is to promote an operator $\hat{A} \in \text{End}\, V$ to a linear operator in the space $\wedge^k V$, $k < N$ (rather than in the top exterior power $\wedge^N V$). We denote this by $\wedge^k \hat{A}^k$:

$$(\wedge^k \hat{A}^k)\mathbf{v}_1 \wedge ... \wedge \mathbf{v}_k = \hat{A}\mathbf{v}_1 \wedge ... \wedge \hat{A}\mathbf{v}_k.$$

This is, of course, a linear map of $\wedge^k \hat{A}^k$ to itself (but not any more a mere multiplication by a scalar!). For instance, in $\wedge^2 V$ we have

$$(\wedge^2 \hat{A}^2)\mathbf{u} \wedge \mathbf{v} = \hat{A}\mathbf{u} \wedge \hat{A}\mathbf{v}.$$

However, this is not the only possibility. We could, for instance, define another map of $\wedge^2 V$ to itself like this,

$$\mathbf{u} \wedge \mathbf{v} \mapsto (\hat{A}\mathbf{u}) \wedge \mathbf{v} + \mathbf{u} \wedge (\hat{A}\mathbf{v}).$$

This map is *linear in* $\hat{A}$ (as well as being a linear map of $\wedge^2 V$ to itself), so I denote this map by $\wedge^2 \hat{A}^1$ to emphasize that it contains $\hat{A}$ only linearly. I call such maps **extensions of** $\hat{A}$ to the exterior power space $\wedge^2 V$ (this is not a standard terminology).

It turns out that operators of this kind play an important role in many results related to determinants. Let us now generalize the examples given above. We denote by $\wedge^m \hat{A}^k$ a linear map $\wedge^m V \to \wedge^m V$ that acts on $\mathbf{v}_1 \wedge ... \wedge \mathbf{v}_m$ by producing a sum of terms with k copies of $\hat{A}$ in each term. For instance,

$$\begin{aligned}
\wedge^2 \hat{A}^1 \left(\mathbf{a} \wedge \mathbf{b}\right) &\equiv \hat{A}\mathbf{a} \wedge \mathbf{b} + \mathbf{a} \wedge \hat{A}\mathbf{b}; \\
\wedge^3 \hat{A}^3 \left(\mathbf{a} \wedge \mathbf{b} \wedge \mathbf{c}\right) &\equiv \hat{A}\mathbf{a} \wedge \hat{A}\mathbf{b} \wedge \hat{A}\mathbf{c}; \\
\wedge^3 \hat{A}^2 \left(\mathbf{a} \wedge \mathbf{b} \wedge \mathbf{c}\right) &\equiv \hat{A}\mathbf{a} \wedge \hat{A}\mathbf{b} \wedge \mathbf{c} + \hat{A}\mathbf{a} \wedge \mathbf{b} \wedge \hat{A}\mathbf{c} \\
&\quad + \mathbf{a} \wedge \hat{A}\mathbf{b} \wedge \hat{A}\mathbf{c}.
\end{aligned}$$

More generally, we can write

$$\begin{aligned}
\wedge^k \hat{A}^k \left(\mathbf{v}_1 \wedge ... \wedge \mathbf{v}_k\right) &= \hat{A}\mathbf{v}_1 \wedge ... \wedge \hat{A}\mathbf{v}_k; \\
\wedge^k \hat{A}^1 \left(\mathbf{v}_1 \wedge ... \wedge \mathbf{v}_k\right) &= \sum_{j=1}^{k} \mathbf{v}_1 \wedge ... \wedge \hat{A}\mathbf{v}_j \wedge ... \wedge \mathbf{v}_k; \\
\wedge^k \hat{A}^m \left(\mathbf{v}_1 \wedge ... \wedge \mathbf{v}_k\right) &= \sum_{\substack{s_1, ..., s_k = 0,1 \\ \sum_j s_j = m}} \hat{A}^{s_1}\mathbf{v}_1 \wedge ... \wedge \hat{A}^{s_k}\mathbf{v}_k.
\end{aligned}$$

In the last line, the sum is over all integers s_j, each being either 0 or 1, so that $\hat{A}^{s_j}$ is either $\hat{1}$ or $\hat{A}$, and the total power of $\hat{A}$ is m.

So far we defined the action of $\wedge^m \hat{A}^k$ only on tensors of the form $\mathbf{v}_1 \wedge ... \wedge \mathbf{v}_m \in \wedge^m V$. Since an arbitrary element of $\wedge^m V$ is a linear combination of such "elementary" tensors, and since we intend $\wedge^m \hat{A}^k$ to be a linear map, we define the action of $\wedge^m \hat{A}^k$ on every element of $\wedge^m V$ using linearity. For example,

$$\wedge^2 \hat{A}^2 \left(\mathbf{a} \wedge \mathbf{b} + \mathbf{c} \wedge \mathbf{d}\right) \equiv \hat{A}\mathbf{a} \wedge \hat{A}\mathbf{b} + \hat{A}\mathbf{c} \wedge \hat{A}\mathbf{d}.$$

By now it should be clear that the extension $\wedge^m \hat{A}^k$ is indeed a linear map $\wedge^m V \to \wedge^m V$. Here is a formal definition.

Definition: For a linear operator $\hat{A}$ in V, the **k-linear extension** of $\hat{A}$ **to the space** $\wedge^m V$ is a linear transformation $\wedge^m V \to \wedge^m V$ denoted by $\wedge^m \hat{A}^k$ and defined by the formula

$$\wedge^m \hat{A}^k \Big(\bigwedge_{j=1}^{m} \mathbf{v}_j \Big) = \sum_{(s_1,...,s_m)} \bigwedge_{j=1}^{m} \hat{A}^{s_j} \mathbf{v}_j, \quad s_j = 0 \text{ or } 1, \ \sum_{j=1}^{m} s_j = k. \tag{3.15}$$

In words: To describe the action of $\wedge^m \hat{A}^k$ on a term $\mathbf{v}_1 \wedge ... \wedge \mathbf{v}_m \in \wedge^m V$, we sum over all possible ways to act with $\hat{A}$ on the various vectors $\mathbf{v}_j$ from the term $\mathbf{v}_1 \wedge ... \wedge \mathbf{v}_m$, where $\hat{A}$ appears exactly k times. The action of $\wedge^m \hat{A}^k$ on a linear combination of terms is by definition the linear combination of the actions on each term. Also by definition we set $\wedge^m \hat{A}^0 \equiv \hat{1}_{\wedge^m V}$ and $\wedge^m \hat{A}^k \equiv \hat{0}_{\wedge^m V}$ for $k < 0$ or $k > m$ or $m > N$. The meaningful values of m and k for $\wedge^m \hat{A}^k$ are thus $0 \le k \le m \le N$.

Example: Let the operator $\hat{A}$ and the vectors $\mathbf{a}, \mathbf{b}, \mathbf{c}$ be such that $\hat{A}\mathbf{a} = 0$, $\hat{A}\mathbf{b} = 2\mathbf{b}$, $\hat{A}\mathbf{c} = \mathbf{b} + \mathbf{c}$. We can then apply the various extensions of the operator $\hat{A}$ to various tensors. For instance,

$$\begin{aligned} \wedge^2 \hat{A}^1 (\mathbf{a} \wedge \mathbf{b}) &= \hat{A}\mathbf{a} \wedge \mathbf{b} + \mathbf{a} \wedge \hat{A}\mathbf{b} = 2\mathbf{a} \wedge \mathbf{b}, \\ \wedge^2 \hat{A}^2 (\mathbf{a} \wedge \mathbf{b}) &= \hat{A}\mathbf{a} \wedge \hat{A}\mathbf{b} = 0, \\ \wedge^3 \hat{A}^2 (\mathbf{a} \wedge \mathbf{b} \wedge \mathbf{c}) &= \mathbf{a} \wedge \hat{A}\mathbf{b} \wedge \hat{A}\mathbf{c} = \mathbf{a} \wedge 2\mathbf{b} \wedge \mathbf{c} = 2(\mathbf{a} \wedge \mathbf{b} \wedge \mathbf{c}) \end{aligned}$$

(in the last line, we dropped terms containing $\hat{A}\mathbf{a}$).

Before we move on to see why the operators $\wedge^m \hat{A}^k$ are useful, let us obtain some basic properties of these operators.

Statement 1: The k-linear extension of $\hat{A}$ is a linear operator in the space $\wedge^m V$.

Proof: To prove the linearity of the map, we need to demonstrate not only that $\wedge^m \hat{A}^k$ maps linear combinations into linear combinations (this is obvious), but also that the result of the action of $\wedge^m \hat{A}^k$ on a tensor $\omega \in \wedge^m V$ does not depend on the particular representation of ω through terms of the form $\mathbf{v}_1 \wedge ... \wedge \mathbf{v}_m$. Thus we need to check that

$$\wedge^m \hat{A}^k \left(\omega \wedge \mathbf{v}_1 \wedge \mathbf{v}_2 \wedge \omega'\right) = -\wedge^m \hat{A}^k \left(\omega \wedge \mathbf{v}_2 \wedge \mathbf{v}_1 \wedge \omega'\right),$$

where ω and ω' are arbitrary tensors such that $\omega \wedge \mathbf{v}_1 \wedge \mathbf{v}_2 \wedge \omega' \in \wedge^m V$. But this property is a simple consequence of the definition of $\wedge^m \hat{A}^k$ which can be verified by explicit computation. ■

Statement 2: For any two operators $\hat{A}, \hat{B} \in \text{End } V$, we have

$$\wedge^m (\hat{A}\hat{B})^m = (\wedge^m \hat{A}^m)(\wedge^m \hat{B}^m).$$

For example,

$$\begin{aligned}\wedge^2 (\hat{A}\hat{B})^2 (\mathbf{u} \wedge \mathbf{v}) &= \hat{A}\hat{B}\mathbf{u} \wedge \hat{A}\hat{B}\mathbf{v} \\ &= \wedge^2 \hat{A}^2 (\hat{B}\mathbf{u} \wedge \hat{B}\mathbf{v}) = \wedge^2 \hat{A}^2 (\wedge^2 \hat{B}^2) (\mathbf{u} \wedge \mathbf{v}).\end{aligned}$$

Proof: This property is a direct consequence of the definition of the operator $\wedge^k \hat{A}^k$:

$$\wedge^k \hat{A}^k (\mathbf{v}_1 \wedge ... \wedge \mathbf{v}_k) = \hat{A}\mathbf{v}_1 \wedge \hat{A}\mathbf{v}_2 \wedge ... \wedge \hat{A}\mathbf{v}_k = \bigwedge_{j=1}^{k} \hat{A}\mathbf{v}_j,$$

therefore

$$\wedge^m (\hat{A}\hat{B})^m (\bigwedge_{j=1}^{k} \mathbf{v}_j) = \bigwedge_{j=1}^{k} \hat{A}\hat{B}\mathbf{v}_j,$$

$$\wedge^m \hat{A}^m \wedge^m \hat{B}^m (\bigwedge_{j=1}^{k} \mathbf{v}_j) = \wedge^m \hat{A}^m (\bigwedge_{j=1}^{k} \hat{B}\mathbf{v}_j) = \bigwedge_{j=1}^{k} \hat{A}\hat{B}\mathbf{v}_j.$$

■

Statement 3: The operator $\wedge^m \hat{A}^k$ is k-linear in $\hat{A}$,

$$\wedge^m (\lambda \hat{A})^k = \lambda^k (\wedge^m \hat{A}^k).$$

For this reason, $\wedge^m \hat{A}^k$ is called a k-linear extension.

Proof: This follows directly from the definition of the operator $\wedge^m \hat{A}^k$. ■

Finally, a formula that will be useful later (you can skip to Sec. 3.8 if you would rather see how $\wedge^m \hat{A}^k$ is used).

Statement 4: The following identity holds for any $\hat{A} \in \text{End } V$ and for any vectors $\{\mathbf{v}_j \,|\, 1 \leq j \leq m\}$ and $\mathbf{u}$,

$$\begin{aligned}\left[\wedge^m \hat{A}^k (\mathbf{v}_1 \wedge ... \wedge \mathbf{v}_m)\right] \wedge \mathbf{u} + \left[\wedge^m \hat{A}^{k-1} (\mathbf{v}_1 \wedge ... \wedge \mathbf{v}_m)\right] \wedge (\hat{A}\mathbf{u}) \\ = \wedge^{m+1} \hat{A}^k (\mathbf{v}_1 \wedge ... \wedge \mathbf{v}_m \wedge \mathbf{u}).\end{aligned}$$

For example,

$$\wedge^2 \hat{A}^2 (\mathbf{u} \wedge \mathbf{v}) \wedge \mathbf{w} + \wedge^2 \hat{A}^1 (\mathbf{u} \wedge \mathbf{v}) \wedge \hat{A}\mathbf{w} = \wedge^3 \hat{A}^2 (\mathbf{u} \wedge \mathbf{v} \wedge \mathbf{w}). \quad (3.16)$$

Proof: By definition, $\wedge^{m+1}\hat{A}^k(\mathbf{v}_1 \wedge ... \wedge \mathbf{v}_m \wedge \mathbf{u})$ is a sum of terms where $\hat{A}$ acts k times on the vectors $\mathbf{v}_j$ and $\mathbf{u}$. We can gather all terms containing $\hat{A}\mathbf{u}$ and separately all terms containing $\mathbf{u}$, and we will get the required expressions. Here is an explicit calculation for the given example:

$$\wedge^2\hat{A}^2(\mathbf{u}\wedge\mathbf{v})\wedge\mathbf{w} = \hat{A}\mathbf{u}\wedge\hat{A}\mathbf{v}\wedge\mathbf{w};$$
$$\wedge^2\hat{A}^1(\mathbf{u}\wedge\mathbf{v})\wedge\hat{A}\mathbf{w} = \left(\hat{A}\mathbf{u}\wedge\mathbf{v}+\mathbf{u}\wedge\hat{A}\mathbf{v}\right)\wedge\hat{A}\mathbf{w}.$$

The formula (3.16) follows.

It should now be clear how the proof proceeds in the general case. A formal proof using Eq. (3.15) is as follows. Applying Eq. (3.15), we need to sum over s_1, ..., s_{m+1}. We can consider terms where $s_{m+1} = 0$ separately from terms where $s_{m+1} = 1$:

$$\begin{aligned}\wedge^{m+1}\hat{A}^k(\mathbf{v}_1\wedge...\wedge\mathbf{v}_m\wedge\mathbf{u}) &= \sum_{(s_1,...,s_m);\sum s_j=k}\Big(\bigwedge_{j=1}^{m}\hat{A}^{s_j}\mathbf{v}_j\Big)\wedge\mathbf{u}\\ &+\sum_{(s_1,...,s_m);\sum s_j=k-1}\Big(\bigwedge_{j=1}^{m}\hat{A}^{s_j}\mathbf{v}_j\Big)\wedge\hat{A}\mathbf{u}\\ &=\left[\wedge^m\hat{A}^k(\mathbf{v}_1\wedge...\wedge\mathbf{v}_m)\right]\wedge\mathbf{u}+\left[\wedge^m\hat{A}^{k-1}(\mathbf{v}_1\wedge...\wedge\mathbf{v}_m)\right]\wedge\hat{A}\mathbf{u}.\end{aligned}$$

■

3.7.1 * Index notation

Let us briefly note how the multilinear action such as $\wedge^m\hat{A}^k$ can be expressed in the index notation.

Suppose that the operator $\hat{A}$ has the index representation A_i^j in a fixed basis. The operator $\wedge^m\hat{A}^k$ acts in the space $\wedge^m V$; tensors ψ in that space are represented in the index notation by totally antisymmetric arrays with m indices, such as $\psi^{i_1...i_m}$. An operator $\hat{B}\in\mathrm{End}\,(\wedge^m V)$ must be therefore represented by an array with $2m$ indices, $B_{i_1...i_m}^{j_1...j_m}$, which is totally antisymmetric with respect to the indices $\{i_s\}$ and separately with respect to $\{j_s\}$.

Let us begin with $\wedge^m\hat{A}^m$ as the simplest case. The action of $\wedge^m\hat{A}^m$ on ψ is written in the index notation as

$$[\wedge^m\hat{A}^m\psi]^{i_1...i_m} = \sum_{j_1,...,j_m=1}^{N} A_{j_1}^{i_1}...A_{j_m}^{i_m}\psi^{j_1...j_m}.$$

This array is totally antisymmetric in i_1, ..., i_m as usual.

Another example is the action of $\wedge^m\hat{A}^1$ on ψ:

$$[\wedge^m\hat{A}^1\psi]^{i_1...i_m} = \sum_{s=1}^{m}\sum_{j=1}^{N} A_j^{i_s}\psi^{i_1...i_{s-1}ji_{s+1}...i_m}.$$

In other words, $\hat{A}$ acts only on the sth index of ψ, and we sum over all s.

In this way, every $\wedge^m\hat{A}^k$ can be written in the index notation, although the expressions become cumbersome.

3.8 Trace

The **trace** of a square matrix A_{jk} is defined as the sum of its diagonal elements, $\mathrm{Tr}A \equiv \sum_{j=1}^{n} A_{jj}$. This definition is quite simple at first sight. However, if this definition is taken as fundamental then one is left with many questions. Suppose A_{jk} is the representation of a linear transformation in a basis; is the number $\mathrm{Tr}A$ independent of the basis? Why is this particular combination of the matrix elements useful? (Why not compute the sum of the elements of A_{jk} along the other diagonal of the square, $\sum_{j=1}^{n} A_{(n+1-j)j}$?)

To clarify the significance of the trace, I will give two other definitions of the trace: one through the canonical linear map $V \otimes V^* \to \mathbb{K}$, and another using the exterior powers construction, quite similar to the definition of the determinant in Sec. 3.3.

Definition Tr1: The trace $\mathrm{Tr}A$ of a tensor $A \equiv \sum_k \mathbf{v}_k \otimes \mathbf{f}_k^* \in V \otimes V^*$ is the number canonically defined by the formula

$$\mathrm{Tr}A = \sum_k \mathbf{f}_k^* (\mathbf{v}_k). \tag{3.17}$$

If we represent the tensor A through the basis tensors $\mathbf{e}_j \otimes \mathbf{e}_k^*$, where $\{\mathbf{e}_j\}$ is some basis and $\{\mathbf{e}_k^*\}$ is its dual basis,

$$A = \sum_{j=1}^{N} \sum_{k=1}^{N} A_{jk} \mathbf{e}_j \otimes \mathbf{e}_k^*,$$

then $\mathbf{e}_k^*(\mathbf{e}_j) = \delta_{ij}$, and it follows that

$$\mathrm{Tr}A = \sum_{j,k=1}^{N} A_{jk} \mathbf{e}_k^*(\mathbf{e}_j) = \sum_{j,k=1}^{N} A_{jk} \delta_{kj} = \sum_{j=1}^{N} A_{jj},$$

in agreement with the traditional definition.

Exercise 1: Show that the trace (according to Definition Tr1) does not depend on the choice of the tensor decomposition $A = \sum_k \mathbf{v}_k \otimes \mathbf{f}_k^*$. ■

Here is another definition of the trace.

Definition Tr2: The **trace** $\mathrm{Tr}\hat{A}$ of an operator $\hat{A} \in \mathrm{End}\, V$ is the number by which any nonzero tensor $\omega \in \wedge^N V$ is multiplied when $\wedge^N \hat{A}^1$ acts on it:

$$(\wedge^N \hat{A}^1)\omega = (\mathrm{Tr}\hat{A})\omega, \quad \forall \omega \in \wedge^N V. \tag{3.18}$$

Alternatively written,

$$\wedge^N \hat{A}^1 = (\mathrm{Tr}\hat{A})\hat{1}_{\wedge^N V}.$$

First we will show that the definition Tr2 is equivalent to the traditional definition of the trace. Recall that, according to the definition of $\wedge^N \hat{A}^1$,

$$\wedge^N \hat{A}^1 (\mathbf{v}_1 \wedge ... \wedge \mathbf{v}_N) = \hat{A}\mathbf{v}_1 \wedge \mathbf{v}_2 \wedge ... \wedge \mathbf{v}_N + ... \\ + \mathbf{v}_1 \wedge ... \wedge \mathbf{v}_{N-1} \wedge \hat{A}\mathbf{v}_N.$$

Statement 1: If $\{\mathbf{e}_j\}$ is any basis in V, $\{\mathbf{e}_j^*\}$ is the dual basis, and a linear operator $\hat{A}$ is represented by a tensor $\hat{A} = \sum_{j,k=1}^{N} A_{jk} \mathbf{e}_j \otimes \mathbf{e}_k^*$, then the trace of $\hat{A}$ computed according to Eq. (3.18) will agree with the formula $\mathrm{Tr}\hat{A} = \sum_{j=1}^{N} A_{jj}$.

Proof: The operator $\hat{A}$ acts on the basis vectors $\{\mathbf{e}_j\}$ as follows,

$$\hat{A}\mathbf{e}_k = \sum_{j=1}^{N} A_{jk} \mathbf{e}_j .$$

Therefore $\mathbf{e}_1 \wedge ... \wedge \hat{A}\mathbf{e}_j \wedge ... \wedge \mathbf{e}_N = A_{jj} \mathbf{e}_1 \wedge ... \wedge \mathbf{e}_N$, and definition (3.18) gives

$$\begin{aligned}(\mathrm{Tr}\hat{A})\, \mathbf{e}_1 \wedge ... \wedge \mathbf{e}_N &= \sum_{j=1}^{N} \mathbf{e}_1 \wedge ... \wedge \hat{A}\mathbf{e}_j \wedge ... \wedge \mathbf{e}_N \\ &= \big(\sum_{j=1}^{N} A_{jj}\big)\, \mathbf{e}_1 \wedge ... \wedge \mathbf{e}_N .\end{aligned}$$

Thus $\mathrm{Tr}\hat{A} = \sum_{j=1}^{N} A_{jj}$. ■

Now we prove some standard properties of the trace.

Statement 2: For any operators $\hat{A}, \hat{B} \in \mathrm{End}\, V$:

(1) $\mathrm{Tr}(\hat{A} + \hat{B}) = \mathrm{Tr}\hat{A} + \mathrm{Tr}\hat{B}$.

(2) $\mathrm{Tr}(\hat{A}\hat{B}) = \mathrm{Tr}(\hat{B}\hat{A})$.

Proof: The formula (3.17) allows one to derive these properties more easily, but I will give proofs using the definition (3.18).

(1) Since

$$\begin{aligned}\mathbf{e}_1 \wedge ... \wedge (\hat{A} + \hat{B})\mathbf{e}_j \wedge ... \wedge \mathbf{e}_N = \mathbf{e}_1 \wedge ... \wedge \hat{A}\mathbf{e}_j \wedge ... \wedge \mathbf{e}_N \\ + \mathbf{e}_1 \wedge ... \wedge \hat{B}\mathbf{e}_j \wedge ... \wedge \mathbf{e}_N ,\end{aligned}$$

from the definition of $\wedge^N \hat{A}^1$ we easily obtain $\wedge^N (\hat{A} + \hat{B})^1 = \wedge^N \hat{A}^1 + \wedge^N \hat{B}^1$.

(2) Since $\wedge^N \hat{A}^1$ and $\wedge^N \hat{B}^1$ are operators in one-dimensional space $\wedge^N V$, they commute, that is

$$(\wedge^N \hat{A}^1)(\wedge^N \hat{B}^1) = (\wedge^N \hat{B}^1)(\wedge^N \hat{A}^1) = (\mathrm{Tr}\hat{A})(\mathrm{Tr}\hat{B})\hat{1}_{\wedge^N V} .$$

Now we explicitly compute the composition $(\wedge^N \hat{A}^1)(\wedge^N \hat{B}^1)$ acting on $\mathbf{e}_1 \wedge .... \wedge \mathbf{e}_N$. First, an example with $N = 2$,

$$\begin{aligned}(\wedge^N \hat{A}^1)(\wedge^N \hat{B}^1)\, (\mathbf{e}_1 \wedge \mathbf{e}_2) &= \wedge^N \hat{A}^1 (\hat{B}\mathbf{e}_1 \wedge \mathbf{e}_2 + \mathbf{e}_1 \wedge \hat{B}\mathbf{e}_2) \\ &= \hat{A}\hat{B}\mathbf{e}_1 \wedge \mathbf{e}_2 + \hat{B}\mathbf{e}_1 \wedge \hat{A}\mathbf{e}_2 \\ &\quad + \hat{A}\mathbf{e}_1 \wedge \hat{B}\mathbf{e}_2 + \mathbf{e}_1 \wedge \hat{A}\hat{B}\mathbf{e}_2 \\ &= \wedge^N (\hat{A}\hat{B})^1 \mathbf{e}_1 \wedge \mathbf{e}_2 + \hat{A}\mathbf{e}_1 \wedge \hat{B}\mathbf{e}_2 + \hat{B}\mathbf{e}_1 \wedge \hat{A}\mathbf{e}_2 .\end{aligned}$$

Now the general calculation:

$$(\wedge^N \hat{A}^1)(\wedge^N \hat{B}^1)\mathbf{e}_1 \wedge .... \wedge \mathbf{e}_N = \sum_{j=1}^{N} \mathbf{e}_1 \wedge ... \wedge \hat{A}\hat{B}\mathbf{e}_j \wedge ... \wedge \mathbf{e}_N$$

$$+\sum_{j=1}^{N} \sum_{\substack{k=1 \\ (k\neq j)}}^{N} \mathbf{e}_1 \wedge ... \wedge \hat{A}\mathbf{e}_j \wedge ... \wedge \hat{B}\mathbf{e}_k \wedge ... \wedge \mathbf{e}_N.$$

The second sum is symmetric in $\hat{A}$ and $\hat{B}$, therefore the identity

$$(\wedge^N \hat{A}^1)(\wedge^N \hat{B}^1)\mathbf{e}_1 \wedge .... \wedge \mathbf{e}_N = (\wedge^N \hat{B}^1)(\wedge^N \hat{A}^1)\mathbf{e}_1 \wedge .... \wedge \mathbf{e}_N$$

entails

$$\sum_{j=1}^{N} \mathbf{e}_1 \wedge ... \wedge \hat{A}\hat{B}\mathbf{e}_j \wedge ... \wedge \mathbf{e}_N = \sum_{j=1}^{N} \mathbf{e}_1 \wedge ... \wedge \hat{B}\hat{A}\mathbf{e}_j \wedge ... \wedge \mathbf{e}_N,$$

that is $\text{Tr}(\hat{A}\hat{B}) = \text{Tr}(\hat{B}\hat{A})$. ■

Exercise 2: The operator $\hat{L}_\mathbf{b}$ acts on the entire exterior algebra $\wedge V$ and is defined by $\hat{L}_\mathbf{b} : \omega \mapsto \mathbf{b} \wedge \omega$, where $\omega \in \wedge V$ and $\mathbf{b} \in V$. Compute the trace of this operator. *Hint:* Use Definition Tr1 of the trace.

Answer: $\text{Tr}\hat{L}_\mathbf{b} = 0$.

Exercise 3: Suppose $\hat{A}\hat{A} = 0$; show that $\text{Tr}\hat{A} = 0$ and $\det \hat{A} = 0$.

Solution: We see that $\det \hat{A} = 0$ because $0 = \det(\hat{A}\hat{A}) = (\det \hat{A})^2$. Now we apply the operator $\wedge^N \hat{A}^1$ to a nonzero tensor $\omega = \mathbf{v}_1 \wedge ... \wedge \mathbf{v}_N \in \wedge^N V$ twice in a row:

$$\begin{aligned}
(\wedge^N \hat{A}^1)(\wedge^N \hat{A}^1)\omega &= (\text{Tr}\hat{A})^2 \omega \\
&= (\wedge^N \hat{A}^1) \sum_{j=1}^{N} \mathbf{v}_1 \wedge ... \wedge \hat{A}\mathbf{v}_j \wedge ... \wedge \mathbf{v}_N \\
&= \sum_{i=1}^{N}\sum_{j=1}^{N} \mathbf{v}_1 \wedge ... \wedge \hat{A}\mathbf{v}_i \wedge ... \wedge \hat{A}\mathbf{v}_j \wedge ... \wedge \mathbf{v}_N \\
&= 2(\wedge^N \hat{A}^2)\omega.
\end{aligned}$$

(In this calculation, we omitted the terms containing $\hat{A}\hat{A}\mathbf{v}_i$ since $\hat{A}\hat{A} = 0$.) Using this trick, we can prove by induction that for $1 \leq k \leq N$

$$(\text{Tr}\hat{A})^k \omega = (\wedge^N \hat{A}^1)^k \omega = k!(\wedge^N \hat{A}^k)\omega.$$

Note that $\wedge^N \hat{A}^N$ multiplies by the determinant of $\hat{A}$, which is zero. Therefore $(\text{Tr}\hat{A})^N = N!(\det \hat{A}) = 0$ and so $\text{Tr}\hat{A} = 0$. ■

3.9 Characteristic polynomial

Definition: The **characteristic polynomial** $Q_{\hat{A}}(x)$ of an operator $\hat{A} \in \text{End } V$ is defined as

$$Q_{\hat{A}}(x) \equiv \det\left(\hat{A} - x\hat{1}_V\right).$$

This is a polynomial of degree N in the variable x.

Example 1: The characteristic polynomial of the operator $a\hat{1}_V$, where $a \in \mathbb{K}$, is

$$Q_{a\hat{1}_V}(x) = (a - x)^N.$$

Setting $a = 0$, we find that the characteristic polynomial of the zero operator $\hat{0}_V$ is simply $(-x)^N$.

Example 2: Consider a **diagonalizable** operator $\hat{A}$, i.e. an operator having a basis $\{\mathbf{v}_1, ..., \mathbf{v}_N\}$ of eigenvectors with eigenvalues $\lambda_1, ..., \lambda_N$ (the eigenvalues are not necessarily all different). This operator can be then written in a tensor form as

$$\hat{A} = \sum_{i=1}^{N} \lambda_i \mathbf{v}_i \otimes \mathbf{v}_i^*,$$

where $\{\mathbf{v}_i^*\}$ is the basis dual to $\{\mathbf{v}_i\}$. The characteristic polynomial of this operator is found from

$$\begin{aligned} \det(\hat{A} - x\hat{1})\mathbf{v}_1 \wedge ... \wedge \mathbf{v}_N &= (\hat{A}\mathbf{v}_1 - x\mathbf{v}_1) \wedge ... \wedge (\hat{A}\mathbf{v}_N - x\mathbf{v}_N) \\ &= (\lambda_1 - x)\,\mathbf{v}_1 \wedge ... \wedge (\lambda_N - x)\,\mathbf{v}_N. \end{aligned}$$

Hence

$$Q_{\hat{A}}(x) = (\lambda_1 - x) ... (\lambda_N - x).$$

Note also that the trace of a diagonalizable operator is equal to the sum of the eigenvalues, $\text{Tr}\,\hat{A} = \lambda_1 + ... + \lambda_N$, and the determinant is equal to the product of the eigenvalues, $\det \hat{A} = \lambda_1\lambda_2...\lambda_N$. This can be easily verified by direct calculations in the eigenbasis of $\hat{A}$.

Exercise 1: If an operator $\hat{A}$ has the characteristic polynomial $Q_{\hat{A}}(x)$ then what is the characteristic polynomial of the operator $a\hat{A}$, where $a \in \mathbb{K}$ is a scalar?

Answer:

$$Q_{a\hat{A}}(x) = a^N Q_{\hat{A}}\left(a^{-1}x\right).$$

Note that the right side of the above formula does *not* actually contain a in the denominator because of the prefactor a^N. ■

The principal use of the characteristic polynomial is to determine the eigenvalues of linear operators. We remind the reader that a polynomial $p(x)$ of degree N has N roots if we count each root with its algebraic multiplicity; the number of different roots may be smaller than N. A root λ has **algebraic multiplicity** k if $p(x)$ contains a factor $(x - \lambda)^k$ but not a factor $(x - \lambda)^{k+1}$. For example, the polynomial

$$p(x) = (x - 3)^2(x - 1) = x^3 - 7x^2 + 15x - 9$$

has two distinct roots, $x = 1$ and $x = 3$, and the root $x = 3$ has multiplicity 2. If we count each root with its multiplicity, we will find that the polynomial $p(x)$ has 3 roots ("not all of them different" as we would say in this case).

Theorem 1: **a)** The set of all the roots of the characteristic polynomial $Q_{\hat{A}}(x)$ is the same as the set of all the eigenvalues of the operator $\hat{A}$.

b) The **geometric multiplicity** of an eigenvalue λ (i.e. the dimension of the space of all eigenvectors with the given eigenvalue λ) is at least 1 but not larger than the algebraic multiplicity of a root λ in the characteristic polynomial.

Proof: **a)** By definition, an eigenvalue of an operator $\hat{A}$ is such a number $\lambda \in \mathbb{K}$ that there exists at least one vector $\mathbf{v} \in V$, $\mathbf{v} \neq 0$, such that $\hat{A}\mathbf{v} = \lambda\mathbf{v}$. This equation is equivalent to $(\hat{A} - \lambda\hat{1}_V)\mathbf{v} = 0$. By Corollary 3.5, there would be no solutions $\mathbf{v} \neq 0$ unless $\det(\hat{A} - \lambda\hat{1}_V) = 0$. It follows that all eigenvalues λ must be roots of the characteristic polynomial. Conversely, if λ is a root then $\det(\hat{A} - \lambda\hat{1}_V) = 0$ and hence the vector equation $(\hat{A} - \lambda\hat{1}_V)\mathbf{v} = 0$ will have at least one nonzero solution $\mathbf{v}$ (see Theorem 2 in Sec. 3.5).

b) Suppose $\{\mathbf{v}_1, ..., \mathbf{v}_k\}$ is a basis in the eigenspace of eigenvalue λ_0. We need to show that λ_0 is a root of $Q_{\hat{A}}(x)$ with multiplicity at least k. We may obtain a basis in the space V as $\{\mathbf{v}_1, ..., \mathbf{v}_k, \mathbf{e}_{k+1}, ..., \mathbf{e}_N\}$ by adding suitable new vectors $\{\mathbf{e}_j\}$, $j = k+1, ..., N$. Now compute the characteristic polynomial:

$$\begin{aligned}
&Q_{\hat{A}}(x)(\mathbf{v}_1 \wedge ... \wedge \mathbf{v}_k \wedge \mathbf{e}_{k+1} \wedge ... \wedge \mathbf{e}_N) \\
&= (\hat{A} - x\hat{1})\mathbf{v}_1 \wedge ... \wedge (\hat{A} - x\hat{1})\mathbf{v}_k \\
&\qquad \wedge (\hat{A} - x\hat{1})\mathbf{e}_{k+1} \wedge ... \wedge (\hat{A} - x\hat{1})\mathbf{e}_N \\
&= (\lambda_0 - x)^k \, \mathbf{v}_1 \wedge ... \wedge \mathbf{v}_k \wedge (\hat{A} - x\hat{1})\mathbf{e}_{k+1} \wedge ... \wedge (\hat{A} - x\hat{1})\mathbf{e}_N .
\end{aligned}$$

It follows that $Q_{\hat{A}}(x)$ contains the factor $(\lambda_0 - x)^k$, which means that λ_0 is a root of $Q_{\hat{A}}(x)$ of multiplicity at least k. ■

Remark: If an operator's characteristic polynomial has a root λ_0 of algebraic multiplicity k, it may or may not have a k-dimensional eigenspace for the eigenvalue λ_0. We only know that λ_0 is an eigenvalue, i.e. that the eigenspace is at least one-dimensional. ■

Theorem 1 shows that all the eigenvalues λ of an operator $\hat{A}$ can be computed as roots of the equation $Q_{\hat{A}}(\lambda) = 0$, which is called the **characteristic equation** for the operator $\hat{A}$.

Now we will demonstrate that the coefficients of the characteristic polynomial $Q_{\hat{A}}(x)$ are related in a simple way to the operators $\wedge^N \hat{A}^k$. First we need an auxiliary calculation to derive an explicit formula for determinants of operators of the form $\hat{A} - \lambda\hat{1}_V$.

Lemma 1: For any $\hat{A} \in \text{End } V$, we have

$$\wedge^N (\hat{A} + \hat{1}_V)^N = \sum_{r=0}^{N} (\wedge^N \hat{A}^r).$$

More generally, for $0 \le q \le p \le N$, we have

$$\wedge^p(\hat{A} + \hat{1}_V)^q = \sum_{r=0}^{q} \binom{p-r}{p-q} (\wedge^p \hat{A}^r). \tag{3.19}$$

Proof: I first give some examples, then prove the most useful case $p = q$, and then show a proof of Eq. (3.19) for arbitrary p and q.

For $p = q = 2$, we compute

$$\begin{aligned}\wedge^2(\hat{A} + \hat{1}_V)^2 \mathbf{a} \wedge \mathbf{b} &= (\hat{A} + \hat{1}_V)\mathbf{a} \wedge (\hat{A} + \hat{1}_V)\mathbf{b} \\ &= \hat{A}\mathbf{a} \wedge \hat{A}\mathbf{b} + \hat{A}\mathbf{a} \wedge \mathbf{b} + \mathbf{a} \wedge \hat{A}\mathbf{b} + \mathbf{a} \wedge \mathbf{b} \\ &= \left[\wedge^2 \hat{A}^2 + \wedge^2 \hat{A}^1 + \wedge^2 \hat{A}^0\right] (\mathbf{a} \wedge \mathbf{b}).\end{aligned}$$

This can be easily generalized to arbitrary $p = q$: The action of the operator $\wedge^p(\hat{A} + \hat{1}_V)^p$ on $\mathbf{e}_1 \wedge ... \wedge \mathbf{e}_p$ is

$$\wedge^p(\hat{A} + \hat{1}_V)^p \mathbf{e}_1 \wedge ... \wedge \mathbf{e}_p = (\hat{A} + \hat{1}_V)\mathbf{e}_1 \wedge ... \wedge (\hat{A} + \hat{1}_V)\mathbf{e}_p,$$

and we can expand the brackets to find first *one* term with p operators $\hat{A}$, then p terms with $(p-1)$ operators $\hat{A}$, etc., and finally one term with no operators $\hat{A}$ acting on the vectors $\mathbf{e}_j$. All terms which contain r operators $\hat{A}$ (with $0 \le r \le p$) are those appearing in the definition of the operator $\wedge^p \hat{A}^r$. Therefore

$$\wedge^p(\hat{A} + \hat{1}_V)^p = \sum_{r=0}^{p} (\wedge^p \hat{A}^r).$$

This is precisely the formula (3.19) because in the particular case $p = q$ the combinatorial coefficient is trivial,

$$\binom{p-r}{p-q} = \binom{p-r}{0} = 1.$$

Now we consider the general case $0 \le q \le p$. First an example: for $p = 2$ and $q = 1$, we compute

$$\begin{aligned}\wedge^2(\hat{A} + \hat{1}_V)^1 \mathbf{a} \wedge \mathbf{b} &= (\hat{A} + \hat{1}_V)\mathbf{a} \wedge \mathbf{b} + \mathbf{a} \wedge (\hat{A} + \hat{1}_V)\mathbf{b} \\ &= 2\mathbf{a} \wedge \mathbf{b} + \hat{A}\mathbf{a} \wedge \mathbf{b} + \mathbf{a} \wedge \hat{A}\mathbf{b} \\ &= \left[\tbinom{2}{1}(\wedge^2 \hat{A}^0) + \tbinom{2}{0}(\wedge^2 \hat{A}^1)\right] \mathbf{a} \wedge \mathbf{b},\end{aligned}$$

since $\binom{2}{1} = 2$ and $\binom{2}{0} = 1$.

To prove the formula (3.19) in the general case, we use induction. The basis of induction consists of the trivial case ($p \ge 0$, $q = 0$) where all operators $\wedge^0 \hat{A}^p$ with $p \ge 1$ are zero operators, and of the case $p = q$, which was already proved. Now we will prove the induction step (p, q) & $(p, q+1) \Rightarrow (p+1, q+1)$. Figure 3.3 indicates why this induction step is sufficient to prove the statement for all $0 \le q \le p \le N$.

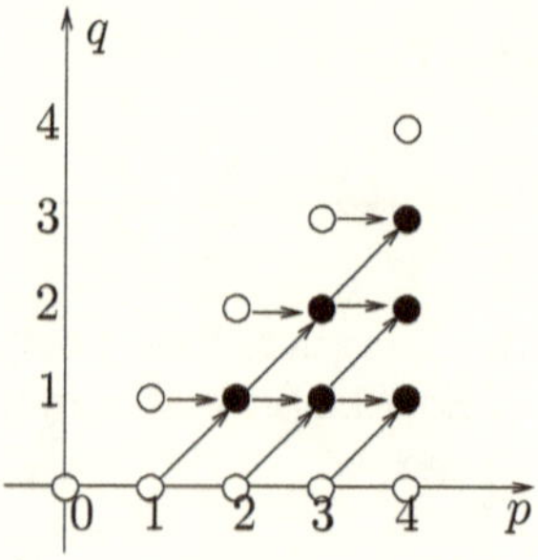

Figure 3.3: Deriving Lemma 1 by induction. White circles correspond to the basis of induction. Black circles are reached by induction steps.

Let $\mathbf{v} \in V$ be an arbitrary vector and $\omega \in \wedge^p V$ be an arbitrary tensor. The induction step is proved by the following chain of equations,

$$\begin{aligned}
&\wedge^{p+1}(\hat{A}+\hat{1}_V)^{q+1}(\mathbf{v}\wedge\omega)\\
&\overset{(1)}{=}(\hat{A}+\hat{1}_V)\mathbf{v}\wedge\left[\wedge^p(\hat{A}+\hat{1}_V)^q\omega\right]+\mathbf{v}\wedge\left[\wedge^p(\hat{A}+\hat{1}_V)^{q+1}\omega\right]\\
&\overset{(2)}{=}\hat{A}\mathbf{v}\wedge\sum_{r=0}^{q}\binom{p-r}{p-q}(\wedge^p\hat{A}^r)\omega+\mathbf{v}\wedge\sum_{r=0}^{q}\binom{p-r}{p-q}(\wedge^p\hat{A}^r)\omega\\
&\quad+\mathbf{v}\wedge\sum_{r=0}^{q+1}\binom{p-r}{p-q-1}(\wedge^p\hat{A}^r)\omega\\
&\overset{(3)}{=}\hat{A}\mathbf{v}\wedge\sum_{k=1}^{q+1}\binom{p-k+1}{p-q}(\wedge^p\hat{A}^{k-1})\omega\\
&\quad+\mathbf{v}\wedge\sum_{r=0}^{q+1}\left[\binom{p-r}{p-q-1}+\binom{p-r}{p-q}\right](\wedge^p\hat{A}^r)\omega\\
&\overset{(4)}{=}\sum_{k=0}^{q+1}\binom{p-k+1}{p-q}\left\{\hat{A}\mathbf{v}\wedge\left[\wedge^p\hat{A}^{k-1}\omega\right]+\mathbf{v}\wedge\left[\wedge^p\hat{A}^k\omega\right]\right\}\\
&\overset{(1)}{=}\sum_{k=0}^{q+1}\binom{p-k+1}{p-q}(\wedge^{p+1}\hat{A}^k)(\mathbf{v}\wedge\omega),
\end{aligned}$$

where $^{(1)}$ is Statement 4 of Sec. 3.7, $^{(2)}$ uses the induction step assumptions for (p, q) and $(p, q+1)$, $^{(3)}$ is the relabeling $r = k - 1$ and rearranging terms (note that the summation over $0 \leq r \leq q$ was formally extended to $0 \leq r \leq q+1$ because the term with $r = q + 1$ vanishes), and $^{(4)}$ is by the binomial identity

$$\binom{n}{m-1}+\binom{n}{m}=\binom{n+1}{m}$$

and a further relabeling $r \to k$ in the preceding summation. ■

Corollary: For any $\hat{A} \in \text{End}\, V$ and $\alpha \in \mathbb{K}$,

$$\wedge^p(\hat{A}+\alpha\hat{1}_V)^q = \sum_{r=0}^{q} \alpha^{q-r}\binom{p-r}{p-q}(\wedge^p \hat{A}^r).$$

Proof: By Statement 3 of Sec. 3.7, $\wedge^p(\alpha\hat{A})^q = \alpha^q(\wedge^p\hat{A}^q)$. Set $\hat{A} = \alpha\hat{B}$, where $\hat{B}$ is an auxiliary operator, and compute

$$\begin{aligned}\wedge^p(\alpha\hat{B}+\alpha\hat{1}_V)^q &= \alpha^q \wedge^p (\hat{B}+\hat{1}_V)^q = \alpha^q \sum_{r=0}^{q}\binom{p-r}{p-q}(\wedge^p\hat{B}^r)\\ &= \sum_{r=0}^{q}\alpha^{q-r}\binom{p-r}{p-q}(\wedge^p(\alpha\hat{B})^r)\\ &= \sum_{r=0}^{q}\alpha^{q-r}\binom{p-r}{p-q}(\wedge^p\hat{A}^r).\end{aligned}$$

■

Theorem 2: The coefficients $q_m(\hat{A})$, $1 \leq m \leq N$ of the characteristic polynomial, defined by

$$Q_{\hat{A}}(\lambda) = (-\lambda)^N + \sum_{k=0}^{N-1}(-1)^k q_{N-k}(\hat{A})\lambda^k,$$

are the numbers corresponding to the operators $\wedge^N\hat{A}^m \in \text{End}(\wedge^N V)$:

$$q_m(\hat{A})\hat{1}_{\wedge^N V} = \wedge^N\hat{A}^m.$$

In particular, $q_N(\hat{A}) = \det\hat{A}$ and $q_1(\hat{A}) = \text{Tr}\hat{A}$. More compactly, the statement can be written as

$$Q_{\hat{A}}(\lambda)\,\hat{1}_{\wedge^N V} = \sum_{k=0}^{N}(-\lambda)^{N-k}(\wedge^N\hat{A}^k).$$

Proof: This is now a consequence of Lemma 1 and its Corollary, where we set $p = q = N$ and obtain

$$\wedge^N(\hat{A}-\lambda\hat{1}_V)^N = \sum_{r=0}^{N}(-\lambda)^{N-r}(\wedge^N\hat{A}^r).$$

■

Exercise 1: Show that the characteristic polynomial of an operator $\hat{A}$ in a *three-dimensional* space V can be written as

$$Q_{\hat{A}}(\lambda) = \det\hat{A} - \tfrac{1}{2}\left[(\text{Tr}\hat{A})^2 - \text{Tr}(\hat{A}^2)\right]\lambda + (\text{Tr}\hat{A})\lambda^2 - \lambda^3.$$

Solution: The first and the third coefficients of $Q_{\hat{A}}(\lambda)$ are, as usual, the determinant and the trace of $\hat{A}$. The second coefficient is equal to $-\wedge^3\hat{A}^2$, so we need to show that

$$\wedge^3\hat{A}^2 = \frac{1}{2}\big[(\text{Tr}\hat{A})^2 - \text{Tr}(\hat{A}^2)\big].$$

We apply the operator $\wedge^3\hat{A}^1$ twice to a tensor $\mathbf{a}\wedge\mathbf{b}\wedge\mathbf{c}$ and calculate:

$$\begin{aligned}
&(\text{Tr}\hat{A})^2\mathbf{a}\wedge\mathbf{b}\wedge\mathbf{c} = (\wedge^3\hat{A}^1)(\wedge^3\hat{A}^1)(\mathbf{a}\wedge\mathbf{b}\wedge\mathbf{c})\\
&= (\wedge^3\hat{A}^1)(\hat{A}\mathbf{a}\wedge\mathbf{b}\wedge\mathbf{c} + \mathbf{a}\wedge\hat{A}\mathbf{b}\wedge\mathbf{c} + \mathbf{a}\wedge\mathbf{b}\wedge\hat{A}\mathbf{c})\\
&= \hat{A}^2\mathbf{a}\wedge\mathbf{b}\wedge\mathbf{c} + 2\hat{A}\mathbf{a}\wedge\hat{A}\mathbf{b}\wedge\mathbf{c} + \mathbf{a}\wedge\hat{A}^2\mathbf{b}\wedge\mathbf{c}\\
&+ 2\hat{A}\mathbf{a}\wedge\mathbf{b}\wedge\hat{A}\mathbf{c} + 2\mathbf{a}\wedge\hat{A}\mathbf{b}\wedge\hat{A}\mathbf{c} + \mathbf{a}\wedge\mathbf{b}\wedge\hat{A}^2\mathbf{c}\\
&= \big[\text{Tr}(\hat{A}^2) + 2\wedge^3\hat{A}^2\big]\mathbf{a}\wedge\mathbf{b}\wedge\mathbf{c}.
\end{aligned}$$

Then the desired formula follows. ■

Exercise 2 (general trace relations): Generalize the result of Exercise 1 to N dimensions:

a) Show that

$$\wedge^N\hat{A}^2 = \tfrac{1}{2}\big[(\text{Tr}\hat{A})^2 - \text{Tr}(\hat{A}^2)\big].$$

b)* Show that all coefficients $\wedge^N\hat{A}^k$ ($k = 1, ..., N$) can be expressed as polynomials in $\text{Tr}\hat{A}$, $\text{Tr}(\hat{A}^2)$, ..., $\text{Tr}(\hat{A}^N)$.

Hint: Define a "mixed" operator $\wedge^N(\hat{A}^n)^j\hat{A}^k$ as a sum of exterior products containing j times $\hat{A}^n$ and k times $\hat{A}$; for example,

$$\begin{aligned}
&\big[\wedge^3(\hat{A}^2)^1\hat{A}^1\big]\mathbf{a}\wedge\mathbf{b}\wedge\mathbf{c} \equiv \hat{A}^2\mathbf{a}\wedge(\hat{A}\mathbf{b}\wedge\mathbf{c} + \mathbf{b}\wedge\hat{A}\mathbf{c})\\
&+ \hat{A}\mathbf{a}\wedge(\hat{A}^2\mathbf{b}\wedge\mathbf{c} + \mathbf{b}\wedge\hat{A}^2\mathbf{c}) + \mathbf{a}\wedge(\hat{A}^2\mathbf{b}\wedge\hat{A}\mathbf{c} + \hat{A}\mathbf{b}\wedge\hat{A}^2\mathbf{c}).
\end{aligned}$$

By applying several operators $\wedge^N\hat{A}^k$ and $\text{Tr}(\hat{A}^k)$ to an exterior product, derive identities connecting these operators and $\wedge^N\hat{A}^k$:

$$\begin{aligned}
(\wedge^N\hat{A}^1)(\wedge^N\hat{A}^k) &= (k+1)\wedge^N\hat{A}^{k+1} + \wedge^N(\hat{A}^2)^1\hat{A}^{k-1},\\
\text{Tr}(\hat{A}^k)\text{Tr}(\hat{A}) &= \text{Tr}(\hat{A}^{k+1}) + \wedge^N(\hat{A}^k)^1\hat{A}^1,
\end{aligned}$$

for $k = 2, ..., N-1$. Using these identities, show by induction that operators of the form $\wedge^N\hat{A}^k$ ($k = 1, ..., N$) can be all expressed through $\text{Tr}\hat{A}$, $\text{Tr}(\hat{A}^2)$, ..., $\text{Tr}(\hat{A}^{N-1})$ as polynomials.

As an example, here is the trace relation for $\wedge^N\hat{A}^3$:

$$\wedge^N\hat{A}^3 = \tfrac{1}{6}(\text{Tr}\hat{A})^3 - \tfrac{1}{2}(\text{Tr}\hat{A})\text{Tr}(\hat{A}^2) + \tfrac{1}{3}\text{Tr}(\hat{A}^3).$$

Note that in three dimensions this formula directly yields the determinant of $\hat{A}$ expressed through traces of powers of $\hat{A}$. Below (Sec. 4.5.3) we will derive a formula for the general trace relation. ■

Since operators in $\wedge^N V$ act as multiplication by a number, it is convenient to omit $\hat{1}_{\wedge^N V}$ and regard expressions such as $\wedge^N\hat{A}^k$ as simply numbers. More formally, there is a canonical isomorphism between $\text{End}\left(\wedge^N V\right)$ and $\mathbb{K}$ (even though there is no canonical isomorphism between $\wedge^N V$ and $\mathbb{K}$).

Exercise 3: Give an explicit formula for the canonical isomorphism: a) between $(\wedge^k V)^*$ and $\wedge^k(V^*)$; b) between End $(\wedge^N V)$ and $\mathbb{K}$.

Answer: a) A tensor $\mathbf{f}_1^* \wedge ... \wedge \mathbf{f}_k^* \in \wedge^k(V^*)$ acts as a linear function on a tensor $\mathbf{v}_1 \wedge ... \wedge \mathbf{v}_k \in \wedge^k V$ by the formula

$$(\mathbf{f}_1^* \wedge ... \wedge \mathbf{f}_k^*)(\mathbf{v}_1 \wedge ... \wedge \mathbf{v}_k) \equiv \det(A_{jk}),$$

where A_{jk} is the square matrix defined by $A_{jk} \equiv \mathbf{f}_j^*(\mathbf{v}_k)$.

b) Since $(\wedge^N V)^*$ is canonically isomorphic to $\wedge^N(V^*)$, an operator $\hat{N} \in$ End $(\wedge^N V)$ can be represented by a tensor

$$\hat{N} = (\mathbf{v}_1 \wedge ... \wedge \mathbf{v}_N) \otimes (\mathbf{f}_1^* \wedge ... \wedge \mathbf{f}_N^*) \in (\wedge^N V) \otimes (\wedge^N V^*).$$

The isomorphism maps $\hat{N}$ into the number $\det(A_{jk})$, where A_{jk} is the square matrix defined by $A_{jk} \equiv \mathbf{f}_j^*(\mathbf{v}_k)$. ■

Exercise 4: Show that an operator $\hat{A} \in$ End V and its canonical transpose operator $\hat{A}^T \in$ End V^* have the same characteristic polynomials.

Hint: Consider the operator $(\hat{A} - x\hat{1}_V)^T$. ■

Exercise 5: Given an operator $\hat{A}$ of rank $r < N$, show that $\wedge^N \hat{A}^k = 0$ for $k \geq r + 1$ but $\wedge^N \hat{A}^r \neq 0$.

Hint: If $\hat{A}$ has rank $r < N$ then $\hat{A}\mathbf{v}_1 \wedge ... \wedge \hat{A}\mathbf{v}_{r+1} = 0$ for any set of vectors $\{\mathbf{v}_1, ..., \mathbf{v}_{r+1}\}$.

3.9.1 Nilpotent operators

There are many operators with the same characteristic polynomial. In particular, there are many operators which have the simplest possible characteristic polynomial, $Q_0(x) = (-x)^N$. Note that the zero operator has this characteristic polynomial. We will now see how to describe all such operators $\hat{A}$ that $Q_{\hat{A}}(x) = (-x)^N$.

Definition: An operator $\hat{A} \in$ End V is **nilpotent** if there exists an integer $p \geq 1$ such that $(\hat{A})^p = \hat{0}$, where $\hat{0}$ is the zero operator and $(\hat{A})^p$ is the p-th power of the operator $\hat{A}$.

Examples: a) The operator defined by the matrix $\begin{pmatrix} 0 & \alpha \\ 0 & 0 \end{pmatrix}$ in some basis $\{\mathbf{e}_1, \mathbf{e}_2\}$ is nilpotent for any number α. This operator can be expressed in tensor form as $\alpha \mathbf{e}_1 \otimes \mathbf{e}_2^*$.

b) In the space of polynomials of degree at most n in the variable x, the linear operator $\frac{d}{dx}$ is nilpotent because the $(n+1)$-th power of this operator will evaluate the $(n+1)$-th derivative, which is zero on any polynomial of degree at most n. ■

Statement: If $\hat{A}$ is a nilpotent operator then $\hat{Q}_{\hat{A}}(x) = (-x)^N$.

Proof: First an example: suppose that $N = 2$ and that $\hat{A}^3 = 0$. By Theorem 2, the coefficients of the characteristic polynomial of the operator $\hat{A}$ correspond to the operators $\wedge^N \hat{A}^k$. We need to show that all these operators are equal to zero.

Consider, for instance, $\wedge^2 \hat{A}^2 = q_2 \hat{1}_{\wedge^2 V}$. This operator raised to the power 3 acts on a tensor $\mathbf{a} \wedge \mathbf{b} \in \wedge^2 V$ as

$$\left(\wedge^2 \hat{A}^2\right)^3 \mathbf{a} \wedge \mathbf{b} = \hat{A}^3 \mathbf{a} \wedge \hat{A}^3 \mathbf{b} = 0$$

since $\hat{A}^3 = 0$. On the other hand,

$$\left(\wedge^2 \hat{A}^2\right)^3 \mathbf{a} \wedge \mathbf{b} = (q_2)^3 \, \mathbf{a} \wedge \mathbf{b}.$$

Therefore $q_2 = 0$. Now consider $\wedge^2 \hat{A}^1$ to the power 3,

$$\left(\wedge^2 \hat{A}^1\right)^3 \mathbf{a} \wedge \mathbf{b} = \hat{A}^2 \mathbf{a} \wedge \hat{A} \mathbf{b} + \hat{A} \mathbf{a} \wedge \hat{A}^2 \mathbf{b}$$

(all other terms vanish because $\hat{A}^3 = 0$). It is clear that the operator $\wedge^2 \hat{A}^1$ to the power 6 vanishes because there will be at least a third power of $\hat{A}$ acting on each vector. Therefore $q_1 = 0$ as well.

Now a general argument. Let p be a positive integer such that $\hat{A}^p = 0$, and consider the (pN)-th power of the operator $\wedge^N \hat{A}^k$ for some $k \geq 1$. We will prove that $(\wedge^N \hat{A}^k)^{pN} = \hat{0}$. Since $\wedge^N \hat{A}^k$ is a multiplication by a number, from $(\wedge^N \hat{A}^k)^{pN} = 0$ it will follow that $\wedge^N \hat{A}^k$ is a zero operator in $\wedge^N V$ for all $k \geq 1$. If all the coefficients q_k of the characteristic polynomial vanish, we will have $Q_{\hat{A}}(x) = (-x)^N$.

To prove that $(\wedge^N \hat{A}^k)^{pN} = \hat{0}$, consider the action of the operator $(\wedge^N \hat{A}^k)^{pN}$ on a tensor $\mathbf{e}_1 \wedge ... \wedge \mathbf{e}_N \in \wedge^N V$. By definition of $\wedge^N \hat{A}^k$, this operator is a sum of terms of the form

$$\hat{A}^{s_1} \mathbf{e}_1 \wedge ... \wedge \hat{A}^{s_N} \mathbf{e}_N,$$

where $s_j = 0$ or $s_j = 1$ are chosen such that $\sum_{j=1}^N s_j = k$. Therefore, the same operator raised to the power pN is expressed as

$$(\wedge^N \hat{A}^k)^{pN} = \sum_{(s_1,...,s_n)} \hat{A}^{s_1} \mathbf{e}_1 \wedge ... \wedge \hat{A}^{s_N} \mathbf{e}_N, \tag{3.20}$$

where now s_j are non-negative integers, $0 \leq s_j \leq pN$, such that $\sum_{j=1}^N s_j = kpN$. It is impossible that all s_j in Eq. (3.20) are less than p, because then we would have $\sum_{j=1}^N s_j < Np$, which would contradict the condition $\sum_{j=1}^N s_j = kpN$ (since $k \geq 1$ by construction). So each term of the sum in Eq. (3.20) contains at least a p-th power of $\hat{A}$. Since $(\hat{A})^p = 0$, each term in the sum in Eq. (3.20) vanishes. Hence $(\wedge^N \hat{A}^k)^{pN} = 0$ as required. ■

Remark: The converse statement is also true: If the characteristic polynomial of an operator $\hat{A}$ is $Q_{\hat{A}}(x) = (-x)^N$ then $\hat{A}$ is nilpotent. This follows easily from the Cayley-Hamilton theorem (see below), which states that $Q_{\hat{A}}(\hat{A}) = 0$, so we obtain immediately $(\hat{A})^N = 0$, i.e. the operator $\hat{A}$ is nilpotent. We find that one cannot distinguish a nilpotent operator from the zero operator by looking only at the characteristic polynomial.

4 Advanced applications

In this chapter we work in an N-dimensional vector space over a number field $\mathbb{K}$.

4.1 The space $\wedge^{N-1}V$

So far we have been using only the top exterior power, $\wedge^N V$. The next-to-top exterior power space, $\wedge^{N-1}V$, has the same dimension as V and is therefore quite useful since it is a space, in some special sense, associated with V. We will now find several important uses of this space.

4.1.1 Exterior transposition of operators

We have seen that a linear operator in the space $\wedge^N V$ is equivalent to multiplication by a number. We can reformulate this statement by saying that the space of linear operators in $\wedge^N V$ is canonically isomorphic to $\mathbb{K}$. Similarly, the space of linear operators in $\wedge^{N-1}V$ is canonically isomorphic to $\operatorname{End} V$, the space of linear operators in V. The isomorphism map will be denoted by the superscript $^{\wedge T}$. We will begin by defining this map explicitly.

Question: What is a nontrivial example of a linear operator in $\wedge^{N-1}V$?

Answer: Any operator of the form $\wedge^{N-1}\hat{A}^p$ with $1 \leq p \leq N-1$ and $\hat{A} \in \operatorname{End} V$. In this book, operators constructed in this way will be the only instance of operators in $\wedge^{N-1}V$.

Definition: If $\hat{X} \in \operatorname{End} V$ is a given linear operator then the **exterior transpose** operator

$$\hat{X}^{\wedge T} \in \operatorname{End}\left(\wedge^{N-1}V\right)$$

is canonically defined by the formula

$$\left(\hat{X}^{\wedge T}\omega\right) \wedge \mathbf{v} \equiv \omega \wedge \hat{X}\mathbf{v},$$

which must hold for all $\omega \in \wedge^{N-1}V$ and all $\mathbf{v} \in V$. If $\hat{Y} \in \operatorname{End}(\wedge^{N-1}V)$ is a linear operator then its exterior transpose $\hat{Y}^{\wedge T} \in \operatorname{End} V$ is defined by the formula

$$\omega \wedge \left(\hat{Y}^{\wedge T}\mathbf{v}\right) \equiv (\hat{Y}\omega) \wedge \mathbf{v}, \quad \forall \omega \in \wedge^{N-1}V, \ \mathbf{v} \in V.$$

We need to check that the definition makes sense, i.e. that the operators defined by these formulas exist and are uniquely defined.

Statement 1: The exterior transpose operators are well-defined, i.e. they exist, are unique, and are linear operators in the respective spaces. The exterior transposition has the linearity property

$$(\hat{A} + \lambda\hat{B})^{\wedge T} = \hat{A}^{\wedge T} + \lambda\hat{B}^{\wedge T}.$$

If $\hat{X} \in \text{End}\, V$ is an exterior transpose of $\hat{Y} \in \text{End}\left(\wedge^{N-1}V\right)$, i.e. $\hat{X} = \hat{Y}^{\wedge T}$, then also conversely $\hat{Y} = \hat{X}^{\wedge T}$.

Proof: We need to show that the formula

$$\left(\hat{X}^{\wedge T}\omega\right) \wedge \mathbf{v} \equiv \omega \wedge \hat{X}\mathbf{v}$$

actually defines an operator $\hat{X}^{\wedge T}$ uniquely when $\hat{X} \in \text{End}\, V$ is a given operator. Let us fix a tensor $\omega \in \wedge^{N-1}V$; to find $\hat{X}^{\wedge T}\omega$ we need to determine a tensor $\psi \in \wedge^{N-1}V$ such that $\psi \wedge \mathbf{v} = \omega \wedge \hat{X}\mathbf{v}$ for all $\mathbf{v} \in V$. When we find such a ψ, we will also show that it is unique; then we will have shown that $\hat{X}^{\wedge T}\omega \equiv \psi$ is well-defined.

An explicit computation of the tensor ψ can be performed in terms of a basis $\{\mathbf{e}_1, ..., \mathbf{e}_N\}$ in V. A basis in the space $\wedge^{N-1}V$ is formed by the set of N tensors of the form $\boldsymbol{\omega}_i \equiv \mathbf{e}_1 \wedge ... \wedge \mathbf{e}_{i-1} \wedge \mathbf{e}_{i+1} \wedge ... \wedge \mathbf{e}_N$, that is, $\boldsymbol{\omega}_i$ is the exterior product of the basis vectors without the vector $\mathbf{e}_i$ ($1 \leq i \leq N$). In the notation of Sec. 2.3.3, we have $\boldsymbol{\omega}_i = *(\mathbf{e}_i)(-1)^{i-1}$. It is sufficient to determine the components of ψ in this basis,

$$\psi = \sum_{i=1}^{N} c_i \boldsymbol{\omega}_i.$$

Taking the exterior product of ψ with $\mathbf{e}_i$, we find that only the term with c_i survives,

$$\psi \wedge \mathbf{e}_i = (-1)^{N-i} c_i \mathbf{e}_1 \wedge ... \wedge \mathbf{e}_N.$$

Therefore, the coefficient c_i is uniquely determined from the condition

$$c_i \mathbf{e}_1 \wedge ... \wedge \mathbf{e}_N = (-1)^{N-i} \psi \wedge \mathbf{e}_i \overset{!}{=} (-1)^{N-i} \omega \wedge \hat{X}\mathbf{e}_i.$$

Since the operator $\hat{X}$ is given, we know all $\hat{X}\mathbf{e}_i$ and can compute $\omega \wedge \hat{X}\mathbf{e}_i \in \wedge^N V$. So we find that every coefficient c_i is uniquely determined.

It is seen from the above formula that each coefficient c_i depends linearly on the operator $\hat{X}$. Therefore the linearity property holds,

$$(\hat{A} + \lambda\hat{B})^{\wedge T} = \hat{A}^{\wedge T} + \lambda\hat{B}^{\wedge T}.$$

The linearity of the operator $\hat{X}^{\wedge T}$ follows straightforwardly from the identity

$$\begin{aligned}
\left(\hat{X}^{\wedge T}(\omega + \lambda\omega')\right) \wedge \mathbf{v} &\overset{!}{=} (\omega + \lambda\omega') \wedge \hat{X}\mathbf{v} \\
&= \omega \wedge \hat{X}\mathbf{v} + \lambda\omega' \wedge \hat{X}\mathbf{v} \\
&\overset{!}{=} (\hat{X}^{\wedge T}\omega) \wedge \mathbf{v} + \lambda(\hat{X}^{\wedge T}\omega') \wedge \mathbf{v}.
\end{aligned}$$

In the same way we prove the existence, the uniqueness, and the linearity of the exterior transpose of an operator from $\text{End}(\wedge^{N-1}V)$. It is then clear that the transpose of the transpose is again the original operator. Details left as exercise. ■

Remark: Note that the space $\wedge^{N-1}V$ is has the same dimension as V but is *not* canonically isomorphic to V. Rather, an element $\psi \in \wedge^{N-1}V$ naturally acts by exterior multiplication on a vector $\mathbf{v} \in V$ and yields a tensor from $\wedge^N V$, i.e. ψ is a linear map $V \rightarrow \wedge^N V$, and we may express this as $\wedge^{N-1}V \cong V^* \otimes \wedge^N V$. Nevertheless, as we will now show, the exterior transpose map allows us to establish that the space of linear operators in $\wedge^{N-1}V$ is canonically isomorphic to the space of linear operators in V. We will use this isomorphism extensively in the following sections. A formal statement follows.

Statement 2: The spaces $\text{End}(\wedge^{N-1}V)$ and $\text{End}\, V$ are canonically isomorphic.

Proof: The map $^{\wedge T}$ between these spaces is one-to-one since no two different operators are mapped to the same operator. If two different operators $\hat{A}, \hat{B}$ had the same exterior transpose, we would have $(\hat{A} - \hat{B})^{\wedge T} = 0$ and yet $\hat{A} - \hat{B} \neq 0$. There exists at least one $\omega \in \wedge^{N-1}V$ and $\mathbf{v} \in V$ such that $\omega \wedge (\hat{A} - \hat{B})\mathbf{v} \neq 0$, and then

$$0 = \left((\hat{A} - \hat{B})^{\wedge T}\omega\right) \wedge \mathbf{v} = \omega \wedge (\hat{A} - \hat{B})\mathbf{v} \neq 0,$$

which is a contradiction. The map $^{\wedge T}$ is linear (Statement 1). Therefore, it is an isomorphism between the vector spaces $\text{End}\left(\wedge^{N-1}V\right)$ and $\text{End}\, V$. ■

A generalization of Statement 1 is the following.

Exercise 1: Show that the spaces $\text{End}(\wedge^k V)$ and $\text{End}(\wedge^{N-k}V)$ are canonically isomorphic ($1 \leq k < N$). Specifically, if $\hat{X} \in \text{End}(\wedge^k V)$ then the linear operator $\hat{X}^{\wedge T} \in \text{End}(\wedge^{N-k}V)$ is uniquely defined by the formula

$$\left(\hat{X}^{\wedge T}\omega_{N-k}\right) \wedge \omega_k \equiv \omega_{N-k} \wedge \hat{X}\omega_k,$$

which must hold for arbitrary tensors $\omega_k \in \wedge^k V, \omega_{N-k} \in \wedge^{N-k}V$.

Remark: It follows that the exterior transpose of $\wedge^N \hat{A}^N \in \text{End}\left(\wedge^N V\right)$ is mapped by the canonical isomorphism to an element of $\text{End}\, \mathbb{K}$, that is, a multiplication by a number. This is precisely the map we have been using in the previous section to define the determinant. In this notation, we have

$$\det \hat{A} \equiv \left(\wedge^N \hat{A}^N\right)^{\wedge T}.$$

Here we identify $\text{End}\, \mathbb{K}$ with $\mathbb{K}$.

Exercise 2: For any operators $\hat{A}, \hat{B} \in \text{End}\left(\wedge^k V\right)$, show that

$$(\hat{A}\hat{B})^{\wedge T} = \hat{B}^{\wedge T}\hat{A}^{\wedge T}.$$

4.1.2 * Index notation

Let us see how the exterior transposition is expressed in the index notation. (Below we will not use the resulting formulas.)

If an operator $\hat{A} \in \text{End}\, V$ is given in the index notation by a matrix A_i^j, the exterior transpose $\hat{A}^{\wedge T} \in \text{End}\left(\wedge^{N-1}V\right)$ is represented by an array $B_{i_1 \ldots i_{N-1}}^{j_1 \ldots j_{N-1}}$, which is totally antisymmetric with respect to its $N-1$ lower and upper indices separately. The action of the operator $\hat{B} \equiv \hat{A}^{\wedge T}$ on a tensor $\psi \in \wedge^{N-1}V$ is written in the index notation as

$$\sum_{i_s} B_{i_1 \ldots i_{N-1}}^{j_1 \ldots j_{N-1}} \psi^{i_1 \ldots i_{N-1}}.$$

(Here we did not introduce any combinatorial factors; the factor $(N-1)!$ will therefore appear at the end of the calculation.)

By definition of the exterior transpose, for any vector $\mathbf{v} \in V$ and for any $\psi \in \wedge^{N-1}V$ we must have

$$(\hat{B}\psi) \wedge \mathbf{v} = \psi \wedge (\hat{A}\mathbf{v}).$$

Using the index representation of the exterior product through the projection operators $\hat{E}$ (see Sec. 2.3.6), we represent the equation above in the the index notation as

$$\sum_{i,i_s,j_s} E_{j_1 \ldots j_{N-1} i}^{k_1 \ldots k_N} (B_{i_1 \ldots i_{N-1}}^{j_1 \ldots j_{N-1}} \psi^{i_1 \ldots i_{N-1}}) v^i = \sum_{j_s,i,j} E_{j_1 \ldots j_{N-1} j}^{k_1 \ldots k_N} \psi^{j_1 \ldots j_{N-1}} (A_i^j v^i).$$

We may simplify this to

$$\sum_{i,i_s,j_s} \varepsilon_{j_1 \ldots j_{N-1} i} (B_{i_1 \ldots i_{N-1}}^{j_1 \ldots j_{N-1}} \psi^{i_1 \ldots i_{N-1}}) v^i = \sum_{i_s,i,j} \varepsilon_{i_1 \ldots i_{N-1} j} \psi^{i_1 \ldots i_{N-1}} (A_i^j v^i),$$

because $E_{j_1 \ldots j_N}^{k_1 \ldots k_N} = \varepsilon_{j_1 \ldots j_N} \varepsilon^{k_1 \ldots k_N}$, and we may cancel the common factor $\varepsilon^{k_1 \ldots k_N}$ whose indices are not being summed over.

Since the equation above should hold for arbitrary $\psi^{i_1 \ldots i_{N-1}}$ and v^i, the equation with the corresponding *free* indices i_s and i should hold:

$$\sum_{j_s} \varepsilon_{j_1 \ldots j_{N-1} i} B_{i_1 \ldots i_{N-1}}^{j_1 \ldots j_{N-1}} = \sum_j \varepsilon_{i_1 \ldots i_{N-1} j} A_i^j. \tag{4.1}$$

This equation can be solved for B as follows. We note that the ε symbol in the left-hand side of Eq. (4.1) has one free index, i. Let us therefore multiply with an additional ε and sum over that index; this will yield the projection operator $\hat{E}$ (see Sec. 2.3.6). Namely, we multiply both sides of Eq. (4.1) with $\varepsilon^{k_1\ldots k_{N-1}i}$ and sum over i:

$$\sum_{j,i} \varepsilon^{k_1\ldots k_{N-1}i}\varepsilon_{i_1\ldots i_{N-1}j}A_i^j = \sum_{j_s,i} \varepsilon^{k_1\ldots k_{N-1}i}\varepsilon_{j_1\ldots j_{N-1}i}B_{i_1\ldots i_{N-1}}^{j_1\ldots j_{N-1}}$$
$$= \sum_{j_s} E_{j_1\ldots j_{N-1}}^{k_1\ldots k_{N-1}} B_{i_1\ldots i_{N-1}}^{j_1\ldots j_{N-1}},$$

where in the last line we used the definition (2.11)–(2.12) of the operator $\hat{E}$. Now we note that the right-hand side is the index representation of the product of the operators $\hat{E}$ and $\hat{B}$ (both operators act in $\wedge^{N-1}V$). The left-hand side is also an operator in $\wedge^{N-1}V$; denoting this operator for brevity by $\hat{X}$, we rewrite the equation as

$$\hat{E}\hat{B} = \hat{X} \in \mathrm{End}\left(\wedge^{N-1}V\right).$$

Using the property

$$\hat{E} = (N-1)!\hat{1}_{\wedge^{N-1}V}$$

(see Exercise in Sec. 2.3.6), we may solve the equation $\hat{E}\hat{B} = \hat{X}$ for $\hat{B}$ as

$$\hat{B} = \frac{1}{(N-1)!}\hat{X}.$$

Hence, the components of $\hat{B} \equiv \hat{A}^{\wedge T}$ are expressed as

$$B_{i_1\ldots i_{N-1}}^{k_1\ldots k_{N-1}} = \frac{1}{(N-1)!}\sum_{j,i} \varepsilon^{k_1\ldots k_{N-1}i}\varepsilon_{i_1\ldots i_{N-1}j}A_i^j.$$

An analogous formula holds for the exterior transpose of an operator in $\wedge^n V$, for any $n = 2, ..., N$. I give the formula without proof and illustrate it by an example.

Statement: If $\hat{A} \in \mathrm{End}\left(\wedge^n V\right)$ is given by its components $A_{i_1\ldots i_n}^{j_1\ldots j_n}$ then the components of $\hat{A}^{\wedge T}$ are

$$\left(\hat{A}^{\wedge T}\right)_{l_1\ldots l_{N-n}}^{k_1\ldots k_{N-n}}$$
$$= \frac{1}{n!(N-n)!}\sum_{j_s,i_s} \varepsilon^{k_1\ldots k_{N-n}i_1\ldots i_n}\varepsilon_{l_1\ldots l_{N-n}j_1\ldots j_n}A_{i_1\ldots i_n}^{j_1\ldots j_n}.$$

Example: Consider the exterior transposition $\hat{A}^{\wedge T}$ of the identity operator $\hat{A} \equiv \hat{1}_{\wedge^2 V}$. The components of the identity operator are given by

$$A_{i_1 i_2}^{j_1 j_2} = \delta_{i_1}^{j_1}\delta_{i_2}^{j_2},$$

so the components of $\hat{A}^{\wedge T}$ are

$$\left(\hat{A}^{\wedge T}\right)_{l_1\ldots l_{N-2}}^{k_1\ldots k_{N-2}} = \frac{1}{2!(N-2)!}\sum_{j_s,i_s} \varepsilon^{k_1\ldots k_{N-2}i_1 i_2}\varepsilon_{l_1\ldots l_{N-2}j_1 j_2}A_{i_1 i_2}^{j_1 j_2}$$
$$= \frac{1}{2!(N-2)!}\sum_{i_1,i_2} \varepsilon^{k_1\ldots k_{N-2}i_1 i_2}\varepsilon_{l_1\ldots l_{N-2}i_1 i_2}.$$

Let us check that this array of components is the same as that representing the operator $\hat{1}_{\wedge^{N-2}V}$. We note that the expression above is the same as

$$\frac{1}{(N-2)!}E^{k_1\ldots k_{N-2}}_{l_1\ldots l_{N-2}},$$

where the numbers $E^{k_1\ldots k_n}_{l_1\ldots l_n}$ are defined by Eqs. (2.11)–(2.12). Since the operator $\hat{E}$ in $\wedge^{N-2}V$ is equal to $(N-2)!\hat{1}_{\wedge^{N-2}V}$, we obtain that

$$\hat{A}^{\wedge T} = \hat{1}_{\wedge^{N-2}V}$$

as required.

4.2 Algebraic complement (adjoint) and beyond

In Sec. 3.3 we defined the determinant and derived various useful properties by considering, essentially, the exterior transpose of $\wedge^N \hat{A}^p$ with $1 \leq p \leq N$ (although we did not introduce this terminology back then). We have just seen that the exterior transposition can be defined more generally — as a map from $\text{End}(\wedge^k V)$ to $\text{End}(\wedge^{N-k}V)$. We will see in this section that the exterior transposition of the operators $\wedge^{N-1}\hat{A}^p$ with $1 \leq p \leq N-1$ yields operators acting in V that are quite useful as well.

4.2.1 Definition of algebraic complement

While we proved that operators like $(\wedge^{N-1}\hat{A}^p)^{\wedge T}$ are well-defined, we still have not obtained any explicit formulas for these operators. We will now compute these operators explicitly because they play an important role in the further development of the theory. It will turn out that every operator of the form $(\wedge^{N-1}\hat{A}^p)^{\wedge T}$ is a *polynomial* in $\hat{A}$ with coefficients that are known if we know the characteristic polynomial of $\hat{A}$.

Example 1: Let us compute $(\wedge^{N-1}\hat{A}^1)^{\wedge T}$. We consider, as a first example, a three-dimensional ($N=3$) vector space V and a linear operator $\hat{A} \in \text{End}\, V$. We are interested in the operator $(\wedge^2\hat{A}^1)^{\wedge T}$. By definition of the exterior transpose,

$$\begin{aligned}\mathbf{a}\wedge\mathbf{b}\wedge(\wedge^2\hat{A}^1)^{\wedge T}\mathbf{c} &= \big((\wedge^2\hat{A}^1)(\mathbf{a}\wedge\mathbf{b})\big)\wedge\mathbf{c}\\ &= \hat{A}\mathbf{a}\wedge\mathbf{b}\wedge\mathbf{c} + \mathbf{a}\wedge\hat{A}\mathbf{b}\wedge\mathbf{c}.\end{aligned}$$

We recognize a fragment of the operator $\wedge^3\hat{A}^1$ and write

$$\begin{aligned}(\wedge^3\hat{A}^1)(\mathbf{a}\wedge\mathbf{b}\wedge\mathbf{c}) &= \hat{A}\mathbf{a}\wedge\mathbf{b}\wedge\mathbf{c} + \mathbf{a}\wedge\hat{A}\mathbf{b}\wedge\mathbf{c} + \mathbf{a}\wedge\mathbf{b}\wedge\hat{A}\mathbf{c}\\ &= (\text{Tr}\,\hat{A})\mathbf{a}\wedge\mathbf{b}\wedge\mathbf{c},\end{aligned}$$

since this operator acts as multiplication by the trace of $\hat{A}$ (Section 3.8). It follows that

$$\begin{aligned}\mathbf{a}\wedge\mathbf{b}\wedge(\wedge^2\hat{A}^1)^{\wedge T}\mathbf{c} &= (\mathrm{Tr}\,\hat{A})\mathbf{a}\wedge\mathbf{b}\wedge\mathbf{c}-\mathbf{a}\wedge\mathbf{b}\wedge\hat{A}\mathbf{c}\\ &= \mathbf{a}\wedge\mathbf{b}\wedge\left((\mathrm{Tr}\,\hat{A})\mathbf{c}-\hat{A}\mathbf{c}\right).\end{aligned}$$

Since this must hold for arbitrary $\mathbf{a},\mathbf{b},\mathbf{c}\in V$, it follows that

$$(\wedge^2\hat{A}^1)^{\wedge T} = (\mathrm{Tr}\,\hat{A})\hat{1}_V - \hat{A}.$$

Thus we have computed the operator $(\wedge^2\hat{A}^1)^{\wedge T}$ in terms of $\hat{A}$ and the trace of $\hat{A}$.

Example 2: Let us now consider the operator $(\wedge^2\hat{A}^2)^{\wedge T}$. We have

$$\mathbf{a}\wedge\mathbf{b}\wedge(\wedge^2\hat{A}^2)^{\wedge T}\mathbf{c} = \left((\wedge^2\hat{A}^2)(\mathbf{a}\wedge\mathbf{b})\right)\wedge\mathbf{c} = \hat{A}\mathbf{a}\wedge\hat{A}\mathbf{b}\wedge\mathbf{c}.$$

We recognize a fragment of the operator $\wedge^3\hat{A}^2$ and write

$$(\wedge^3\hat{A}^2)(\mathbf{a}\wedge\mathbf{b}\wedge\mathbf{c}) = \hat{A}\mathbf{a}\wedge\hat{A}\mathbf{b}\wedge\mathbf{c}+\mathbf{a}\wedge\hat{A}\mathbf{b}\wedge\hat{A}\mathbf{c}+\hat{A}\mathbf{a}\wedge\mathbf{b}\wedge\hat{A}\mathbf{c}.$$

Therefore,

$$\begin{aligned}\mathbf{a}\wedge\mathbf{b}\wedge(\wedge^2\hat{A}^2)^{\wedge T}\mathbf{c} &= (\wedge^3\hat{A}^2)(\mathbf{a}\wedge\mathbf{b}\wedge\mathbf{c})\\ &\quad -(\mathbf{a}\wedge\hat{A}\mathbf{b}+\hat{A}\mathbf{a}\wedge\mathbf{b})\wedge\hat{A}\mathbf{c}\\ &\overset{(1)}{=} (\wedge^3\hat{A}^2)(\mathbf{a}\wedge\mathbf{b}\wedge\mathbf{c})-\mathbf{a}\wedge\mathbf{b}\wedge(\wedge^2\hat{A}^1)^{\wedge T}\hat{A}\mathbf{c}\\ &= \mathbf{a}\wedge\mathbf{b}\wedge(\wedge^3\hat{A}^2-(\wedge^2\hat{A}^1)^{\wedge T}\hat{A})\mathbf{c},\end{aligned}$$

where $^{(1)}$ used the definition of the operator $(\wedge^2\hat{A}^1)^{\wedge T}$. It follows that

$$\begin{aligned}(\wedge^2\hat{A}^2)^{\wedge T} &= (\wedge^3\hat{A}^2)\hat{1}_V-(\wedge^2\hat{A}^1)^{\wedge T}\hat{A}\\ &= (\wedge^3\hat{A}^2)\hat{1}_V-(\mathrm{Tr}\,\hat{A})\hat{A}+\hat{A}\hat{A}.\end{aligned}$$

Thus we have expressed the operator $(\wedge^2\hat{A}^2)^{\wedge T}$ as a *polynomial in* $\hat{A}$. Note that $\wedge^3\hat{A}^2$ is the second coefficient of the characteristic polynomial of $\hat{A}$.

Exercise 1: Consider a three-dimensional space V, a linear operator $\hat{A}$, and show that

$$(\wedge^2\hat{A}^2)^{\wedge T}\hat{A}\mathbf{v} = (\det\hat{A})\mathbf{v},\quad \forall\mathbf{v}\in V.$$

Hint: Consider $\mathbf{a}\wedge\mathbf{b}\wedge(\wedge^2\hat{A}^2)^{\wedge T}\hat{A}\mathbf{c} = \hat{A}\mathbf{a}\wedge\hat{A}\mathbf{b}\wedge\hat{A}\mathbf{c}$. ■

These examples are straightforwardly generalized. We will now express every operator of the form $(\wedge^{N-1}\hat{A}^p)^{\wedge T}$ as a polynomial in $\hat{A}$. For brevity, we introduce the notation

$$\hat{A}_{(k)} \equiv (\wedge^{N-1}\hat{A}^{N-k})^{\wedge T},\quad 1\leq k\leq N-1.$$

Lemma 1: For any operator $\hat{A} \in \text{End}\, V$ and for an integer p, $1 \le p \le N$, the following formula holds as an identity of operators in V:

$$(\wedge^{N-1}\hat{A}^{p-1})^{\wedge T}\hat{A} + (\wedge^{N-1}\hat{A}^{p})^{\wedge T} = (\wedge^N \hat{A}^p)\hat{1}_V.$$

Here, in order to provide a meaning for this formula in cases $p = 1$ and $p = N$, we define $\wedge^{N-1}\hat{A}^N \equiv \hat{0}$ and $\wedge^{N-1}\hat{A}^0 \equiv \hat{1}$. In the shorter notation, this is

$$\hat{A}_{(k)}\hat{A} + \hat{A}_{(k-1)} = (\wedge^N \hat{A}^{N-k+1})\hat{1}_V.$$

Note that $\wedge^N \hat{A}^{N-k+1} \equiv q_{k-1}$, where q_j are the coefficients of the characteristic polynomial of $\hat{A}$ (see Sec. 3.9).

Proof: We use Statement 4 in Sec. 3.7 with $\omega \equiv \mathbf{v}_1 \wedge ... \wedge \mathbf{v}_{N-1}$, $m \equiv N-1$ and $k \equiv p$:

$$(\wedge^{N-1}\hat{A}^p\omega) \wedge \mathbf{u} + (\wedge^{N-1}\hat{A}^{p-1}\omega) \wedge (\hat{A}\mathbf{u}) = \wedge^N \hat{A}^p (\omega \wedge \mathbf{u}).$$

This holds for $1 \le p \le N-1$. Applying the definition of the exterior transpose, we find

$$\omega \wedge (\wedge^{N-1}\hat{A}^p)^{\wedge T}\mathbf{u} + \omega \wedge (\wedge^{N-1}\hat{A}^{p-1})^{\wedge T}\hat{A}\mathbf{u} = (\wedge^N \hat{A}^p)\omega \wedge \mathbf{u}.$$

Since this holds for all $\omega \in \wedge^{N-1}V$ and $\mathbf{u} \in V$, we obtain the required formula,

$$(\wedge^{N-1}\hat{A}^p)^{\wedge T} + \omega \wedge (\wedge^{N-1}\hat{A}^{p-1})^{\wedge T}\hat{A} = (\wedge^N \hat{A}^p)\hat{1}_V.$$

It remains to verify the case $p = N$. In that case we compute directly,

$$\begin{aligned}(\wedge^{N-1}\hat{A}^{N-1}\omega) \wedge (\hat{A}\mathbf{u}) &= \hat{A}\mathbf{v}_1 \wedge ... \wedge \hat{A}\mathbf{v}_{N-1} \wedge \hat{A}\mathbf{u} \\ &= \wedge^N \hat{A}^N (\omega \wedge \mathbf{u}).\end{aligned}$$

Hence,

$$(\wedge^{N-1}\hat{A}^{N-1})^{\wedge T}\hat{A} = (\wedge^N \hat{A}^N)\hat{1}_V \equiv (\det \hat{A})\hat{1}_V.$$

■

Remark: In these formulas we interpret the operators $\wedge^N \hat{A}^p \in \text{End}\left(\wedge^N V\right)$ as simply numbers multiplying some operators. This is justified since $\wedge^N V$ is one-dimensional, and linear operators in it act as multiplication by numbers. In other words, we implicitly use the canonical isomorphism $\text{End}\left(\wedge^N V\right) \cong \mathbb{K}$. ■

Exercise 2: Use induction in p (for $1 \le p \le N-1$) and Lemma 1 to express $\hat{A}_{(k)}$ explicitly as polynomials in $\hat{A}$:

$$\hat{A}_{(N-p)} \equiv (\wedge^{N-1}\hat{A}^p)^{\wedge T} = \sum_{k=0}^{p} (-1)^k (\wedge^N \hat{A}^{p-k})(\hat{A})^k.$$

Hint: Start applying Lemma 1 with $p = 1$ and $\hat{A}_{(N)} \equiv \hat{1}$. ■

Using the coefficients $q_k \equiv \wedge^N \hat{A}^{N-k}$ of the characteristic polynomial, the result of Exercise 2 can be rewritten as

$$\begin{aligned}
\left(\wedge^{N-1}\hat{A}^1\right)^{\wedge T} &\equiv \hat{A}_{(N-1)} = q_{N-1}\hat{1}_V - \hat{A}, \\
\left(\wedge^{N-1}\hat{A}^2\right)^{\wedge T} &\equiv \hat{A}_{(N-2)} = q_{N-2}\hat{1}_V - q_{N-1}\hat{A} + (\hat{A})^2, \\
&\ldots\ldots, \\
\left(\wedge^{N-1}\hat{A}^{N-1}\right)^{\wedge T} &\equiv \hat{A}_{(1)} = q_1\hat{1}_V + q_2(-\hat{A}) + ... \\
&\qquad + q_{N-1}(-\hat{A})^{N-2} + (-\hat{A})^{N-1}.
\end{aligned}$$

Note that the characteristic polynomial of $\hat{A}$ is

$$Q_{\hat{A}}(\lambda) = q_0 + q_1(-\lambda) + ... + q_{N-1}(-\lambda)^{N-1} + (-\lambda)^N.$$

Thus the operators denoted by $\hat{A}_{(k)}$ are computed as suitable "fragments"' of the characteristic polynomial into which $\hat{A}$ is substituted instead of λ.

Exercise 3:* Using the definition of exterior transpose for general exterior powers (Exercise 1 in Sec. 4.1.1), show that for $1 \leq k \leq N-1$ and $1 \leq p \leq k$ the following identity holds,

$$\sum_{q=0}^{p} \left(\wedge^{N-k}\hat{A}^{p-q}\right)^{\wedge T}(\wedge^k \hat{A}^q) = (\wedge^N \hat{A}^p)\hat{1}_{\wedge^k V}.$$

Deduce that the operators $\left(\wedge^{N-k}\hat{A}^p\right)^{\wedge T}$ can be expressed as polynomials in the (mutually commuting) operators $\wedge^k \hat{A}^j$ $(1 \leq j \leq k)$.

Hints: Follow the proof of Statement 4 in Sec. 3.7. The idea is to apply both sides to $\omega_k \wedge \omega_{N-k}$, where $\omega_k \equiv \mathbf{v}_1 \wedge ... \wedge \mathbf{v}_k$ and $\omega_{N-k} = \mathbf{v}_{N-k+1} \wedge ... \wedge \mathbf{v}_N$. Since $\wedge^N \hat{A}^p$ acts on $\omega_k \wedge \omega_{N-k}$ by distributing p copies of $\hat{A}$ among the N vectors $\mathbf{v}_j$, one needs to show that the same terms will occur when one first distributes q copies of $\hat{A}$ among the first k vectors and $p-q$ copies of $\hat{A}$ among the last $N-k$ vectors, and then sums over all q from 0 to p. Once the identity is proved, one can use induction to express the operators $\left(\wedge^{N-k}\hat{A}^p\right)^{\wedge T}$. For instance, the identity with $k=2$ and $p=1$ yields

$$\left(\wedge^{N-2}\hat{A}^0\right)^{\wedge T}(\wedge^2\hat{A}^1) + \left(\wedge^{N-2}\hat{A}^1\right)^{\wedge T}(\wedge^2\hat{A}^0) = (\wedge^N\hat{A}^1)\hat{1}_{\wedge^k V}.$$

Therefore

$$\left(\wedge^{N-2}\hat{A}^1\right)^{\wedge T} = (\mathrm{Tr}\hat{A})\hat{1}_{\wedge^k V} - \wedge^2\hat{A}^1.$$

Similarly, with $k=2$ and $p=2$ we find

$$\begin{aligned}
\left(\wedge^{N-2}\hat{A}^2\right)^{\wedge T} &= (\wedge^N\hat{A}^2)\hat{1}_{\wedge^k V} - \left(\wedge^{N-2}\hat{A}^1\right)^{\wedge T}(\wedge^2\hat{A}^1) - \wedge^2\hat{A}^2 \\
&= (\wedge^N\hat{A}^2)\hat{1}_{\wedge^k V} - (\mathrm{Tr}\hat{A})(\wedge^2\hat{A}^1) + (\wedge^2\hat{A}^1)^2 - \wedge^2\hat{A}^2.
\end{aligned}$$

It follows by induction that all the operators $\left(\wedge^{N-k}\hat{A}^p\right)^{\wedge T}$ are expressed as polynomials in $\wedge^k\hat{A}^j$. ■

At the end of the proof of Lemma 1 we have obtained a curious relation,

$$(\wedge^{N-1}\hat{A}^{N-1})^{\wedge T}\hat{A} = (\det \hat{A})\hat{1}_V.$$

If $\det \hat{A} \neq 0$, we may divide by it and immediately find the following result.
Lemma 2: If $\det \hat{A} \neq 0$, the inverse operator satisfies

$$\hat{A}^{-1} = \frac{1}{\det \hat{A}}(\wedge^{N-1}\hat{A}^{N-1})^{\wedge T}.$$

Thus we are able to express the inverse operator $\hat{A}^{-1}$ as a *polynomial* in $\hat{A}$. If $\det \hat{A} = 0$ then the operator $\hat{A}$ has no inverse, but the operator $(\wedge^{N-1}\hat{A}^{N-1})^{\wedge T}$ is still well-defined and sufficiently useful to deserve a special name.
Definition: The **algebraic complement** (also called the **adjoint**) of $\hat{A}$ is the operator

$$\tilde{\hat{A}} \equiv (\wedge^{N-1}\hat{A}^{N-1})^{\wedge T} \in \operatorname{End} V.$$

Exercise 4: Compute the algebraic complement of the operator $\hat{A} = \mathbf{a} \otimes \mathbf{b}^*$, where $\mathbf{a} \in V$ and $\mathbf{b} \in V^*$, and V is an N-dimensional space ($N \geq 2$).

Answer: Zero if $N \geq 3$. For $N = 2$ we use Example 1 to compute

$$(\wedge^1\hat{A}^1)^{\wedge T} = (\operatorname{Tr} \hat{A})\hat{1} - \hat{A} = \mathbf{b}^*(\mathbf{a})\hat{1} - \mathbf{a} \otimes \mathbf{b}^*.$$

Exercise 5: For the operator $\hat{A} = \mathbf{a} \otimes \mathbf{b}^*$ in N-dimensional space, as in Exercise 4, show that $(\wedge^{N-1}\hat{A}^p)^{\wedge T} = 0$ for $p \geq 2$.

4.2.2 Algebraic complement of a matrix

The algebraic complement is usually introduced in terms of matrix determinants. Namely, one takes a matrix A_{ij} and deletes the column number k and the row number l. Then one computes the determinant of the resulting matrix and multiplies by $(-1)^{k+l}$. The result is the element B_{kl} of the matrix that is the algebraic complement of A_{ij}. I will now show that our definition is equivalent to this one, if we interpret matrices as coefficients of linear operators in a basis.
Statement: Let $\hat{A} \in \operatorname{End} V$ and let $\{\mathbf{e}_j\}$ be a basis in V. Let A_{ij} be the matrix of the operator $\hat{A}$ in this basis. Let $\hat{B} = (\wedge^{N-1}\hat{A}^{N-1})^{\wedge T}$ and let B_{kl} be the matrix of $\hat{B}$ in the same basis. Then B_{kl} is equal to $(-1)^{k+l}$ times the determinant of the matrix obtained from A_{ij} by deleting the column number k and the row number l.

Proof: Given an operator $\hat{B}$, the matrix element B_{kl} in the basis $\{\mathbf{e}_j\}$ can be computed as the coefficient in the following relation (see Sec. 2.3.3),

$$B_{kl}\mathbf{e}_1 \wedge ... \wedge \mathbf{e}_N = \mathbf{e}_1 \wedge ... \wedge \mathbf{e}_{k-1} \wedge (\hat{B}\mathbf{e}_l) \wedge \mathbf{e}_{k+1} \wedge ... \wedge \mathbf{e}_N.$$

Since $\hat{B} = (\wedge^{N-1}\hat{A}^{N-1})^{\wedge T}$, we have

$$B_{kl}\mathbf{e}_1 \wedge ... \wedge \mathbf{e}_N = \hat{A}\mathbf{e}_1 \wedge ... \wedge \hat{A}\mathbf{e}_{k-1} \wedge \mathbf{e}_l \wedge \hat{A}\mathbf{e}_{k+1} \wedge ... \wedge \hat{A}\mathbf{e}_N.$$

Now the right side can be expressed as the determinant of another operator, call it $\hat{X}$,

$$\begin{aligned} B_{kl}\mathbf{e}_1 \wedge ... \wedge \mathbf{e}_N &= (\det \hat{X})\mathbf{e}_1 \wedge ... \wedge \mathbf{e}_N \\ &= \hat{X}\mathbf{e}_1 \wedge ... \wedge \hat{X}\mathbf{e}_{k-1} \wedge \hat{X}\mathbf{e}_k \wedge \hat{X}\mathbf{e}_{k+1} \wedge ... \wedge \hat{X}\mathbf{e}_N, \end{aligned}$$

if we define $\hat{X}$ as an operator such that $\hat{X}\mathbf{e}_k \equiv \mathbf{e}_l$ while on other basis vectors $\hat{X}\mathbf{e}_j \equiv \hat{A}\mathbf{e}_j$ ($j \neq k$). Having defined $\hat{X}$ in this way, we have $B_{kl} = \det \hat{X}$.

We can now determine the matrix X_{ij} representing $\hat{X}$ in the basis $\{\mathbf{e}_j\}$. By the definition of the matrix representation of operators,

$$\hat{A}\mathbf{e}_j = \sum_{i=1}^{N} A_{ij}\mathbf{e}_i, \quad \hat{X}\mathbf{e}_j = \sum_{i=1}^{N} X_{ij}\mathbf{e}_i, \quad 1 \leq j \leq N.$$

It follows that $X_{ij} = A_{ij}$ for $j \neq k$ while $X_{ik} = \delta_{il}$ ($1 \leq i \leq N$), which means that the entire k-th column in the matrix A_{ij} has been replaced by a column containing zeros except for a single nonzero element $X_{lk} = 1$.

It remains to show that the determinant of the matrix X_{ij} is equal to $(-1)^{k+l}$ times the determinant of the matrix obtained from A_{ij} by deleting column k and row l. We may move in the matrix X_{ij} the k-th column to the first column and the l-th row to the first row, without changing the order of any other rows and columns. This produces the sign factor $(-1)^{k+l}$ but otherwise does not change the determinant. The result is

$$\begin{aligned} B_{kl} = \det \hat{X} &= (-1)^{k+l} \det \begin{vmatrix} 1 & X_{12} & ... & X_{1N} \\ 0 & * & * & * \\ \vdots & * & * & * \\ 0 & * & * & * \end{vmatrix} \\ &= (-1)^{k+l} \det \begin{vmatrix} * & * & * \\ * & * & * \\ * & * & * \end{vmatrix}, \end{aligned}$$

where the stars represent the matrix obtained from A_{ij} by deleting column k and row l, and the numbers X_{12}, ..., X_{1N} do not enter the determinant. This is the result we needed. ■

Exercise 5:* Show that the matrix representation of the algebraic complement can be written through the Levi-Civita symbol ε as

$$\tilde{A}^i_k = \frac{1}{(N-1)!} \sum_{i_2,...,i_N} \sum_{k_2,...,k_N} \varepsilon_{kk_2...k_N} \varepsilon^{ii_2...i_N} A^{k_2}_{i_2} ... A^{k_N}_{i_N}.$$

Hint: See Sections 3.4.1 and 4.1.2.

4.2.3 Further properties and generalizations

In our approach, the algebraic complement $\tilde{\hat{A}}$ of an operator $\hat{A}$ comes from considering the set of $N-1$ operators

$$\hat{A}_{(k)} \equiv \left(\wedge^{N-1}\hat{A}^{N-k}\right)^{\wedge T}, \quad 1 \leq k \leq N-1.$$

(For convenience we might define $\hat{A}_{(N)} \equiv \hat{1}_V$.)

The operators $\hat{A}_{(k)}$ can be expressed as polynomials in $\hat{A}$ through the identity (Lemma 1 in Sec. 4.2.1)

$$\hat{A}_{(k)}\hat{A} + \hat{A}_{(k-1)} = q_{k-1}\hat{1}, \quad q_j \equiv \wedge^N \hat{A}^{N-j}.$$

The numbers q_j introduced here are the coefficients of the characteristic polynomial of $\hat{A}$; for instance, $\det \hat{A} \equiv q_0$ and $\mathrm{Tr}\hat{A} \equiv q_{N-1}$. It follows by induction (Exercise 2 in Sec. 4.2.1) that

$$\begin{aligned}\hat{A}_{(N-k)} &= q_{N-k}\hat{1} - q_{N-k+1}\hat{A} + ... \\ &\quad + q_{N-1}(-\hat{A})^{k-1} + (-\hat{A})^k.\end{aligned}$$

The algebraic complement is $\tilde{\hat{A}} \equiv \hat{A}_1$, but it appears natural to study the properties of all the operators $\hat{A}_{(k)}$. (The operators $\hat{A}_{(k)}$ do not seem to have an established name for $k \geq 2$.)

Statement 1: The coefficients of the characteristic polynomial of the algebraic complement, $\tilde{\hat{A}}$, are

$$\wedge^N \tilde{\hat{A}}^k = (\det \hat{A})^{k-1}(\wedge^N \hat{A}^{N-k}) \equiv q_0^{k-1} q_k.$$

For instance,

$$\begin{aligned}\mathrm{Tr}\,\tilde{\hat{A}} &= \wedge^N \tilde{\hat{A}}^1 = q_1 = \wedge^N \hat{A}^{N-1}, \\ \det \tilde{\hat{A}} &= \wedge^N \tilde{\hat{A}}^N = q_0^{N-1} q_N = (\det \hat{A})^{N-1}.\end{aligned}$$

Proof: Let us first assume that $\det \hat{A} \equiv q_0 \neq 0$. We use the property $\hat{A}\tilde{\hat{A}} = q_0\hat{1}$ (Lemma 2 in Sec. 4.2.1) and the multiplicativity of determinants to find

$$\begin{aligned}\det(\tilde{\hat{A}} - \lambda\hat{1})q_0 &= \det(q_0\hat{1} - \lambda\hat{A}) = (-\lambda)^N \det(\hat{A} - \frac{q_0}{\lambda}\hat{1}) \\ &= (-\lambda^N)Q_{\hat{A}}(\frac{q_0}{\lambda}),\end{aligned}$$

hence the characteristic polynomial of $\tilde{\hat{A}}$ is

$$\begin{aligned}Q_{\tilde{\hat{A}}}(\lambda) \equiv \det(\tilde{\hat{A}} - \lambda\hat{1}) &= \frac{(-\lambda^N)}{q_0} Q_{\hat{A}}(\frac{q_0}{\lambda}) \\ &= \frac{(-\lambda)^N}{q_0}\left[\left(-\frac{q_0}{\lambda}\right)^N + q_{N-1}\left(-\frac{q_0}{\lambda}\right)^{N-1} + ... + q_0\right] \\ &= (-\lambda)^N + q_1(-\lambda)^{N-1} + q_2 q_0\,(-\lambda)^{N-2} + ... + q_0^{N-1}.\end{aligned}$$

This agrees with the required formula.

It remains to prove the case $q_0 \equiv \det \hat{A} = 0$. Although this result could be achieved as a limit of nonzero q_0 with $q_0 \to 0$, it is instructive to see a direct proof without using the assumption $q_0 \neq 0$ or taking limits.

Consider a basis $\{\mathbf{v}_j\}$ in V and the expression

$$(\wedge^N \tilde{\hat{A}}^k)\mathbf{v}_1 \wedge ... \wedge \mathbf{v}_N.$$

This expression contains $\binom{N}{k}$ terms of the form

$$\tilde{\hat{A}}\mathbf{v}_1 \wedge ... \wedge \tilde{\hat{A}}\mathbf{v}_k \wedge \mathbf{v}_{k+1} \wedge ... \wedge \mathbf{v}_N,$$

where $\tilde{\hat{A}}$ is applied only to k vectors. Using the definition of $\tilde{\hat{A}}$, we can rewrite such a term as follows. First, we use the definition of $\tilde{\hat{A}}$ to write

$$\tilde{\hat{A}}\mathbf{v}_1 \wedge \psi = \mathbf{v}_1 \wedge (\wedge^{N-1}\hat{A}^{N-1})\psi,$$

for any $\psi \in \wedge^{N-1}V$. In our case, we use

$$\psi \equiv \tilde{\hat{A}}\mathbf{v}_2 \wedge ... \wedge \tilde{\hat{A}}\mathbf{v}_k \wedge \mathbf{v}_{k+1} \wedge ... \wedge \mathbf{v}_N$$

and find

$$\tilde{\hat{A}}\mathbf{v}_1 \wedge \psi = \mathbf{v}_1 \wedge \hat{A}\tilde{\hat{A}}\mathbf{v}_2 \wedge ... \wedge \hat{A}\tilde{\hat{A}}\mathbf{v}_k \wedge \hat{A}\mathbf{v}_{k+1} \wedge ... \wedge \hat{A}\mathbf{v}_N.$$

By assumption $q_0 = 0$, hence $\hat{A}\tilde{\hat{A}} = 0 = \tilde{\hat{A}}\hat{A}$ (since $\tilde{\hat{A}}$, being a polynomial in $\hat{A}$, commutes with $\hat{A}$) and thus

$$(\wedge^N \tilde{\hat{A}}^k)\mathbf{v}_1 \wedge ... \wedge \mathbf{v}_N = 0, \quad k \geq 2.$$

For $k = 1$ we find

$$\tilde{\hat{A}}\mathbf{v}_1 \wedge \psi = \mathbf{v}_1 \wedge \hat{A}\mathbf{v}_2 \wedge ... \wedge \hat{A}\mathbf{v}_N.$$

Summing N such terms, we obtain the same expression as that in the definition of $\wedge^N \hat{A}^{N-1}$, hence

$$(\wedge^N \tilde{\hat{A}}^1)\mathbf{v}_1 \wedge ... \wedge \mathbf{v}_N = \wedge^N \hat{A}^{N-1}\mathbf{v}_1 \wedge ... \wedge \mathbf{v}_N.$$

This concludes the proof for the case $\det \hat{A} = 0$. ■

Exercise:* Suppose that $\hat{A}$ has the **simple** eigenvalue $\lambda = 0$ (i.e. this eigenvalue has multiplicity 1). Show that the algebraic complement, $\tilde{\hat{A}}$, has rank 1, and that the image of $\tilde{\hat{A}}$ is the one-dimensional subspace Span $\{\mathbf{v}\}$.

Hint: An operator has rank 1 if its image is one-dimensional. The eigenvalue $\lambda = 0$ has multiplicity 1 if $\wedge^N \hat{A}^{N-1} \neq 0$. Choose a basis consisting of the eigenvector $\mathbf{v}$ and $N-1$ other vectors $\mathbf{u}_2$, ..., $\mathbf{u}_N$. Show that

$$\tilde{\hat{A}}\mathbf{v} \wedge \mathbf{u}_2 \wedge ... \wedge \mathbf{u}_N = \wedge^N \hat{A}^{N-1}(\mathbf{v} \wedge \mathbf{u}_2 \wedge ... \wedge \mathbf{u}_N) \neq 0,$$

while

$$\mathbf{v} \wedge \mathbf{u}_2 \wedge ... \wedge \tilde{\hat{A}}\mathbf{u}_j \wedge ... \wedge \mathbf{u}_N = 0, \quad 2 \leq j \leq N.$$

Consider other expressions, such as

$$\tilde{\hat{A}}\mathbf{v} \wedge \mathbf{v} \wedge \mathbf{u}_3 \wedge ... \wedge \mathbf{u}_N \text{ or } \tilde{\hat{A}}\mathbf{u}_j \wedge \mathbf{v} \wedge \mathbf{u}_3 \wedge ... \wedge \mathbf{u}_N,$$

and finally deduce that the image of $\tilde{\hat{A}}$ is precisely the one-dimensional subspace Span $\{\mathbf{v}\}$. ■

Now we will demonstrate a useful property of the operators $\hat{A}_{(k)}$.

Statement 2: The trace of $\hat{A}_{(k)}$ satisfies

$$\frac{\mathrm{Tr}\hat{A}_{(k)}}{k} = \wedge^N \hat{A}^{N-k} \equiv q_k.$$

Proof: Consider the action of $\wedge^N \hat{A}^{N-k}$ on a basis tensor $\omega \equiv \mathbf{v}_1 \wedge ... \wedge \mathbf{v}_N$; the result is a sum of $\binom{N}{N-k}$ terms,

$$\wedge^N \hat{A}^{N-k}\omega = \hat{A}\mathbf{v}_1 \wedge ... \wedge \hat{A}\mathbf{v}_{N-k} \wedge \mathbf{v}_{N-k+1} \wedge ... \wedge \mathbf{v}_N + \text{(permutations)}.$$

Consider now the action of $\mathrm{Tr}\hat{A}_{(k)}$ on ω,

$$\mathrm{Tr}\hat{A}_{(k)}\omega = \wedge^N [\hat{A}_{(k)}]^1 \omega = \sum_{j=1}^{N} \mathbf{v}_1 \wedge ... \wedge \hat{A}_{(k)}\mathbf{v}_j \wedge ... \wedge \mathbf{v}_N.$$

Using the definition of $\hat{A}_{(k)}$, we rewrite

$$\mathbf{v}_1 \wedge ... \wedge \hat{A}_{(k)}\mathbf{v}_j \wedge ... \wedge \mathbf{v}_N = \hat{A}\mathbf{v}_1 \wedge ... \wedge \hat{A}\mathbf{v}_{N-k} \wedge \mathbf{v}_{N-k+1} \wedge ... \wedge \mathbf{v}_j \wedge ... \wedge \mathbf{v}_N + \text{(permutations not including } \hat{A}\mathbf{v}_j).$$

After summing over j, we will obtain all the same terms as were present in the expression for $\wedge^N \hat{A}^{N-k}\omega$, but each term will occur several times. We can show that each term will occur exactly k times. For instance, the term

$$\hat{A}\mathbf{v}_1 \wedge ... \wedge \hat{A}\mathbf{v}_{N-k} \wedge \mathbf{v}_{N-k+1} \wedge ... \wedge \mathbf{v}_j \wedge ... \wedge \mathbf{v}_N$$

will occur k times in the expression for $\mathrm{Tr}\hat{A}_{(k)}\omega$ because it will be generated once by each of the terms

$$\mathbf{v}_1 \wedge ... \wedge \hat{A}_{(k)}\mathbf{v}_j \wedge ... \wedge \mathbf{v}_N$$

with $N-k+1 \leq j \leq N$. The same argument holds for every other term. Therefore

$$\mathrm{Tr}\hat{A}_{(k)}\omega = k\,(\wedge^N \hat{A}^{N-k})\omega = kq_k\omega.$$

Since this holds for any $\omega \in \wedge^N V$, we obtain the required statement. ■

Remark: We have thus computed the trace of every operator $\hat{A}_{(k)}$, as well as the characteristic polynomial of $\hat{A}_{(1)} \equiv \tilde{\hat{A}}$. Computing the entire characteristic polynomial of each $\hat{A}_k$ is certainly possible but will perhaps lead to cumbersome expressions. ■

An interesting application of Statement 2 is the following algorithm for computing the characteristic polynomial of an operator.[1] This algorithm is more economical compared with the computation of $\det(\hat{A} - \lambda\hat{1})$ via permutations, and requires only operator (or matrix) multiplications and the computation of a trace.

Statement 3: (Leverrier's algorithm) The coefficients $\wedge^N \hat{A}^k \equiv q_{N-k}$ $(1 \leq k \leq N)$ of the characteristic polynomial of an operator $\hat{A}$ can be computed together with the operators $\hat{A}_{(j)}$ by starting with $\hat{A}_{(N)} \equiv \hat{1}_V$ and using the descending recurrence relation for $j = N-1, ..., 0$:

$$\begin{aligned} q_j &= \frac{1}{N-j} \mathrm{Tr}\,[\hat{A}\hat{A}_{(j+1)}], \\ \hat{A}_{(j)} &= q_j \hat{1} - \hat{A}\hat{A}_{(j+1)}. \end{aligned} \tag{4.2}$$

At the end of the calculation, we will have

$$q_0 = \det \hat{A}, \quad \hat{A}_{(1)} = \tilde{\hat{A}}, \quad \hat{A}_{(0)} = 0.$$

Proof: At the beginning of the recurrence, we have

$$j = N-1, \quad q_{N-1} = \frac{1}{N-j} \mathrm{Tr}\,[\hat{A}\hat{A}_{(j+1)}] = \mathrm{Tr}\hat{A},$$

which is correct. The recurrence relation (4.2) for $\hat{A}_{(j)}$ coincides with the result of Lemma 1 in Sec. 4.2.1 and thus yields at each step j the correct operator $\hat{A}_{(j)}$ — as long as q_j was computed correctly at that step. So it remains to verify that q_j is computed correctly. Taking the trace of Eq. (4.2) and using $\mathrm{Tr}\,\hat{1} = N$, we get

$$\mathrm{Tr}\,[A\hat{A}_{(j+1)}] = Nq_j - \mathrm{Tr}\hat{A}_{(j)}.$$

We now substitute for $\mathrm{Tr}\hat{A}_{(j)}$ the result of Statement 2 and find

$$\mathrm{Tr}\,[A\hat{A}_{(j+1)}] = Nq_j - jq_j = (N-j)\,q_j.$$

Thus q_j is also computed correctly from the previously known $\hat{A}_{(j+1)}$ at each step j. ■

Remark: This algorithm provides another illustration for the "trace relations" (see Exercises 1 and 2 in Sec. 3.9), i.e. for the fact that the coefficients q_j of the characteristic polynomial of $\hat{A}$ can be expressed as polynomials in the traces of $\hat{A}$ and its powers. These expressions will be obtained in Sec. 4.5.3.

---

[1] I found this algorithm in an online note by W. Kahan, *"Jordan's normal form"* (downloaded from `http://www.cs.berkeley.edu/~wkahan/MathH110/jordan.pdf` on October 6, 2009). Kahan attributes this algorithm to Leverrier, Souriau, Frame, and Faddeev.

4.3 Cayley-Hamilton theorem and beyond

The characteristic polynomial of an operator $\hat{A}$ has roots λ that are eigenvalues of $\hat{A}$. It turns out that we can substitute $\hat{A}$ as an operator into the characteristic polynomial, and the result is the zero operator, as if $\hat{A}$ were one of its eigenvalues. In other words, $\hat{A}$ satisfies (as an operator) its own characteristic equation.

Theorem 1 (Cayley-Hamilton): If $Q_{\hat{A}}(\lambda) \equiv \det(\hat{A} - \lambda \hat{1}_V)$ is the characteristic polynomial of the operator $\hat{A}$ then $Q_{\hat{A}}(\hat{A}) = \hat{0}_V$.

Proof: The coefficients of the characteristic polynomial are $\wedge^N \hat{A}^m$. When we substitute the operator $\hat{A}$ into $Q_{\hat{A}}(\lambda)$, we obtain the operator

$$Q_{\hat{A}}(\hat{A}) = (\det \hat{A})\hat{1}_V + (\wedge^N \hat{A}^{N-1})(-\hat{A}) + ... + (-\hat{A})^N.$$

We note that this expression is similar to that for the algebraic complement of $\hat{A}$ (see Exercise 2 in Sec. 4.2.1), so

$$\begin{aligned} Q_{\hat{A}}(\hat{A}) &= (\det \hat{A})\hat{1}_V + (\wedge^N \hat{A}^{N-1} + ... + (-\hat{A})^{N-1})(-\hat{A}) \\ &= (\det \hat{A})\hat{1}_V - (\wedge^{N-1} \hat{A}^{N-1})^{\wedge T} \hat{A} = \hat{0}_V \end{aligned}$$

by Lemma 1 in Sec. 4.2.1. Hence $Q_{\hat{A}}(\hat{A}) = \hat{0}_V$ for any operator $\hat{A}$. ■

Remark: While it is true that the characteristic polynomial vanishes on $\hat{A}$, it is not necessarily the simplest such polynomial. A polynomial of a lower degree may vanish on $\hat{A}$. A trivial example of this is given by an operator $\hat{A} = \alpha\hat{1}$, that is, the identity operator times a constant α. The characteristic polynomial of $\hat{A}$ is $Q_{\hat{A}}(\lambda) = (\alpha - \lambda)^N$. In agreement with the Cayley-Hamilton theorem, $(\alpha\hat{1} - \hat{A})^N = \hat{0}$. However, the simpler polynomial $p(\lambda) = \lambda - \alpha$ also has the property $p(\hat{A}) = \hat{0}$. We will look into this at the end of Sec. 4.6. ■

We have derived the Cayley-Hamilton theorem by considering the exterior transpose of $\wedge^{N-1} \hat{A}^{N-1}$. A generalization is found if we similarly use the operators of the form $(\wedge^a \hat{A}^b)^{\wedge T}$.

Theorem 2 (Cayley-Hamilton in $\wedge^k V$): For any operator $\hat{A}$ in V and for $1 \leq k \leq N, 1 \leq p \leq N$, the following identity holds,

$$\sum_{q=0}^{p} (\wedge^{N-k} \hat{A}^{p-q})^{\wedge T} (\wedge^k \hat{A}^q) = (\wedge^N \hat{A}^p) \hat{1}_{\wedge^k V}. \tag{4.3}$$

In this identity, we set $\wedge^k \hat{A}^0 \equiv \hat{1}_{\wedge^k V}$ and $\wedge^k \hat{A}^r \equiv 0$ for $r > k$. Explicit expressions can be derived for all operators $(\wedge^{N-k} \hat{A}^p)^{\wedge T}$ as polynomials in the (mutually commuting) operators $\wedge^k \hat{A}^j$, $1 \leq j \leq k$. (See Exercise 3 in Sec. 4.2.1.) Hence, there exist k identically vanishing operator-valued polynomials involving $\wedge^k \hat{A}^j$. (In the ordinary Cayley-Hamilton theorem, we have $k = 1$ and a single polynomial $Q_{\hat{A}}(\hat{A})$ that identically vanishes as an operator in $V \equiv \wedge^1 V$.) The coefficients of those polynomials will be known functions of $\hat{A}$. One can also obtain an identically vanishing polynomial in $\wedge^k \hat{A}^1$.

Proof: Let us fix k and first write Eq. (4.3) for $1 \leq p \leq N-k$. These $N-k$ equations are all of the form

$$\left(\wedge^{N-k}\hat{A}^p\right)^{\wedge T} + [...] = (\wedge^N \hat{A}^p)\hat{1}_{\wedge^k V}, \quad 1 \leq p \leq N-k.$$

In the p-th equation, the omitted terms in square brackets contain only the operators $\left(\wedge^{N-k}\hat{A}^r\right)^{\wedge T}$ with $r < p$ and $\wedge^k \hat{A}^q$ with $1 \leq q \leq k$. Therefore, these equations can be used to express $\left(\wedge^{N-k}\hat{A}^p\right)^{\wedge T}$ for $1 \leq p \leq N-k$ through the operators $\wedge^k \hat{A}^q$ explicitly as polynomials. Substituting these expressions into Eq. (4.3), we obtain k identically vanishing polynomials in the k operators $\wedge^k \hat{A}^q$ (with $1 \leq q \leq k$). These polynomials can be considered as a system of polynomial equations in the variables $\hat{\alpha}_q \equiv \wedge^k \hat{A}^q$. (As an exercise, you may verify that all the operators $\hat{\alpha}_q$ commute.) A system of polynomial equations may be reduced to a single polynomial equation in one of the variables, say $\hat{\alpha}_1$. (The technique for doing this in practice, called the "Gröbner basis," is complicated and beyond the scope of this book.) ■

The following two examples illustrate Theorem 2 in three and four dimensions.

Example 1: Suppose V is a three-dimensional space ($N = 3$) and an operator $\hat{A}$ is given. The ordinary Cayley-Hamilton theorem is obtained from Theorem 2 with $k = 1$,

$$q_0 - q_1\hat{A} + q_2\hat{A}^2 - \hat{A}^3 = 0,$$

where $q_j \equiv \wedge^N \hat{A}^{N-j}$ are the coefficients of the characteristic polynomial of $\hat{A}$. The generalization of the Cayley-Hamilton theorem is obtained with $k = 2$ (the only remaining case $k = 3$ will not yield interesting results).

We write the identity (4.3) for $k = 2$ and $p = 1, 2, 3$. Using the properties $\wedge^k \hat{A}^{k+j} = 0$ (with $j > 0$) and $\wedge^k \hat{A}^0 = \hat{1}$, we get the following three identities of operators in $\wedge^2 V$:

$$\begin{aligned}\left(\wedge^1\hat{A}^1\right)^{\wedge T} + \wedge^2\hat{A}^1 &= q_2\hat{1}_{\wedge^2 V},\\ \left(\wedge^1\hat{A}^1\right)^{\wedge T}(\wedge^2\hat{A}^1) + \wedge^2\hat{A}^2 &= q_1\hat{1}_{\wedge^2 V},\\ \left(\wedge^1\hat{A}^1\right)^{\wedge T}(\wedge^2\hat{A}^2) &= q_0\hat{1}_{\wedge^2 V}.\end{aligned}$$

Let us denote for brevity $\hat{\alpha}_1 \equiv \wedge^2 \hat{A}^1$ and $\hat{\alpha}_2 \equiv \wedge^2 \hat{A}^2$. Expressing $\left(\wedge^1 \hat{A}^1\right)^{\wedge T}$ through $\hat{\alpha}_1$ from the first line above and substituting into the last two lines, we find

$$\begin{aligned}\hat{\alpha}_2 &= q_1\hat{1} - q_2\hat{\alpha}_1 + \hat{\alpha}_1^2,\\ (q_2\hat{1} - \hat{\alpha}_1)\hat{\alpha}_2 &= q_0\hat{1}.\end{aligned}$$

We can now express $\hat{\alpha}_2$ through $\hat{\alpha}_1$ and substitute into the last equation to find

$$\hat{\alpha}_1^3 - 2q_2\hat{\alpha}_1^2 + (q_1 + q_2^2)\hat{\alpha}_1 - (q_1 q_2 - q_0)\hat{1} = 0.$$

Thus, the generalization of the Cayley-Hamilton theorem in $\wedge^2 V$ yields an identically vanishing polynomial in $\wedge^2 \hat{A}^1 \equiv \hat{\alpha}_1$ with coefficients that are expressed through q_j.

Question: Is this the characteristic polynomial of $\hat{\alpha}_1$?

Answer: I do not know! It could be since it has the correct degree. However, not every polynomial $p(x)$ such that $p(\hat{\alpha}) = 0$ for some operator $\hat{\alpha}$ is the characteristic polynomial of $\hat{\alpha}$.

Example 2: Let us now consider the case $N = 4$ and $k = 2$. We use Eq. (4.3) with $p = 1, 2, 3, 4$ and obtain the following four equations,

$$\begin{aligned}
(\wedge^2 \hat{A}^1)^{\wedge T} + \wedge^2 \hat{A}^1 &= (\wedge^4 \hat{A}^1)\hat{1}_{\wedge^2 V},\\
(\wedge^2 \hat{A}^2)^{\wedge T} + (\wedge^2 \hat{A}^1)^{\wedge T}(\wedge^2 \hat{A}^1) + \wedge^2 \hat{A}^2 &= (\wedge^4 \hat{A}^2)\hat{1}_{\wedge^2 V},\\
(\wedge^2 \hat{A}^2)^{\wedge T}(\wedge^2 \hat{A}^1) + (\wedge^2 \hat{A}^1)^{\wedge T}(\wedge^2 \hat{A}^2) &= (\wedge^4 \hat{A}^3)\hat{1}_{\wedge^2 V},\\
(\wedge^2 \hat{A}^2)^{\wedge T}(\wedge^2 \hat{A}^2) &= (\wedge^4 \hat{A}^4)\hat{1}_{\wedge^2 V}.
\end{aligned}$$

Let us denote, as before, $q_j = \wedge^4 \hat{A}^{4-j}$ (with $0 \leq j \leq 3$) and $\hat{\alpha}_r \equiv \wedge^2 \hat{A}^r$ (with $r = 1, 2$). Using the first two equations above, we can then express $(\wedge^2 \hat{A}^r)^{\wedge T}$ through $\hat{\alpha}_r$ and substitute into the last two equations. We obtain

$$\begin{aligned}
(\wedge^2 \hat{A}^1)^{\wedge T} &= q_3 \hat{1} - \hat{\alpha}_1,\\
(\wedge^2 \hat{A}^2)^{\wedge T} &= q_2 \hat{1} + \hat{\alpha}_1^2 - q_3 \hat{\alpha}_1 - \hat{\alpha}_2,
\end{aligned}$$

and finally

$$\begin{aligned}
(q_2 \hat{1} + \hat{\alpha}_1^2 - q_3 \hat{\alpha}_1 - \hat{\alpha}_2)\hat{\alpha}_1 + (q_3 \hat{1} - \hat{\alpha}_1)\hat{\alpha}_2 &= q_1 \hat{1},\\
(q_2 \hat{1} + \hat{\alpha}_1^2 - q_3 \hat{\alpha}_1 - \hat{\alpha}_2)\hat{\alpha}_2 &= q_0 \hat{1}.
\end{aligned}$$

One cannot express $\hat{\alpha}_2$ directly through $\hat{\alpha}_1$ using these last equations. However, one can show (for instance, using a computer algebra program[2]) that there exists an identically vanishing polynomial of degree 6 in $\hat{\alpha}_1$, namely $p(\hat{\alpha}_1) = 0$ with

$$\begin{aligned}
p(x) \equiv\; & x^6 - 3q_3 x^5 + \left(2q_2 + 3q_3^2\right) x^4 - \left(4q_2 q_3 + q_3^3\right) x^3\\
& + \left(q_2^2 - 4q_0 + q_1 q_3 + 2q_2 q_3^2\right) x^2 - \left(q_1 q_3^2 + q_2^2 q_3 - 4q_0 q_3\right) x\\
& + q_1 q_2 q_3 - q_0 q_3^2 - q_1^2.
\end{aligned}$$

The coefficients of $p(x)$ are known functions of the coefficients q_j of the characteristic polynomial of $\hat{A}$. Note that the space $\wedge^2 V$ has dimension 6 in this example; the polynomial $p(x)$ has the same degree.

Question: In both examples we found an identically vanishing polynomial in $\wedge^k \hat{A}^1$. Is there a general formula for the coefficients of this polynomial?

Answer: I do not know!

[2]This can be surely done by hand, but I have not yet learned the Gröbner basis technique necessary to do this, so I cannot show the calculation here.

4.4 Functions of operators

We will now consider some calculations with operators.

Let $\hat{A} \in \operatorname{End} V$. Since linear operators can be multiplied, it is straightforward to evaluate $\hat{A}\hat{A} \equiv \hat{A}^2$ and other powers of $\hat{A}$, as well as arbitrary polynomials in $\hat{A}$. For example, the operator $\hat{A}$ can be substituted instead of x into the polynomial $p(x) = 2 + 3x + 4x^2$; the result is the operator $\hat{2} + 3\hat{A} + 4\hat{A}^2 \equiv p(\hat{A})$.

Exercise: For a linear operator $\hat{A}$ and an arbitrary polynomial $p(x)$, show that $p(\hat{A})$ has the same eigenvectors as $\hat{A}$ (although perhaps with different eigenvalues). ■

Another familiar function of $\hat{A}$ is the inverse operator, $\hat{A}^{-1}$. Clearly, we can evaluate a polynomial in $\hat{A}^{-1}$ as well (if $\hat{A}^{-1}$ exists). It is interesting to ask whether we can evaluate an arbitrary function of $\hat{A}$; for instance, whether we can raise $\hat{A}$ to a non-integer power, or compute $\exp(\hat{A})$, $\ln(\hat{A})$, $\cos(\hat{A})$. Generally, can we substitute $\hat{A}$ instead of x in an arbitrary function $f(x)$ and evaluate an operator-valued function $f(\hat{A})$? If so, how to do this in practice?

4.4.1 Definitions. Formal power series

The answer is that *sometimes* we can. There are two situations when $f(\hat{A})$ makes sense, i.e. can be defined and has reasonable properties.

The first situation is when $\hat{A}$ is **diagonalizable**, i.e. there exists a basis $\{\mathbf{e}_i\}$ such that every basis vector is an eigenvector of $\hat{A}$,

$$\hat{A}\mathbf{e}_i = \lambda_i \mathbf{e}_i.$$

In this case, we simply define $f(\hat{A})$ as the linear operator that acts on the basis vectors as follows,

$$f(\hat{A})\mathbf{e}_i \equiv f(\lambda_i)\mathbf{e}_i.$$

Definition 1: Given a function $f(x)$ and a diagonalizable linear operator

$$\hat{A} = \sum_{i=1}^{N} \lambda_i \mathbf{e}_i \otimes \mathbf{e}_i^*,$$

the function $f(\hat{A})$ is the linear operator defined by

$$f(\hat{A}) \equiv \sum_{i=1}^{N} f(\lambda_i)\, \mathbf{e}_i \otimes \mathbf{e}_i^*,$$

provided that $f(x)$ is well-defined at the points $x = \lambda_i$, $i = 1, ..., N$.

This definition might appear to be "cheating" since we simply substituted the eigenvalues into $f(x)$, rather than evaluate the operator $f(\hat{A})$ in some "natural" way. However, the result is reasonable since we, in effect, define $f(\hat{A})$ separately in each eigenspace $\operatorname{Span}\{\mathbf{e}_i\}$ where $\hat{A}$ acts as multiplication by λ_i. It is natural to define $f(\hat{A})$ in each eigenspace as multiplication by $f(\lambda_i)$.

The second situation is when $f(x)$ is an **analytic function**, that is, a function represented by a power series

$$f(x) = \sum_{n=0}^{\infty} c_n x^n,$$

such that the series converges to the value $f(x)$ for some x. Further, we need this series to converge for a sufficiently wide range of values of x such that all eigenvalues of $\hat{A}$ are within that range. Then one can show that the operator-valued series

$$f(\hat{A}) = \sum_{n=0}^{\infty} c_n (\hat{A})^n$$

converges. The technical details of this proof are beyond the scope of this book; one needs to define the limit of a sequence of operators and other notions studied in functional analysis. Here is a simple argument that gives a condition for convergence. Suppose that the operator $\hat{A}$ is diagonalizable and has eigenvalues λ_i and the corresponding eigenvectors $\mathbf{v}_i$ ($i = 1, ..., N$) such that $\{\mathbf{v}_i\}$ is a basis and $\hat{A}$ has a tensor representation

$$\hat{A} = \sum_{i=1}^{N} \lambda_i \mathbf{v}_i \otimes \mathbf{v}_i^*.$$

Note that

$$\hat{A}^n = \left[\sum_{i=1}^{N} \lambda_i \mathbf{v}_i \otimes \mathbf{v}_i^* \right]^n = \sum_{i=1}^{N} \lambda_i^n \mathbf{v}_i \otimes \mathbf{v}_i^*$$

due to the property of the dual basis, $\mathbf{v}_i^*(\mathbf{v}_j) = \delta_{ij}$. So if the series $\sum_{n=0}^{\infty} c_n x^n$ converges for every eigenvalue $x = \lambda_i$ of the operator $\hat{A}$ then the tensor-valued series also converges and yields a new tensor

$$\begin{aligned}\sum_{n=0}^{\infty} c_n (\hat{A})^n &= \sum_{n=0}^{\infty} c_n \sum_{i=1}^{N} \lambda_i^n \mathbf{v}_i \otimes \mathbf{v}_i^* \\ &= \sum_{i=1}^{N} \left[\sum_{n=0}^{\infty} c_n \lambda^n \right] \mathbf{v}_i \otimes \mathbf{v}_i^*.\end{aligned}$$

This argument indicates at least one case where the operator-valued power series surely converges.

Instead of performing an in-depth study of operator-valued power series, I will restrict myself to considering "formal power series" containing a parameter t, that is, infinite power series in t considered without regard for convergence. Let us discuss this idea in more detail.

By definition, a **formal power series** (FPS) is an infinite sequence of numbers $(c_0, c_1, c_2, ...)$. This sequence, however, is written as if it were a power series in a parameter t,

$$c_0 + c_1 t + c_2 t^2 + ... = \sum_{n=0}^{\infty} c_n t^n .$$

It appears that we need to calculate the sum of the above series. However, while we manipulate an FPS, we *do not* assign any value to t and thus do not have to consider the issue of convergence of the resulting infinite series. Hence, we work with an FPS as with an algebraic expression containing a variable t, an expression that we do not evaluate (although we may simplify it). These expressions can be manipulated term by term, so that, for example, the sum and the product of two FPS are always defined; the result is another FPS. Thus, the notation for FPS should be understood as a convenient shorthand that simplifies working with FPS, rather than an actual sum of an infinite series. At the same time, the notation for FPS makes it easy to evaluate the actual infinite series when the need arises. Therefore, any results obtained using FPS will hold whenever the series converges.

Now I will use the formal power series to define $f(t\hat{A})$.

Definition 2: Given an analytic function $f(x)$ shown above and a linear operator $\hat{A}$, the function $f(t\hat{A})$ denotes the operator-valued formal power series

$$f(t\hat{A}) \equiv \sum_{n=0}^{\infty} c_n (\hat{A})^n t^n .$$

(According to the definition of formal power series, the variable t is a parameter that does not have a value and serves only to label the terms of the series.)

One can define the derivative of a formal power series, *without* using the notion of a limit (and without discussing convergence).

Definition 3: The **derivative** ∂_t of a formal power series $\sum_k a_k t^k$ is another formal power series defined by

$$\partial_t \Big(\sum_{k=0}^{\infty} a_k t^k \Big) = \sum_{k=0}^{\infty} (k+1)\, a_{k+1} t^k .$$

This definition gives us the usual properties of the derivative. For instance, it is obvious that ∂_t is a linear operator in the space of formal power series. Further, we have the important distributive property:

Statement 1: The Leibniz rule,

$$\partial_t \left[f(t) g(t) \right] = \left[\partial_t f(t) \right] g(t) + f(t) \left[\partial_t g(t) \right] ,$$

holds for formal power series.

Proof: Since ∂_t is a linear operation, it is sufficient to check that the Leibniz rule holds for single terms, $f(t) = t^a$ and $g(t) = t^b$. Details left as exercise. ■

This definition of $f(t\hat{A})$ has reasonable and expected properties, such as:

Exercise: For an analytic function $f(x)$, show that

$$f(\hat{A})\hat{A} = \hat{A}f(\hat{A})$$

and that

$$\frac{d}{dt}f(t\hat{A}) = \hat{A}f'(\hat{A})$$

for an analytic function $f(x)$. Here both sides are interpreted as formal power series. Deduce that $f(\hat{A})g(\hat{A}) = g(\hat{A})f(\hat{A})$ for any two analytic functions $f(x)$ and $g(x)$.

Hint: Linear operations with formal power series must be performed term by term (by definition). So it is sufficient to consider a single term in $f(x)$, such as $f(x) = x^a$. ■

Now we can show that the two definitions of the operator-valued function $f(\hat{A})$ agree when both are applicable.

Statement 2: If $f(x)$ is an analytic function and $\hat{A}$ is a diagonalizable operator then the two definitions agree, i.e. for $f(x) = \sum_{n=0}^{\infty} c_n x^n$ and $\hat{A} = \sum_{i=1}^{N} \lambda_i \mathbf{e}_i \otimes \mathbf{e}_i^*$ we have the equality of formal power series,

$$\sum_{n=0}^{\infty} c_n (t\hat{A})^n = \sum_{i=1}^{N} f(t\lambda_i)\, \mathbf{e}_i \otimes \mathbf{e}_i^*. \tag{4.4}$$

Proof: It is sufficient to prove that the terms multiplying t^n coincide for each n. We note that the square of $\hat{A}$ is

$$\begin{aligned}\left(\sum_{i=1}^{N} \lambda_i \mathbf{e}_i \otimes \mathbf{e}_i^*\right)^2 &= \left(\sum_{i=1}^{N} \lambda_i \mathbf{e}_i \otimes \mathbf{e}_i^*\right)\left(\sum_{j=1}^{N} \lambda_j \mathbf{e}_j \otimes \mathbf{e}_j^*\right) \\ &= \sum_{i=1}^{N} \lambda_i^2 \mathbf{e}_i \otimes \mathbf{e}_i^*\end{aligned}$$

because $\mathbf{e}_i^*(\mathbf{e}_j) = \delta_{ij}$. In this way we can compute any power of $\hat{A}$. Therefore, the term in the left side of Eq. (4.4) is

$$c_n t^n (\hat{A})^n = c_n t^n \left(\sum_{i=1}^{N} \lambda_i \mathbf{e}_i \otimes \mathbf{e}_i^*\right)^n = c_n t^n \sum_{i=1}^{N} \lambda_i^n \mathbf{e}_i \otimes \mathbf{e}_i^*,$$

which coincides with the term at t^n in the right side. ■

4.4.2 Computations: Sylvester's method

Now that we know when an operator-valued function $f(\hat{A})$ is defined, how can we actually compute the operator $f(\hat{A})$? The first definition requires us to diagonalize $\hat{A}$ (this is already a lot of work since we need to determine every eigenvector). Moreover, Definition 1 does not apply when $\hat{A}$ is non-diagonalizable. On the other hand, Definition 2 requires us to evaluate infinitely many terms of a power series. Is there a simpler way?

There is a situation when $f(\hat{A})$ can be computed without such effort. Let us first consider a simple example where the operator $\hat{A}$ happens to be a projector, $(\hat{A})^2 = \hat{A}$. In this case, any power of $\hat{A}$ is again equal to $\hat{A}$. It is then easy to compute a power series in $\hat{A}$:

$$\sum_{n=0}^{\infty} c_n (\hat{A})^n = c_0 \hat{1} + \big(\sum_{n=1}^{\infty} c_n\big) \hat{A}.$$

In this way we can compute any analytic function of $\hat{A}$ (as long as the series $\sum_{n=1}^{\infty} c_n$ converges). For example,

$$\begin{aligned}
\cos \hat{A} &= \hat{1} - \frac{1}{2!}(\hat{A})^2 + \frac{1}{4!}(\hat{A})^4 - ... = \hat{1} - \frac{1}{2!}\hat{A} + \frac{1}{4!}\hat{A} - ... \\
&= (1 - \frac{1}{2!} + \frac{1}{4!} - ...)\hat{A} + \hat{1} - \hat{A} \\
&= [(\cos 1) - 1]\,\hat{A} + \hat{1}.
\end{aligned}$$

Remark: In the above computation, we obtained a formula that *expresses the end result through* $\hat{A}$. We have that formula even though we do not know an explicit form of the operator $\hat{A}$ — not even the dimension of the space where $\hat{A}$ acts or whether $\hat{A}$ is diagonalizable. We do not need to know any eigenvectors of $\hat{A}$. We only use the given fact that $\hat{A}^2 = \hat{A}$, and we are still able to find a useful result. If such an operator $\hat{A}$ is given explicitly, we can substitute it into the formula

$$\cos \hat{A} = [(\cos 1) - 1]\,\hat{A} + \hat{1}$$

to obtain an explicit expression for $\cos \hat{A}$. Note also that the result is a formula *linear* in $\hat{A}$.

Exercise 1: a) Given that $(\hat{P})^2 = \hat{P}$, express $(\lambda\hat{1} - \hat{P})^{-1}$ and $\exp \hat{P}$ through $\hat{P}$. Assume that $|\lambda| > 1$ so that the Taylor series for $f(x) = (\lambda - x)^{-1}$ converges for $x = 1$.

b) It is known only that $(\hat{A})^2 = \hat{A} + 2$. Determine the possible eigenvalues of $\hat{A}$. Show that any analytic function of $\hat{A}$ can be reduced to the form $\alpha\hat{1} + \beta\hat{A}$ with some suitable coefficients α and β. Express $(\hat{A})^3$, $(\hat{A})^4$, and $\hat{A}^{-1}$ as linear functions of $\hat{A}$.

Hint: Write $\hat{A}^{-1} = \alpha\hat{1} + \beta\hat{A}$ with unknown α, β. Write $\hat{A}\hat{A}^{-1} = \hat{1}$ and simplify to determine α and β.

Exercise 2: The operator $\hat{A}$ is such that $\hat{A}^3 + \hat{A} = 0$. Compute $\exp(\lambda\hat{A})$ as a quadratic polynomial of $\hat{A}$ (here λ is a fixed number). ■

Let us now consider a more general situation. Suppose we know the characteristic polynomial $Q_{\hat{A}}(\lambda)$ of $\hat{A}$. The characteristic polynomial has the form

$$Q_{\hat{A}}(\lambda) = (-\lambda)^N + \sum_{k=0}^{N-1} (-1)^k q_{N-k} \lambda^k,$$

where q_i ($i = 1, ..., N$) are known coefficients. The Cayley-Hamilton theorem indicates that $\hat{A}$ satisfies the polynomial identity,

$$(\hat{A})^N = -\sum_{k=0}^{N-1} q_{N-k} (-1)^{N-k} (\hat{A})^k.$$

It follows that any power of $\hat{A}$ larger than $N-1$ can be expressed as a linear combination of smaller powers of $\hat{A}$. Therefore, a power series in $\hat{A}$ can be reduced to a polynomial $p(\hat{A})$ of degree not larger than $N-1$. The task of computing an arbitrary function $f(\hat{A})$ is then reduced to the task of determining the N coefficients of $p(x) \equiv p_0 + ... + p_{N-1}x^{n-1}$. Once the coefficients of that polynomial are found, the function can be evaluated as $f(\hat{A}) = p(\hat{A})$ for any operator $\hat{A}$ that has the given characteristic polynomial.

Determining the coefficients of the polynomial $p(\hat{A})$ might appear to be difficult because one can get rather complicated formulas when one converts an arbitrary power of $\hat{A}$ to smaller powers. This work can be avoided if the eigenvalues of $\hat{A}$ are known, by using the **method of Sylvester**, which I will now explain.

The present task is to calculate $f(\hat{A})$ — equivalently, the polynomial $p(\hat{A})$ — when the characteristic polynomial $Q_{\hat{A}}(\lambda)$ is known. The characteristic polynomial has order N and hence has N (complex) roots, counting each root with its multiplicity. The eigenvalues λ_i of the operator $\hat{A}$ are roots of its characteristic polynomial, and there exists *at least one* eigenvector $\mathbf{v}_i$ for each λ_i (Theorem 1 in Sec. 3.9). Knowing the characteristic polynomial $Q_{\hat{A}}(\lambda)$, we may determine its roots λ_i.

Let us first assume that the roots λ_i ($i = 1, ..., N$) are *all different*. Then we have N different eigenvectors $\mathbf{v}_i$. The set $\{\mathbf{v}_i \,|\, i = 1, ..., N\}$ is linearly independent (Statement 1 in Sec. 3.6.1) and hence is a basis in V; that is, $\hat{A}$ is diagonalizable. We will not actually need to determine the eigenvectors $\mathbf{v}_i$; it will be sufficient that they exist. Let us now apply the function $f(\hat{A})$ to each of these N eigenvectors: we must have

$$f(\hat{A})\mathbf{v}_i = f(\lambda_i)\mathbf{v}_i.$$

On the other hand, we may express

$$f(\hat{A})\mathbf{v}_i = p(\hat{A})\mathbf{v}_i = p(\lambda_i)\mathbf{v}_i.$$

Since the set $\{\mathbf{v}_i\}$ is linearly independent, the vanishing linear combination

$$\sum_{i=1}^{N} [f(\lambda_i) - p(\lambda_i)]\, \mathbf{v}_i = 0$$

must have all vanishing coefficients; hence we obtain a system of N equations for N unknowns $\{p_0, ..., p_{N-1}\}$:

$$p_0 + p_1\lambda_i + ... + p_{N-1}\lambda_i^{N-1} = f(\lambda_i), \quad i = 1, ..., N.$$

Note that this system of equations has the Vandermonde matrix (Sec. 3.6). Since by assumption all λ_i's are different, the determinant of this matrix is nonzero, therefore the solution $\{p_0, ..., p_{N-1}\}$ exists and is unique. The polynomial $p(x)$ is the interpolating polynomial for $f(x)$ at the points $x = \lambda_i$ $(i = 1, ..., N)$.

We have proved the following theorem:

Theorem 1: If the roots $\{\lambda_1, ..., \lambda_N\}$ of the characteristic polynomial of $\hat{A}$ are all different, a function of $\hat{A}$ can be computed as $f(\hat{A}) = p(\hat{A})$, where $p(x)$ is the interpolating polynomial for $f(x)$ at the N points $\{\lambda_1, ..., \lambda_N\}$.

Exercise 3: It is given that the operator $\hat{A}$ has the characteristic polynomial $Q_{\hat{A}}(\lambda) = \lambda^2 - \lambda + 6$. Determine the eigenvalues of $\hat{A}$ and calculate $\exp(\hat{A})$ as a linear expression in $\hat{A}$.

If we know that an operator $\hat{A}$ satisfies a certain operator equation, say $(\hat{A})^2 - \hat{A} + 6 = 0$, then it is not necessary to know the characteristic polynomial in order to compute functions $f(\hat{A})$. It can be that the characteristic polynomial has a high order due to many repeated eigenvalues; however, as far as analytic functions are concerned, all that matters is the possibility to reduce high powers of $\hat{A}$ to low powers. This possibility can be provided by a polynomial of a lower degree than the characteristic polynomial.

In the following theorem, we will determine $f(\hat{A})$ knowing only *some* polynomial $Q(x)$ for which $p(\hat{A}) = 0$.

Theorem 2: Suppose that a linear operator $\hat{A}$ and a polynomial $Q(x)$ are such that $Q(\hat{A}) = 0$, and assume that the equation $Q(\lambda) = 0$ has all distinct roots λ_i $(i = 1, ..., n)$, where n is not necessarily equal to the dimension N of the vector space. Then an analytic function $f(\hat{A})$ can be computed as

$$f(\hat{A}) = p(\hat{A}),$$

where $p(x)$ is the interpolating polynomial for the function $f(x)$ at the points $x = \lambda_i$ $(i = 1, ..., n)$.

Proof: The polynomial $p(x)$ is defined uniquely by substituting x^k with $k \geq n$ through lower powers of x in the series for $f(x)$, using the equation $p(x) = 0$. Consider the operator $\hat{A}_1$ that acts as multiplication by λ_1. This operator satisfies $p(\hat{A}_1) = 0$, and so $f(\hat{A}_1)$ is simplified to the same polynomial $p(\hat{A}_1)$. Hence we must have $f(\hat{A}_1) = p(\hat{A}_1)$. However, $f(\hat{A}_1)$ is simply the operator of multiplication by $f(\lambda_1)$. Hence, $p(x)$ must be equal to $f(x)$ when evaluated at $x = \lambda_1$. Similarly, we find that $p(\lambda_i) = f(\lambda_i)$ for $i = 1, ..., n$. The interpolating polynomial for $f(x)$ at the points $x = \lambda_i$ $(i = 1, ..., n)$ is unique and has degree $n - 1$. Therefore, this polynomial must be equal to $p(x)$. ■

It remains to develop a procedure for the case when *not all* roots λ_i of the polynomial $Q(\lambda)$ are different. To be specific, let us assume that $\lambda_1 = \lambda_2$ and that all other eigenvalues are different. In this case we will first solve an auxiliary problem where $\lambda_2 = \lambda_1 + \varepsilon$ and then take the limit $\varepsilon \to 0$. The equations determining the coefficients of the polynomial $p(x)$ are

$$p(\lambda_1) = f(\lambda_1), \quad p(\lambda_1 + \varepsilon) = f(\lambda_1 + \varepsilon), \; p(\lambda_3) = f(\lambda_3), \; ...$$

Subtracting the first equation from the second and dividing by ε, we find

$$\frac{p(\lambda_1+\varepsilon)-p(\lambda_1)}{\varepsilon}=\frac{f(\lambda_1+\varepsilon)-f(\lambda_1)}{\varepsilon}.$$

In the limit $\varepsilon \to 0$ this becomes

$$p'(\lambda_1)=f'(\lambda_1).$$

Therefore, the polynomial $p(x)$ is determined by the requirements that

$$p(\lambda_1)=f(\lambda_1),\ p'(\lambda_1)=f'(\lambda_1),\ p(\lambda_3)=f(\lambda_3),\ ...$$

If *three* roots coincide, say $\lambda_1=\lambda_2=\lambda_3$, we introduce two auxiliary parameters ε_2 and ε_3 and first obtain the three equations

$$p(\lambda_1)=f(\lambda_1),\ p(\lambda_1+\varepsilon_2)=f(\lambda_1+\varepsilon_2),$$
$$p(\lambda_1+\varepsilon_2+\varepsilon_3)=f(\lambda_1+\varepsilon_2+\varepsilon_3).$$

Subtracting the equations and taking the limit $\varepsilon_2 \to 0$ as before, we find

$$p(\lambda_1)=f(\lambda_1),\ p'(\lambda_1)=f'(\lambda_1),\ p'(\lambda_1+\varepsilon_3)=f'(\lambda_1+\varepsilon_3).$$

Subtracting now the second equation from the third and taking the limit $\varepsilon_3 \to 0$, we find $p''(\lambda_1)=f''(\lambda_1)$. Thus we have proved the following.

Theorem 3: If a linear operator $\hat{A}$ satisfies a polynomial operator equation $Q(\hat{A})=0$, such that the equation $Q(\lambda)=0$ has roots λ_i ($i=1,...,n$) with multiplicities m_i,

$$Q(\lambda)=\text{const}\cdot(\lambda-\lambda_1)^{m_1}...(\lambda-\lambda_n)^{m_n},$$

an analytic function $f(\hat{A})$ can be computed as

$$f(\hat{A})=p(\hat{A}),$$

where $p(x)$ is the polynomial determined by the conditions

$$p(\lambda_i)=f(\lambda_i),\ p'(\lambda_i)=f'(\lambda_i),\ ...,$$
$$\left.\frac{d^{m_i-1}p(x)}{dx^{m_i-1}}\right|_{x=\lambda_i}=\left.\frac{d^{m_i-1}f(x)}{dx^{m_i-1}}\right|_{x=\lambda_i},\quad i=1,...,n.$$

Theorems 1 to 3, which comprise Sylvester's method, allow us to compute functions of an operator when only the eigenvalues are known, without determining any eigenvectors and without assuming that the operator is diagonalizable.

4.4.3 * Square roots of operators

In the previous section we have seen that functions of operators can be sometimes computed explicitly. However, our methods work either for diagonalizable operators $\hat{A}$ or for functions $f(x)$ given by a power series that converges for every eigenvalue of the operator $\hat{A}$. If these conditions are not met, functions of operators may not exist or may not be uniquely defined. As an example where these problems arise, we will briefly consider the task of computing the square root of a given operator.

Given an operator $\hat{A}$ we would like to define its square root as an operator $\hat{B}$ such that $\hat{B}^2 = \hat{A}$. For a diagonalizable operator $\hat{A} = \sum_{i=1}^{N} \lambda_i \mathbf{e}_i \otimes \mathbf{e}_i^*$ (where $\{\mathbf{e}_i\}$ is an eigenbasis and $\{\mathbf{e}_i^*\}$ is the dual basis) we can easily find a suitable $\hat{B}$ by writing

$$\hat{B} \equiv \sum_{i=1}^{N} \sqrt{\lambda_i} \mathbf{e}_i \otimes \mathbf{e}_i^*.$$

Note that the numeric square root $\sqrt{\lambda_i}$ has an ambiguous sign; so with each possible choice of sign for each $\sqrt{\lambda_i}$, we obtain a possible choice of $\hat{B}$. (Depending on the problem at hand, there might be a natural way of fixing the signs; for instance, if all λ_i are positive then it might be useful to choose also all $\sqrt{\lambda_i}$ as positive.) The ambiguity of signs is expected; what is unexpected is that there could be many other operators $\hat{B}$ satisfying $\hat{B}^2 = \hat{A}$, as the following example shows.

Example 1: Let us compute the square root of the identity operator in a two-dimensional space. We look for $\hat{B}$ such that $\hat{B}^2 = \hat{1}$. Straightforward solutions are $\hat{B} = \pm\hat{1}$. However, consider the following operator,

$$\hat{B} \equiv \begin{pmatrix} a & b \\ c & -a \end{pmatrix}, \quad \hat{B}^2 = \begin{pmatrix} a^2 + bc & 0 \\ 0 & a^2 + bc \end{pmatrix} = \left(a^2 + bc\right)\hat{1}.$$

This $\hat{B}$ satisfies $\hat{B}^2 = \hat{1}$ for any $a, b, c \in \mathbb{C}$ as long as $a^2 + bc = 1$. The square root is quite ambiguous for the identity operator! ■

We will now perform a simple analysis of square roots of operators in two- and three-dimensional spaces using the Cayley-Hamilton theorem.

Let us assume that $\hat{B}^2 = \hat{A}$, where $\hat{A}$ is a given operator, and denote for brevity $a \equiv \mathrm{Tr}\hat{A}$ and $b \equiv \mathrm{Tr}\hat{B}$ (where a is given but b is still unknown). In two dimensions, any operator $\hat{B}$ satisfies the characteristic equation

$$\hat{B}^2 - (\mathrm{Tr}\hat{B})\hat{B} + (\det \hat{B})\hat{1} = 0.$$

Taking the trace of this equation, we can express the determinant as

$$\det \hat{B} = \frac{1}{2}(\mathrm{Tr}\hat{B})^2 - \frac{1}{2}\mathrm{Tr}(\hat{B}^2)$$

and hence

$$b\hat{B} = \hat{A} + \frac{b^2 - a}{2}\hat{1}. \qquad (4.5)$$

This equation will yield an explicit formula for $\hat{B}$ through $\hat{A}$ if we only determine the value of the constant b such that $b \neq 0$. Squaring the above equation and taking the trace, we find

$$b^4 - 2b^2 a + c = 0, \quad c \equiv 2\mathrm{Tr}(\hat{A}^2) - a^2 = a^2 - 4\det\hat{A}.$$

Hence, we obtain up to *four* possible solutions for b,

$$b = \pm\sqrt{a \pm \sqrt{a^2 - c}} = \pm\sqrt{\mathrm{Tr}\hat{A} \pm 2\sqrt{\det\hat{A}}}. \tag{4.6}$$

Each value of b such that $b \neq 0$ yield possible operators $\hat{B}$ through Eq. (4.5). Denoting by $s_1 = \pm 1$ and $s_2 = \pm 1$ the two free choices of signs in Eq. (4.6), we may write the general solution (assuming $b \neq 0$) as

$$\hat{B} = s_1 \frac{\hat{A} + s_2\sqrt{\det\hat{A}}\,\hat{1}}{\sqrt{\mathrm{Tr}\hat{A} + 2s_2\sqrt{\det\hat{A}}}}. \tag{4.7}$$

It is straightforward to verify (using the Cayley-Hamilton theorem for $\hat{A}$) that every such $\hat{B}$ indeed satisfies $\hat{B}^2 = \hat{A}$.

Note also that $\hat{B}$ is expressed as a *linear* polynomial in $\hat{A}$. Due to the Cayley-Hamilton theorem, any analytic function of $\hat{A}$ reduces to a linear polynomial in the two-dimensional case. Hence, we can view Eq. (4.7) as a formula yielding the *analytic* solutions of the equation $\hat{B}^2 = \hat{A}$.

If $b = 0$ is a solution of Eq. (4.6) then we must consider the possibility that solutions $\hat{B}$ with $b \equiv \mathrm{Tr}\,\hat{B} = 0$ may exist. In that case, Eq. (4.5) indicates that $\hat{A}$ plus a multiple of $\hat{1}$ must be equal to the zero operator. Note that Eq. (4.5) is a *necessary* consequence of $\hat{B}^2 = \hat{A}$, obtained only by assuming that $\hat{B}$ exists. Hence, when $\hat{A}$ is *not* proportional to the identity operator, no solutions $\hat{B}$ with $\mathrm{Tr}\,\hat{B} = 0$ can exist. On the other hand, if $\hat{A}$ *is* proportional to $\hat{1}$, solutions with $\mathrm{Tr}\,\hat{B} = 0$ exist but the present method does not yield these solutions. (Note that this method can only yield solutions $\hat{B}$ that are linear combinations of the operator $\hat{A}$ and the identity operator!) It is easy to see that the operators from Example 1 fall into this category, with $\mathrm{Tr}\hat{B} = 0$. There are no other solutions except those shown in Example 1 because in that example we have obtained all possible traceless solutions.

Another interesting example is found when $\hat{A}$ is a nilpotent (but nonzero).

Example 2: Consider a nilpotent operator $\hat{A}_1 = \begin{pmatrix} 0 & 1 \\ 0 & 0 \end{pmatrix}$. In that case, both the trace and the determinant of $\hat{A}_1$ are equal to zero; it follows that $b = 0$ is the only solution of Eq. (4.6). However, $\hat{A}_1$ is not proportional to the identity operator. Hence, a square root of $\hat{A}_1$ *does not exist*.

Remark: This problem with the nonexistence of the square root is not the same as the nonexistence of $\sqrt{-1}$ within real numbers; the square root of $\hat{A}_1$

does not exist even if we allow complex numbers! The reason is that the existence of $\sqrt{\hat{A}_1}$ would be *algebraically inconsistent* (because it would contradict the Cayley-Hamilton theorem). ■

Let us summarize our results so far. In two dimensions, the general calculation of a square root of a given operator $\hat{A}$ proceeds as follows: If $\hat{A}$ is proportional to the identity operator, we have various solutions of the form shown in Example 1. (Not every one of these solutions may be relevant for the problem at hand, but they exist.) If $\hat{A}$ is not proportional to the identity operator, we solve Eq. (4.6) and obtain up to four possible values of b. If the only solution is $b = 0$, the square root of $\hat{A}$ does not exist. Otherwise, every *nonzero* value of b yields a solution $\hat{B}$ according to Eq. (4.5), and there are no other solutions.

Example 3: We would like to determine a square root of the operator

$$\hat{A} = \begin{pmatrix} 1 & 3 \\ 0 & 4 \end{pmatrix}.$$

We compute $\det \hat{A} = 4$ and $a = \mathrm{Tr}\hat{A} = 5$. Hence Eq. (4.6) gives four nonzero values,

$$b = \pm\sqrt{5 \pm 4} = \{\pm 1, \pm 3\}.$$

Substituting these values of b into Eq. (4.5) and solving for $\hat{B}$, we compute the four possible square roots

$$\hat{B} = \pm \begin{pmatrix} 1 & 1 \\ 0 & 2 \end{pmatrix}, \quad \hat{B} = \pm \begin{pmatrix} -1 & 3 \\ 0 & 2 \end{pmatrix}.$$

Since $b = 0$ is not a solution, while $\hat{A} \neq \lambda \hat{1}$, there are no other square roots.

Exercise 1: Consider a diagonalizable operator represented in a certain basis by the matrix

$$\hat{A} = \begin{pmatrix} \lambda^2 & 0 \\ 0 & \mu^2 \end{pmatrix},$$

where λ and μ are any complex numbers, possibly zero, such that $\lambda^2 \neq \mu^2$. Use Eqs. (4.5)–(4.6) to show that the possible square roots are

$$\hat{B} = \begin{pmatrix} \pm\lambda & 0 \\ 0 & \pm\mu \end{pmatrix}.$$

and that there are no other square roots. ■

Exercise 2: Obtain all possible square roots of the zero operator in two dimensions. ■

Let us now consider a given operator $\hat{A}$ in a *three*-dimensional space and assume that there exists $\hat{B}$ such that $\hat{B}^2 = \hat{A}$. We will be looking for a formula expressing $\hat{B}$ as a polynomial in $\hat{A}$. As we have seen, this will certainly not give *every* possible solution $\hat{B}$, but we do expect to get the interesting solutions that can be expressed as *analytic* functions of $\hat{A}$.

As before, we denote $a \equiv \mathrm{Tr}\hat{A}$ and $b \equiv \mathrm{Tr}\hat{B}$. The Cayley-Hamilton theorem for $\hat{B}$ together with Exercise 1 in Sec. 3.9 (page 141) yields a simplified equation,

$$\begin{aligned} 0 &= \hat{B}^3 - b\hat{B}^2 + s\hat{B} - (\det \hat{B})\hat{1} \\ &= (\hat{A} + s\hat{1})\hat{B} - b\hat{A} - (\det \hat{B})\hat{1}, \\ s &\equiv \frac{b^2 - a}{2}. \end{aligned} \tag{4.8}$$

Note that $\det \hat{B} = \pm\sqrt{\det \hat{A}}$ and hence can be considered known. Moving $\hat{B}$ to another side in Eq. (4.8) and squaring the resulting equation, we find

$$(\hat{A}^2 + 2s\hat{A} + s^2\hat{1})\hat{A} = (b\hat{A} + (\det \hat{B})\hat{1})^2.$$

Expanding the brackets and using the Cayley-Hamilton theorem for $\hat{A}$ in the form

$$\hat{A}^3 - a\hat{A}^2 + p\hat{A} - (\det \hat{A})\hat{1} = 0,$$

where the coefficient p can be expressed as

$$p = \frac{1}{2}(a^2 - \mathrm{Tr}(\hat{A}^2)),$$

we obtain after simplifications

$$(s^2 - p - 2b \det \hat{B})\hat{A} = 0.$$

This yields a fourth-order polynomial equation for b,

$$\left(\frac{b^2 - a}{2}\right)^2 - p - 2b \det \hat{B} = 0.$$

This equation can be solved, in principle. Since $\det \hat{B}$ has up to *two* possible values, $\det \hat{B} = \pm\sqrt{\det \hat{A}}$, we can then determine *up to eight* possible values of b (and the corresponding values of s).

Now we use a trick to express $\hat{B}$ as a function of $\hat{A}$. We rewrite Eq. (4.8) as

$$\hat{A}\hat{B} = -s\hat{B} + b\hat{A} + (\det \hat{B})\hat{1}$$

and multiply both sides by $\hat{B}$, substituting $\hat{A}\hat{B}$ back into the equation,

$$\begin{aligned} \hat{A}^2 + s\hat{A} &= b\hat{A}\hat{B} + (\det \hat{B})\hat{B} \\ &= b[-s\hat{B} + b\hat{A} + (\det \hat{B})\hat{1}] + (\det \hat{B})\hat{B}. \end{aligned}$$

The last line yields

$$\hat{B} = \frac{1}{(\det \hat{B}) - sb}[\hat{A}^2 + (s - b^2)\hat{A} - b(\det \hat{B})\hat{1}].$$

This is the final result, provided that the denominator $(\det \hat{B} - sb)$ does not vanish. In case this denominator vanishes, the present method cannot yield a formula for $\hat{B}$ in terms of $\hat{A}$.

Exercise 3:* Verify that the square root of a diagonalizable operator,

$$\hat{A} = \begin{pmatrix} p^2 & 0 & 0 \\ 0 & q^2 & 0 \\ 0 & 0 & r^2 \end{pmatrix},$$

where $p^2, q^2, r^2 \in \mathbb{C}$ are all different, can be determined using this approach, which yields the eight possibilities

$$\hat{B} = \begin{pmatrix} \pm p & 0 & 0 \\ 0 & \pm q & 0 \\ 0 & 0 & \pm r \end{pmatrix}.$$

Hint: Rather than trying to solve the fourth-order equation for b directly (a cumbersome task), one can just verify, by substituting into the equation, that the eight values $b = \pm p \pm q \pm r$ (with all the possible choices of signs) are roots of that equation.

Exercise 4:*[3] It is given that a three-dimensional operator $\hat{A}$ satisfies

$$\mathrm{Tr}\,(\hat{A}^2) = \frac{1}{2}(\mathrm{Tr}\,\hat{A})^2, \quad \det \hat{A} \neq 0.$$

Show that there exists $\hat{B}$, unique up to a sign, such that $\mathrm{Tr}\,\hat{B} = 0$ and $\hat{B}^2 = \hat{A}$.

Answer:

$$\hat{B} = \pm \frac{1}{\sqrt{\det \hat{A}}} \big[\hat{A}^2 - \frac{1}{2}(\mathrm{Tr}\,\hat{A})\hat{A}\big].$$

4.5 Formulas of Jacobi and Liouville

Definition: The **Liouville formula** is the identity

$$\det(\exp \hat{A}) = \exp(\mathrm{Tr}\hat{A}), \tag{4.9}$$

where $\hat{A}$ is a linear operator and $\exp \hat{A}$ is defined by the power series,

$$\exp \hat{A} \equiv \sum_{n=0}^{\infty} \frac{1}{n!}(\hat{A})^n.$$

Example: Consider a **diagonalizable** operator $\hat{A}$ (an operator such that there exists an eigenbasis $\{\mathbf{e}_i \,|\, i = 1, ..., N\}$) and denote by λ_i the eigenvalues, so that $\hat{A}\mathbf{e}_i = \lambda_i \mathbf{e}_i$. (The eigenvalues λ_i are not necessarily all different.) Then we have $(\hat{A})^n \mathbf{e}_i = \lambda_i^n \mathbf{e}_i$ and therefore

$$(\exp \hat{A})\mathbf{e}_i = \sum_{n=0}^{\infty} \frac{1}{n!}(\hat{A})^n \mathbf{e}_i = \sum_{n=0}^{\infty} \frac{1}{n!}\lambda_i^n \mathbf{e}_i = e^{\lambda_i}\mathbf{e}_i.$$

[3] This is motivated by the article by R. Capovilla, J. Dell, and T. Jacobson, *Classical and Quantum Gravity* 8 (1991), pp. 59–73; see p. 63 in that article.

The trace of $\hat{A}$ is $\mathrm{Tr}\hat{A} = \sum_{i=1}^{N} \lambda_i$ and the determinant is $\det \hat{A} = \prod_{i=1}^{N} \lambda_i$. Hence we can easily verify the Liouville formula,

$$\det(\exp \hat{A}) = e^{\lambda_1}...e^{\lambda_N} = \exp(\lambda_1 + ... + \lambda_n) = \exp(\mathrm{Tr}\hat{A}).$$

However, the Liouville formula is valid also for non-diagonalizable operators. ■

The formula (4.9) is useful in several areas of mathematics and physics. A proof of Eq. (4.9) for matrices can be given through the use of the Jordan canonical form of the matrix, which is a powerful but complicated construction that actually is not needed to derive the Liouville formula. We will derive it using operator-valued differential equations for power series. A useful by-product is a formula for the derivative of the determinant.

Theorem 1 (Liouville's formula): For an operator $\hat{A}$ in a finite-dimensional space V,

$$\det \exp(t\hat{A}) = \exp(t\mathrm{Tr}\hat{A}). \tag{4.10}$$

Here both sides are understood as **formal power series** in the variable t, e.g.

$$\exp(t\hat{A}) \equiv \sum_{n=0}^{\infty} \frac{t^n}{n!}(\hat{A})^n,$$

i.e. an infinite series considered without regard for convergence (Sec. 4.4).

Remark: Although we establish Theorem 1 only in the sense of equality of formal power series, the result is useful because both sides of Eq. (4.10) will be equal whenever both series converge. Since the series for $\exp(x)$ converges for all x, one expects that Eq. (4.10) has a wide range of applicability. In particular, it holds for any operator in finite dimensions. ■

The idea of the proof will be to represent both sides of Eq. (4.10) as power series in t satisfying some differential equation. First we figure out how to solve differential equations for formal power series. Then we will guess a suitable differential equation that will enable us to prove the theorem.

Lemma 1: The operator-valued function $\hat{F}(t) \equiv \exp(t\hat{A})$ is the unique solution of the differential equation

$$\partial_t \hat{F}(t) = \hat{F}(t)\,\hat{A}, \quad \hat{F}\,(t=0) = \hat{1}_V,$$

where both sides of the equation are understood as formal power series.

Proof: The initial condition means that

$$\hat{F}(t) = \hat{1} + \hat{F}_1 t + \hat{F}_2 t^2 + ...,$$

where $\hat{F}_1$, $\hat{F}_2$, ..., are some operators. Then we equate terms with equal powers of t in the differential equation, which yields $\hat{F}_{j+1} = \frac{1}{j}\hat{F}_j\hat{A}$, $j = 1, 2, ...$, and so we obtain the desired exponential series. ■

Lemma 2: If $\phi(t)$ and $\psi(t)$ are power series in t with coefficients from $\wedge^m V$ and $\wedge^n V$ respectively, then the Leibniz rule holds,

$$\partial_t\,(\phi \wedge \psi) = (\partial_t \phi) \wedge \psi + \phi \wedge (\partial_t \psi)\,.$$

Proof: Since the derivative of formal power series, as defined above, is a linear operation, it is sufficient to verify the statement in the case when $\phi = t^a\omega_1$ and $\psi = t^b\omega_2$. Then we find

$$\partial_t \left(\phi \wedge \psi\right) = (a+b)\, t^{a+b-1}\omega_1 \wedge \omega_2,$$
$$(\partial_t\phi) \wedge \psi + \phi \wedge (\partial_t\psi) = at^{a-1}\omega_1 \wedge t^b\omega_2 + t^a\omega_1 \wedge bt^{b-1}\omega_2.$$

■

Lemma 3: The inverse to a formal power series $\phi(t)$ exists (as a formal power series) if and only if $\phi(0) \neq 0$.

Proof: The condition $\phi(0) \neq 0$ means that we can express $\phi(t) = \phi(0)+t\psi(t)$ where $\psi(t)$ is another power series. Then we can use the identity of formal power series,

$$1 = (1+x)\left[\sum_{n=0}^{\infty} (-1)^n x^n\right],$$

to express $1/\phi(t)$ as a formal power series,

$$\frac{1}{\phi(t)} = \frac{1}{\phi(0)+t\psi(t)} = \sum_{n=0}^{\infty} (-1)^n \left[\phi(0)\right]^{-n-1} \left[t\psi(t)\right]^n.$$

Since each term $[t\psi(t)]^n$ is expanded into a series that starts with t^n, we can compute each term of $1/\phi(t)$ by adding finitely many other terms, i.e. the above equation does specify a well-defined formal power series. ■

Corollary: If $\hat{A}(t)$ is an operator-valued formal power series, the inverse to $\hat{A}(t)$ exists (as a formal power series) if and only if $\det \hat{A}(0) \neq 0$.

The next step towards guessing the differential equation is to compute the derivative of a determinant.

Lemma 4 (Jacobi's formula): If $\hat{A}(t)$ is an operator-valued formal power series such that the inverse $\hat{A}^{-1}(t)$ exists, we have

$$\partial_t \det \hat{A}(t) = (\det \hat{A})\mathrm{Tr}\,[\hat{A}^{-1}\partial_t\hat{A}] = \mathrm{Tr}\,[(\det \hat{A})\hat{A}^{-1}\partial_t\hat{A}]. \qquad (4.11)$$

If the inverse does not exist, we need to replace $\det \hat{A} \cdot \hat{A}^{-1}$ in Eq. (4.11) by the algebraic complement,

$$\tilde{\hat{A}} \equiv \left(\wedge^{N-1}\hat{A}^{N-1}\right)^{\wedge T}$$

(see Sec. 4.2.1), so that we obtain the formula of Jacobi,

$$\partial_t \det \hat{A} = \mathrm{Tr}\,[\tilde{\hat{A}}\,\partial_t\hat{A}].$$

Proof of Lemma 4: A straightforward calculation using Lemma 2 gives

$$\left(\partial_t \det \hat{A}(t)\right)\mathbf{v}_1 \wedge ... \wedge \mathbf{v}_N = \partial_t[\hat{A}\mathbf{v}_1 \wedge ... \wedge \hat{A}\mathbf{v}_N]$$
$$= \sum_{k=1}^{N} \hat{A}\mathbf{v}_1 \wedge ... \wedge (\partial_t\hat{A})\mathbf{v}_k \wedge ... \wedge \hat{A}\mathbf{v}_N.$$

Now we use the definition of the algebraic complement operator to rewrite

$$\hat{A}\mathbf{v}_1 \wedge ... \wedge (\partial_t \hat{A})\mathbf{v}_k \wedge ... \wedge \hat{A}\mathbf{v}_N = \mathbf{v}_1 \wedge ... \wedge (\tilde{\hat{A}}\, \partial_t \hat{A}\mathbf{v}_k) \wedge ... \wedge \mathbf{v}_N .$$

Hence

$$\begin{aligned}(\partial_t \det \hat{A})\mathbf{v}_1 \wedge ... \wedge \mathbf{v}_N &= \sum_{k=1}^{N} \mathbf{v}_1 \wedge ... \wedge (\tilde{\hat{A}}\, \partial_t \hat{A}\mathbf{v}_k) \wedge ... \wedge \mathbf{v}_N \\ &= \wedge^N (\tilde{\hat{A}}\, \partial_t \hat{A})^1 \mathbf{v}_1 \wedge ... \wedge \mathbf{v}_N \\ &= \mathrm{Tr}\,[\tilde{\hat{A}}\, \partial_t \hat{A}]\mathbf{v}_1 \wedge ... \wedge \mathbf{v}_N .\end{aligned}$$

Therefore $\partial_t \det \hat{A} = \mathrm{Tr}\,[\tilde{\hat{A}}\, \partial_t \hat{A}]$. When $\hat{A}^{-1}$ exists, we may express $\tilde{\hat{A}}$ through the inverse matrix, $\tilde{\hat{A}} = (\det \hat{A})\hat{A}^{-1}$, and obtain Eq. (4.11).

Proof of Theorem 1: It follows from Lemma 3 that $\hat{F}^{-1}(t)$ exists since $\hat{F}(0) = \hat{1}$, and it follows from Lemma 4 that the operator-valued function $\hat{F}(t) = \exp(t\hat{A})$ satisfies the differential equation

$$\partial_t \det \hat{F}(t) = \det \hat{F}(t) \cdot \mathrm{Tr}[\hat{F}^{-1}\partial_t \hat{F}].$$

From Lemma 1, we have $\hat{F}^{-1}\partial_t \hat{F} = \hat{F}^{-1}\hat{F}\hat{A} = \hat{A}$, therefore

$$\partial_t \det \hat{F}(t) = \det \hat{F}(t) \cdot \mathrm{Tr}\hat{A}.$$

This is a differential equation for the number-valued formal power series $f(t) \equiv \det \hat{F}(t)$, with the initial condition $f(0) = 1$. The solution (which we may still regard as a formal power series) is

$$f(t) = \exp(t\mathrm{Tr}\hat{A}).$$

Therefore

$$\det \hat{F}(t) \equiv \det \exp(t\hat{A}) = \exp(t\mathrm{Tr}\hat{A}).$$

■

Exercise 1: (generalized Liouville's formula) If $\hat{A} \in \mathrm{End}\, V$ and $p \leq N \equiv \dim V$, show that

$$\wedge^p (\exp t\hat{A})^p = \exp\big(t(\wedge^p \hat{A}^1)\big),$$

where both sides are understood as formal power series of operators in $\wedge^p V$. (The Liouville formula is a special case with $p = N$.)

Exercise 2:* (Sylvester's theorem) For any two linear maps $\hat{A} : V \to W$ and $\hat{B} : W \to V$, we have well-defined composition maps $\hat{A}\hat{B} \in \mathrm{End}\, W$ and $\hat{B}\hat{A} \in \mathrm{End}\, V$. Then

$$\det(\hat{1}_V + \hat{B}\hat{A}) = \det(\hat{1}_W + \hat{A}\hat{B}).$$

Note that the operators at both sides act in different spaces.

Hint: Introduce a real parameter t and consider the functions $f(t) \equiv \det(1+t\hat{A}\hat{B})$, $g(t) \equiv \det(1+t\hat{B}\hat{A})$. These functions are polynomials of finite degree in t. Consider the differential equation for these functions; show that $f(t)$ satisfies

$$\frac{df}{dt} = f(t)\mathrm{Tr}\,[\hat{A}\hat{B}(1+t\hat{A}\hat{B})^{-1}],$$

and similarly for g. Expand in series in t and use the identities $\mathrm{Tr}\,(\hat{A}\hat{B}) = \mathrm{Tr}\,(\hat{B}\hat{A})$, $\mathrm{Tr}\,(\hat{A}\hat{B}\hat{A}\hat{B}) = \mathrm{Tr}\,(\hat{B}\hat{A}\hat{B}\hat{A})$, etc. Then show that f and g are solutions of the same differential equation, with the same conditions at $t = 0$. Therefore, show that these functions are identical as formal power series. Since f and g are actually polynomials in t, they must be equal.

4.5.1 Derivative of characteristic polynomial

Jacobi's formula expresses the derivative of the determinant, $\partial_t \det \hat{A}$, in terms of the derivative $\partial_t \hat{A}$ of the operator $\hat{A}$. The determinant is the last coefficient q_0 of the characteristic polynomial of $\hat{A}$. It is possible to obtain similar formulas for the derivatives of all other coefficients of the characteristic polynomial.
Statement: The derivative of the coefficient

$$q_k \equiv \wedge^N \hat{A}^{N-k}$$

of the characteristic polynomial of $\hat{A}$ is expressed (for $0 \leq k \leq N-1$) as

$$\partial_t q_k = \mathrm{Tr}\,\big[(\wedge^{N-1}\hat{A}^{N-k-1})^{\wedge T}\partial_t \hat{A}\big].$$

Note that the first operator in the brackets is the one we denoted by $\hat{A}_{(k+1)}$ in Sec. 4.2.3, so we can write

$$\partial_t q_k = \mathrm{Tr}\,[\hat{A}_{(k+1)}\partial_t \hat{A}].$$

Proof: We apply the operator $\partial_t(\wedge^N \hat{A}^{N-k})$ to the tensor $\omega \equiv \mathbf{v}_1 \wedge ... \wedge \mathbf{v}_N$, where $\{\mathbf{v}_j\}$ is a basis. We assume that the vectors $\mathbf{v}_j$ do not depend on t, so we can compute

$$\big[\partial_t(\wedge^N \hat{A}^{N-k})\big]\omega = \partial_t\big[\wedge^N \hat{A}^{N-k}\omega\big].$$

The result is a sum of terms such as

$$\hat{A}\mathbf{v}_1 \wedge ... \wedge \hat{A}\mathbf{v}_{N-k-1} \wedge \partial_t \hat{A}\mathbf{v}_{N-k} \wedge \mathbf{v}_{N-k+1} \wedge ... \wedge \mathbf{v}_N$$

and other terms obtained by permuting the vectors $\mathbf{v}_j$ (without introducing any minus signs!). The total number of these terms is equal to $N\binom{N-1}{N-k-1}$, since we need to choose a single vector to which $\partial_t \hat{A}$ will apply, and then $(N-k-1)$ vectors to which $\hat{A}$ will apply, among the $(N-1)$ remaining vectors. Now consider the expression

$$\mathrm{Tr}\,\big[(\wedge^{N-1}\hat{A}^{N-k-1})^{\wedge T}\partial_t \hat{A}\big]\omega.$$

This expression is the sum of terms such as

$$\hat{A}_{(k+1)}\partial_t\hat{A}\mathbf{v}_1 \wedge \mathbf{v}_2 \wedge ... \wedge \mathbf{v}_N$$

and other terms with permuted vectors $\mathbf{v}_j$. There will be N such terms, since we choose one vector out of N to apply the operator $\hat{A}_{(k+1)}\partial_t\hat{A}$. Using the definition of $\hat{A}_{(k+1)}$, we write

$$\begin{aligned}
&\hat{A}_{(k+1)}\partial_t\hat{A}\mathbf{v}_1 \wedge \mathbf{v}_2 \wedge ... \wedge \mathbf{v}_N \\
&\quad = \partial_t\hat{A}\mathbf{v}_1 \wedge \left[\wedge^{N-1}\hat{A}^{N-k-1}\right](\mathbf{v}_2 \wedge ... \wedge \mathbf{v}_N) \\
&\quad = \partial_t\hat{A}\mathbf{v}_1 \wedge \hat{A}\mathbf{v}_2 \wedge ... \wedge \hat{A}\mathbf{v}_{N-k} \wedge \mathbf{v}_{N-k+1} \wedge ... \wedge \mathbf{v}_N + ...,
\end{aligned}$$

where in the last line we omitted all other permutations of the vectors. (There will be $\binom{N-1}{N-k-1}$ such permutations.) It follows that the tensor expressions

$$\partial_t q_k \omega \equiv \partial_t(\wedge^N \hat{A}^{N-k})\omega$$

and $\mathrm{Tr}\,[\hat{A}_{(k+1)}\partial_t\hat{A}]\omega$ consist of the same terms; thus they are equal,

$$\partial_t q_k \omega = \mathrm{Tr}\,[\hat{A}_{(k+1)}\partial_t\hat{A}]\omega.$$

Since this holds for any $\omega \in \wedge^N V$, we obtain the required statement. ■

Exercise: Assuming that $\hat{A}(t)$ is invertible, derive a formula for the derivative of the algebraic complement, $\partial_t\tilde{\hat{A}}$.

Hint: Compute ∂_t of both sides of the identity $\tilde{\hat{A}}\hat{A} = (\det \hat{A})\hat{1}$.

Answer:

$$\partial_t\tilde{\hat{A}} = \frac{\mathrm{Tr}\,[\tilde{\hat{A}}\partial_t\hat{A}]\tilde{\hat{A}} - \tilde{\hat{A}}(\partial_t\hat{A})\tilde{\hat{A}}}{\det \hat{A}}.$$

Remark: Since $\tilde{\hat{A}}$ is a polynomial in $\hat{A}$,

$$\tilde{\hat{A}} = q_1 - q_2\hat{A} + ... + q_{N-1}(-\hat{A})^{N-2} + (-\hat{A})^{N-1},$$

all derivatives of $\tilde{\hat{A}}$ may be expressed directly as polynomials in $\hat{A}$ and derivatives of $\hat{A}$, even when $\hat{A}$ is not invertible. Explicit expressions not involving $\hat{A}^{-1}$ are cumbersome — for instance, the derivative of a polynomial in $\hat{A}$ will contain expressions like

$$\partial_t(\hat{A}^3) = (\partial_t\hat{A})\hat{A}^2 + \hat{A}(\partial_t\hat{A})\hat{A} + \hat{A}^2\partial_t\hat{A}.$$

Nevertheless, these expressions can be derived using the known formulas for $\partial_t q_k$ and $\hat{A}_{(k)}$. ■

4.5.2 Derivative of a simple eigenvalue

Suppose an operator $\hat{A}$ is a function of a parameter t; we will consider $\hat{A}(t)$ as a formal power series (FPS). Then the eigenvectors and the eigenvalues of $\hat{A}$ are also functions of t. We can obtain a simple formula for the derivative of an eigenvalue λ if it is an eigenvalue of multiplicity 1. It will be sufficient to know the eigenvalue λ and the algebraic complement of $\hat{A} - \lambda\hat{1}$; we do not need to know any eigenvectors of $\hat{A}$ explicitly, nor the other eigenvalues.

Statement: Suppose $\hat{A}(t)$ is an operator-valued formal power series and $\lambda(0)$ is a simple eigenvalue, i.e. an eigenvalue of $\hat{A}(0)$ having multiplicity 1. We also assume that there exists an FPS $\lambda(t)$ and a vector-valued FPS $\mathbf{v}(t)$ such that $\hat{A}\mathbf{v} = \lambda\mathbf{v}$ in the sense of formal power series. Then the following identity of FPS holds,

$$\partial_t\lambda = \frac{\mathrm{Tr}\,(\tilde{\hat{B}}\partial_t\hat{A})}{\wedge^N\hat{B}^{N-1}} = \frac{\mathrm{Tr}\,(\tilde{\hat{B}}\partial_t\hat{A})}{\mathrm{Tr}\,\tilde{\hat{B}}},$$
$$\hat{B}(t) \equiv \hat{A}(t) - \lambda(t)\hat{1}_V.$$

The number

$$\mathrm{Tr}\tilde{\hat{B}}(0) \equiv \wedge^N\hat{B}^{N-1}\Big|_{t=0} \neq 0$$

if and only if $\lambda(0)$ is a simple eigenvalue.

Proof: We consider the derivative ∂_t of the identity $\det\hat{B} = 0$:

$$0 = \partial_t\det\hat{B} = \mathrm{Tr}\,(\tilde{\hat{B}}\partial_t\hat{B}) = \mathrm{Tr}\,[\tilde{\hat{B}}(\partial_t\hat{A} - \hat{1}\partial_t\lambda)]$$
$$= \mathrm{Tr}\,(\tilde{\hat{B}}\partial_t\hat{A}) - (\mathrm{Tr}\,\tilde{\hat{B}})\partial_t\lambda.$$

We have from Statement 1 in Sec. 4.2.3 the relation

$$\mathrm{Tr}\,\tilde{\hat{B}} = \wedge^N\hat{B}^{N-1}$$

for any operator $\hat{B}$. Since (by assumption) $\mathrm{Tr}\tilde{\hat{B}}(t) \neq 0$ at $t = 0$, we may divide by $\mathrm{Tr}\tilde{\hat{B}}(t)$ because $1/\mathrm{Tr}\tilde{\hat{B}}(t)$ is a well-defined FPS (Lemma 3 in Sec. 4.5). Hence, we have

$$\partial_t\lambda = \frac{\mathrm{Tr}\,(\tilde{\hat{B}}\partial_t\hat{A})}{\mathrm{Tr}\,\tilde{\hat{B}}} = \frac{\mathrm{Tr}\,(\tilde{\hat{B}}\partial_t\hat{A})}{\wedge^N\hat{B}^{N-1}}.$$

The condition $\wedge^N\hat{B}^{N-1} \neq 0$ is equivalent to

$$\frac{\partial}{\partial\mu}Q_{\hat{B}}(\mu) \neq 0 \quad \text{at } \mu = 0,$$

which is the same as the condition that $\mu = 0$ is a simple zero of the characteristic polynomial of $\hat{B} \equiv \hat{A} - \lambda\hat{1}$. ■

Remark: If $\hat{A}(t)$, say, at $t=0$ has an eigenvalue $\lambda(0)$ of multiplicity higher than 1, the formula derived in Statement 1 does not apply, and the analysis requires knowledge of the eigenvectors. For example, the eigenvalue $\lambda(0)$ could have multiplicity 2 because there are two eigenvalues $\lambda_1(t)$ and $\lambda_2(t)$, corresponding to different eigenvectors, which are accidentally equal at $t=0$. One cannot compute $\partial_t\lambda$ without specifying which of the two eigenvalues, $\lambda_1(t)$ or $\lambda_2(t)$, needs to be considered, i.e. without specifying the corresponding eigenvectors $\mathbf{v}_1(t)$ or $\mathbf{v}_2(t)$. Here I do not consider these more complicated situations but restrict attention to the case of a simple eigenvalue.

4.5.3 General trace relations

We have seen in Sec. 3.9 (Exercises 1 and 2) that the coefficients of the characteristic polynomial of an operator $\hat{A}$ can be expressed by algebraic formulas through the N traces $\mathrm{Tr}\hat{A}$, ..., $\mathrm{Tr}(\hat{A}^N)$, and we called these formulas "trace relations." We will now compute the coefficients in the trace relations in the general case.

We are working with a given operator $\hat{A}$ in an N-dimensional space.

Statement: We denote for brevity $q_k \equiv \wedge^N \hat{A}^k$ and $t_k \equiv \mathrm{Tr}(\hat{A}^k)$, where $k = 1, 2, ...$, and set $q_k \equiv 0$ for $k > N$. Then all q_k can be expressed as polynomials in t_k, and these polynomials are equal to the coefficients at x^k of the formal power series

$$G(x) = \exp\left[t_1 x - t_2\frac{x^2}{2} + ... + (-1)^{n-1} t_n \frac{x^n}{n} + ...\right] \equiv \sum_{k=1}^{\infty} x^k q_k$$

by collecting the powers of the formal variable x up to the desired order.

Proof: Consider the expression $\det(\hat{1}+x\hat{A})$ as a formal power series in x. By the Liouville formula, we have the following identity of formal power series,

$$\begin{aligned}
\ln\det(\hat{1}+x\hat{A}) &= \mathrm{Tr}\left[\ln(\hat{1}+x\hat{A})\right] \\
&= \mathrm{Tr}\left[x\hat{A} - \frac{x^2}{2}\hat{A}^2 + ... + (-1)^{n-1}\frac{x^n}{n}\hat{A}^n + ...\right] \\
&= xt_1 - \frac{x^2}{2}t_2 + ... + (-1)^{n-1} t_n \frac{x^n}{n} + ...,
\end{aligned}$$

where we substituted the power series for the logarithm function and used the notation $t_k \equiv \mathrm{Tr}(\hat{A}^k)$. Therefore, we have

$$\det(\hat{1}+x\hat{A}) = \exp G(x)$$

as the identity of formal power series. On the other hand, $\det(\hat{1}+x\hat{A})$ is actually a *polynomial* of degree N in x, i.e. a formal power series that has all zero coefficients from x^{N+1} onwards. The coefficients of this polynomial are found by using $x\hat{A}$ instead of $\hat{A}$ in Lemma 1 of Sec. 3.9:

$$\det(\hat{1}+x\hat{A}) = 1 + q_1 x + ... + q_N x^N.$$

Therefore, the coefficient at x^k in the formal power series $\exp G(x)$ is indeed equal to q_k for $k = 1, ..., N$. (The coefficients at x^k for $k > N$ are all zero!) ■

Example: Expanding the given series up to terms of order x^4, we find after some straightforward calculations

$$G(x) = t_1 x + \frac{t_1^2 - t_2}{2} x^2 + \left[\frac{t_1^3}{6} - \frac{t_1 t_2}{2} + \frac{t_3}{3} \right] x^3$$
$$+ \left[\frac{t_1^4}{24} - \frac{t_1^2 t_2}{4} + \frac{t_2^2}{8} + \frac{t_1 t_3}{3} - \frac{t_4}{4} \right] x^4 + O(x^5).$$

Replacing t_j with $\mathrm{Tr}(\hat{A}^j)$ and collecting the terms at the k-th power of x, we obtain the k-th trace relation. For example, the trace relation for $k = 4$ is

$$\wedge^N \hat{A}^4 = \frac{1}{24}(\mathrm{Tr}\hat{A})^4 - \frac{1}{4}\mathrm{Tr}(\hat{A}^2)(\mathrm{Tr}\hat{A})^2 + \frac{1}{8}\left[\mathrm{Tr}(\hat{A}^2)\right]^2$$
$$+ \frac{1}{3}\mathrm{Tr}(\hat{A}^3)\mathrm{Tr}\hat{A} - \frac{1}{4}\mathrm{Tr}(\hat{A}^4).$$

Note that this formula is valid for all N, even for $N < 4$; in the latter case, $\wedge^N \hat{A}^4 = 0$.

4.6 Jordan canonical form

We have seen in Sec. 3.9 that the eigenvalues of a linear operator are the roots of the characteristic polynomial, and that there exists *at least one* eigenvector corresponding to each eigenvalue. In this section we will assume that the total number of roots of the characteristic polynomial, counting the algebraic multiplicity, is equal to N (the dimension of the space). This is the case, for instance, when the field $\mathbb{K}$ is that of the complex numbers ($\mathbb{C}$); otherwise not all polynomials will have roots belonging to $\mathbb{K}$.

The dimension of the eigenspace corresponding to an eigenvalue λ (the **geometric multiplicity**) is not larger than the algebraic multiplicity of the root λ in the characteristic polynomial (Theorem 1 in Sec. 3.9). The geometric multiplicity is in any case not less than 1 because at least one eigenvector exists (Theorem 2 in Sec. 3.5.1). However, it may happen that the algebraic multiplicity of an eigenvalue λ is larger than 1 but the geometric multiplicity is strictly smaller than the algebraic multiplicity. For example, an operator given in some basis by the matrix

$$\begin{pmatrix} 0 & 1 \\ 0 & 0 \end{pmatrix}$$

has only one eigenvector corresponding to the eigenvalue $\lambda = 0$ of algebraic multiplicity 2. Note that this has nothing to do with missing real roots of algebraic equations; this operator has only one eigenvector even if we allow complex eigenvectors. In this case, the operator is not diagonalizable because there are insufficiently many eigenvectors to build a basis. The theory of the

Jordan canonical form explains the structure of the operator in this case and finds a suitable basis that contains all the eigenvectors and also some additional vectors (called the **root vectors**), such that the given operator has a particularly simple form when expressed through that basis. This form is block-diagonal and consists of **Jordan cells**, which are square matrices such as

$$\begin{pmatrix} \lambda & 1 & 0 \\ 0 & \lambda & 1 \\ 0 & 0 & \lambda \end{pmatrix},$$

and similarly built matrices of higher dimension.

To perform the required analysis, it is convenient to consider each eigenvalue of a given operator separately and build the required basis gradually. Since the procedure is somewhat long, we will organize it by steps. The result of the procedure will be a construction of a basis (the **Jordan basis**) in which the operator $\hat{A}$ has the Jordan canonical form.

Step 0: Set up the initial basis. Let $\hat{A} \in \text{End}\, V$ be a linear operator having the eigenvalues $\lambda_1,...,\lambda_n$, and let us consider the first eigenvalue λ_1; suppose λ_1 has algebraic multiplicity m. If the geometric multiplicity of λ_1 is also equal to m, we can choose a linearly independent set of m basis eigenvectors $\{\mathbf{v}_1, ..., \mathbf{v}_m\}$ and continue to work with the next eigenvalue λ_2. If the geometric multiplicity of λ_1 is less than m, we can only choose a set of $r < m$ basis eigenvectors $\{\mathbf{v}_1, ..., \mathbf{v}_r\}$.

In either case, we have found a set of eigenvectors with eigenvalue λ_1 that spans the entire eigenspace. We can repeat Step 0 for every eigenvalue λ_i and obtain the spanning sets of eigenvectors. The resulting set of eigenvectors can be completed to a basis in V. At the end of Step 0, we have a basis $\{\mathbf{v}_1, ..., \mathbf{v}_k, \mathbf{u}_{k+1}, ..., \mathbf{u}_N\}$, where the vectors $\mathbf{v}_i$ are eigenvectors of $\hat{A}$ and the vectors $\mathbf{u}_i$ are chosen arbitrarily — as long as the result is a basis in V. By construction, any eigenvector of $\hat{A}$ is a linear combination of the $\mathbf{v}_i$'s. If the eigenvectors $\mathbf{v}_i$ are sufficiently numerous as to make a basis in V without any $\mathbf{u}_i$'s, the operator $\hat{A}$ is diagonalizable and its Jordan basis is the eigenbasis; the procedure is finished. We need to proceed with the next steps only in the case when the eigenvectors $\mathbf{v}_i$ do not yet span the entire space V, so the Jordan basis is not yet determined.

Step 1: Determine a root vector. We will now concentrate on an eigenvalue λ_1 for which the geometric multiplicity r is less than the algebraic multiplicity m. At the previous step, we have found a basis containing all the eigenvectors needed to span every eigenspace. The basis presently has the form $\{\mathbf{v}_1, ..., \mathbf{v}_r, \mathbf{u}_{r+1}, ..., \mathbf{u}_N\}$, where $\{\mathbf{v}_i \,|\, 1 \leq i \leq r\}$ span the eigenspace of the eigenvalue λ_1, and $\{\mathbf{u}_i \,|\, r+1 \leq i \leq N\}$ are either eigenvectors of $\hat{A}$ corresponding to other eigenvalues, or other basis vectors. Without loss of generality, we may assume that $\lambda_1 = 0$ (otherwise we need to consider temporarily the operator $\hat{A} - \lambda_1 \hat{1}_V$, which has all the same eigenvectors as $\hat{A}$). Since the operator $\hat{A}$ has eigenvalue 0 with algebraic multiplicity m, the characteristic polynomial has the form $Q_{\hat{A}}(\lambda) = \lambda^m \tilde{q}(\lambda)$, where $\tilde{q}(\lambda)$ is some other polynomial. Since the coefficients of the characteristic polynomial are proportional to the operators $\wedge^N \hat{A}^k$ for $1 \leq k \leq N$, we find that

$$\wedge^N \hat{A}^{N-m} \neq 0, \text{ while } \wedge^N \hat{A}^{N-k} = 0, \quad 0 \leq k < m.$$

In other words, we have found that several operators of the form $\wedge^N \hat{A}^{N-k}$ vanish. Let us now try to obtain some information about the vectors $\mathbf{u}_i$ by considering the action of these operators on the N-vector

$$\omega \equiv \mathbf{v}_1 \wedge ... \wedge \mathbf{v}_r \wedge \mathbf{u}_{r+1} \wedge ... \wedge \mathbf{u}_N.$$

The result must be zero; for instance, we have

$$(\wedge^N \hat{A}^N)\omega = \hat{A}\mathbf{v}_1 \wedge ... = 0$$

since $\hat{A}\mathbf{v}_1 = 0$. We do not obtain any new information by considering the operator $\wedge^N \hat{A}^N$ because the application of $\wedge^N \hat{A}^N$ on ω acts with $\hat{A}$ on $\mathbf{v}_i$, which immediately yields zero. A nontrivial result can be obtained only if we do not act with $\hat{A}$ on *any* of the r eigenvectors $\mathbf{v}_i$. Thus, we turn to considering the operators $\wedge^N \hat{A}^{N-k}$ with $k \geq r$; these operators involve sufficiently few powers of $\hat{A}$ so that $\wedge^N \hat{A}^{N-k}\omega$ may avoid containing any terms $\hat{A}\mathbf{v}_i$.

The first such operator is

$$0 \stackrel{!}{=} (\wedge^N \hat{A}^{N-r})\omega = \mathbf{v}_1 \wedge ... \wedge \mathbf{v}_r \wedge \hat{A}\mathbf{u}_{r+1} \wedge ... \wedge \hat{A}\mathbf{u}_N.$$

It follows that the set $\{\mathbf{v}_1, ..., \mathbf{v}_r, \hat{A}\mathbf{u}_{r+1}, ..., \hat{A}\mathbf{u}_N\}$ is linearly dependent, so there exists a vanishing linear combination

$$\sum_{i=1}^{r} c_i \mathbf{v}_i + \sum_{i=r+1}^{N} c_i \hat{A}\mathbf{u}_i = 0 \tag{4.12}$$

with at least some $c_i \neq 0$. Let us define the vectors

$$\tilde{\mathbf{v}} \equiv \sum_{i=1}^{r} c_i \mathbf{v}_i, \quad \mathbf{x} \equiv - \sum_{i=r+1}^{N} c_i \mathbf{u}_i,$$

so that Eq. (4.12) is rewritten as $\hat{A}\mathbf{x} = \tilde{\mathbf{v}}$. Note that $\mathbf{x} \neq 0$, for otherwise we would have $\sum_{i=1}^{r} c_i \mathbf{v}_i = 0$, which contradicts the linear independence of the set $\{\mathbf{v}_1, ..., \mathbf{v}_r\}$. Further, the vector $\tilde{\mathbf{v}}$ cannot be equal to zero, for otherwise we would have $\hat{A}\mathbf{x} = 0$, so there would exist an additional eigenvector $\mathbf{x} \neq 0$ that is not a linear combination of $\mathbf{v}_i$, which is impossible since (by assumption) the set $\{\mathbf{v}_1, ..., \mathbf{v}_r\}$ spans the entire subspace of all eigenvectors with eigenvalue 0. Therefore, $\tilde{\mathbf{v}} \neq 0$, so at least one of the coefficients $\{c_i \,|\, 1 \leq i \leq r\}$ is nonzero. Without loss of generality, we assume that $c_1 \neq 0$. Then we can replace $\mathbf{v}_1$ by $\tilde{\mathbf{v}}$ in the basis; the set $\{\tilde{\mathbf{v}}, \mathbf{v}_2, ..., \mathbf{v}_r, \mathbf{u}_{r+1}, ..., \mathbf{u}_N\}$ is still a basis because

$$\begin{aligned} \tilde{\mathbf{v}} \wedge \mathbf{v}_2 \wedge ... \wedge \mathbf{v}_r &= (c_1\mathbf{v}_1 + ...) \wedge \mathbf{v}_2 \wedge ... \wedge \mathbf{v}_r \\ &= c_1 \mathbf{v}_1 \wedge \mathbf{v}_2 \wedge ... \wedge \mathbf{v}_r \neq 0. \end{aligned}$$

Similarly, at least one of the coefficients $\{c_i \,|\, r+1 \leq i \leq N\}$ is nonzero. We would like to replace one of the $\mathbf{u}_i$'s in the basis by $\mathbf{x}$; it is possible to replace $\mathbf{u}_i$ by $\mathbf{x}$ as long as $c_i \neq 0$. However, we do not wish to remove from the basis any of the eigenvectors corresponding to other eigenvalues; so we need to choose the index i such that $\mathbf{u}_i$ is not one of the other eigenvectors and at the same time $c_i \neq 0$. This choice is possible; for were it impossible, the vector $\mathbf{x}$ were a linear combination of other eigenvectors of $\hat{A}$ (all having nonzero eigenvalues), so $\hat{A}\mathbf{x}$ is again a linear combination of those eigenvectors, which contradicts the equations $\hat{A}\mathbf{x} = \tilde{\mathbf{v}}$ and $\hat{A}\tilde{\mathbf{v}} = 0$ because $\tilde{\mathbf{v}}$ is linearly independent of all other eigenvectors. Therefore, we can choose a vector $\mathbf{u}_i$ that is not an eigenvector and such that $\mathbf{x}$ can be replaced by $\mathbf{u}_i$. Without loss of generality, we may assume that this vector is $\mathbf{u}_{r+1}$. The new basis, $\{\tilde{\mathbf{v}}, \mathbf{v}_2, ..., \mathbf{v}_r, \mathbf{x}, \mathbf{u}_{r+2}, ..., \mathbf{u}_N\}$ is still linearly independent because

$$\tilde{\omega} \equiv \tilde{\mathbf{v}} \wedge \mathbf{v}_2 \wedge ... \wedge \mathbf{v}_r \wedge \mathbf{x} \wedge \mathbf{u}_{r+2} ... \wedge \mathbf{u}_N \neq 0$$

due to $c_{r+1} \neq 0$. Renaming now $\tilde{\mathbf{v}} \to \mathbf{v}_1$, $\mathbf{x} \to \mathbf{x}_1$, and $\tilde{\omega} \to \omega$, we obtain a new basis $\{\mathbf{v}_1, ..., \mathbf{v}_r, \mathbf{x}_1, \mathbf{u}_{r+2}, ..., \mathbf{u}_N\}$ such that $\mathbf{v}_i$ are eigenvectors ($\hat{A}\mathbf{v}_i = 0$) and $\hat{A}\mathbf{x}_1 = \mathbf{v}_1$. The vector $\mathbf{x}_1$ is called a **root vector** of order 1 corresponding to the given eigenvalue $\lambda_1 = 0$. Eventually the Jordan basis will contain all the root vectors as well as all the eigenvectors for each eigenvalue. So our goal is to determine all the root vectors.

Example 1: The operator $\hat{A} = \mathbf{e}_1 \otimes \mathbf{e}_2^*$ in a two-dimensional space has an eigenvector $\mathbf{e}_1$ with eigenvalue 0 and a root vector $\mathbf{e}_2$ (of order 1) so that $\hat{A}\mathbf{e}_2 = \mathbf{e}_1$ and $\hat{A}\mathbf{e}_1 = 0$. The matrix representation of $\hat{A}$ in the basis $\{\mathbf{e}_1, \mathbf{e}_2\}$ is

$$\hat{A} = \begin{pmatrix} 0 & 1 \\ 0 & 0 \end{pmatrix}.$$

Step 2: Determine other root vectors. If $r+1 = m$ then we are finished with the eigenvalue λ_1; there are no more operators $\wedge^N \hat{A}^{N-k}$ that vanish, and we cannot extract any more information. Otherwise $r+1 < m$, and we will continue by considering the operator $\wedge^N \hat{A}^{N-r-1}$, which vanishes as well:

$$0 = (\wedge^N \hat{A}^{N-r-1})\omega = \mathbf{v}_1 \wedge ... \wedge \mathbf{v}_r \wedge \mathbf{x}_1 \wedge \hat{A}\mathbf{u}_{r+2} \wedge ... \wedge \hat{A}\mathbf{u}_N.$$

(Note that $\mathbf{v}_1 \wedge \hat{A}\mathbf{x}_1 = 0$, so in writing $(\wedge^N \hat{A}^{N-r-1})\omega$ we omit the terms where $\hat{A}$ acts on $\mathbf{v}_i$ or on $\mathbf{x}_1$ and write only the term where the operators $\hat{A}$ act on the $N-r-1$ vectors $\mathbf{u}_i$.) As before, it follows that there exists a vanishing linear combination

$$\sum_{i=1}^{r} c_i \mathbf{v}_i + c_{r+1}\mathbf{x}_1 + \sum_{i=r+2}^{N} c_i \hat{A}\mathbf{u}_i = 0. \qquad (4.13)$$

We introduce the auxiliary vectors

$$\tilde{\mathbf{v}} \equiv \sum_{i=1}^{r} c_i \mathbf{v}_i, \quad \mathbf{x} \equiv -\sum_{i=r+2}^{N} c_i \mathbf{u}_i,$$

and rewrite Eq. (4.13) as

$$\hat{A}\mathbf{x} = c_{r+1}\mathbf{x}_1 + \tilde{\mathbf{v}}. \tag{4.14}$$

As before, we find that $\mathbf{x} \neq 0$. There are now two possibilities: either $c_{r+1} = 0$ or $c_{r+1} \neq 0$. If $c_{r+1} = 0$ then $\mathbf{x}$ is another root vector of order 1. As before, we show that one of the vectors $\mathbf{v}_i$ (but not $\mathbf{v}_1$) may be replaced by $\tilde{\mathbf{v}}$, and one of the vectors $\mathbf{u}_i$ (but not one of the other eigenvectors or root vectors) may be replaced by $\mathbf{x}$. After renaming the vectors ($\tilde{\mathbf{v}} \to \mathbf{v}_i$ and $\mathbf{x} \to \mathbf{x}_2$), the result is a new basis

$$\{\mathbf{v}_1, ..., \mathbf{v}_r, \mathbf{x}_1, \mathbf{x}_2, \mathbf{u}_{r+3}, ..., \mathbf{u}_N\}, \tag{4.15}$$

such that $\hat{A}\mathbf{x}_1 = \mathbf{v}_1$ and $\hat{A}\mathbf{x}_2 = \mathbf{v}_2$. It is important to keep the information that $\mathbf{x}_1$ and $\mathbf{x}_2$ are root vectors of order 1.

The other possibility is that $c_{r+1} \neq 0$. Without loss of generality, we may assume that $c_{r+1} = 1$ (otherwise we divide Eq. (4.14) by c_{r+1} and redefine $\mathbf{x}$ and $\tilde{\mathbf{v}}$). In this case $\mathbf{x}$ is a root vector of order 2; according to Eq. (4.14), acting with $\hat{A}$ on $\mathbf{x}$ yields a root vector of order 1 and a linear combination of some eigenvectors. We will modify the basis again in order to simplify the action of $\hat{A}$; namely, we redefine $\tilde{\mathbf{x}}_1 \equiv \mathbf{x}_1 + \tilde{\mathbf{v}}$ so that $\hat{A}\mathbf{x} = \tilde{\mathbf{x}}_1$. The new vector $\tilde{\mathbf{x}}_1$ is still a root vector of order 1 because it satisfies $\hat{A}\tilde{\mathbf{x}}_1 = \mathbf{v}_1$, and the vector $\mathbf{x}_1$ in the basis may be replaced by $\tilde{\mathbf{x}}_1$. As before, one of the $\mathbf{u}_i$'s can be replaced by $\mathbf{x}$. Renaming $\tilde{\mathbf{x}}_1 \to \mathbf{x}_1$ and $\mathbf{x} \to \mathbf{x}_2$, we obtain the basis

$$\{\mathbf{v}_1, ..., \mathbf{v}_r, \mathbf{x}_1, \mathbf{x}_2, \mathbf{u}_{r+3}, ..., \mathbf{u}_N\},$$

where now we record that $\mathbf{x}_2$ is a root vector of order 2.

The procedure of determining the root vectors can be continued in this fashion until all the root vectors corresponding to the eigenvalue 0 are found. The end result will be a basis of the form

$$\{\mathbf{v}_1, ..., \mathbf{v}_r, \mathbf{x}_1, ..., \mathbf{x}_{m-r}, \mathbf{u}_{m+1}, ..., \mathbf{u}_N\},$$

where $\{\mathbf{v}_i\}$ are eigenvectors, $\{\mathbf{x}_i\}$ are root vectors of various orders, and $\{\mathbf{u}_i\}$ are the vectors that do not belong to this eigenvalue.

Generally, a root vector of order k for the eigenvalue $\lambda_1 = 0$ is a vector $\mathbf{x}$ such that $(\hat{A})^k\mathbf{x} = 0$. However, we have constructed the root vectors such that they come in "chains," for example $\hat{A}\mathbf{x}_2 = \mathbf{x}_1$, $\hat{A}\mathbf{x}_1 = \mathbf{v}_1$, $\hat{A}\mathbf{v}_1 = 0$. Clearly, this is the simplest possible arrangement of basis vectors. There are at most r chains for a given eigenvalue because each eigenvector $\mathbf{v}_i$ ($i = 1, ..., r$) may have an associated chain of root vectors. Note that the root chains for an eigenvalue $\lambda \neq 0$ have the form $\hat{A}\mathbf{v}_1 = \lambda\mathbf{v}_1$, $\hat{A}\mathbf{x}_1 = \lambda\mathbf{x}_1 + \mathbf{v}_1$, $\hat{A}\mathbf{x}_2 = \lambda\mathbf{x}_2 + \mathbf{x}_1$, etc.

Example 2: An operator given by the matrix

$$\hat{A} = \begin{pmatrix} 20 & 1 & 0 \\ 0 & 20 & 1 \\ 0 & 0 & 20 \end{pmatrix}$$

has an eigenvector $\mathbf{e}_1$ with eigenvalue $\lambda = 20$ and the root vectors $\mathbf{e}_2$ (of order 1) and $\mathbf{e}_3$ (of order 2) since $\hat{A}\mathbf{e}_1 = 20\mathbf{e}_1$, $\hat{A}\mathbf{e}_2 = 20\mathbf{e}_2 + \mathbf{e}_1$, and $\hat{A}\mathbf{e}_3 = 20\mathbf{e}_3 + \mathbf{e}_2$. A tensor representation of $\hat{A}$ is

$$\hat{A} = \mathbf{e}_1 \otimes (20\mathbf{e}_1^* + \mathbf{e}_2^*) + \mathbf{e}_2 \otimes (20\mathbf{e}_2^* + \mathbf{e}_3^*) + 20\mathbf{e}_3 \otimes \mathbf{e}_3^*.$$

Step 3: Proceed to other eigenvalues. At Step 2, we determined all the root vectors for one eigenvalue λ_1. The eigenvectors and the root vectors belonging to a given eigenvalue λ_1 span a subspace called the **Jordan cell** for that eigenvalue. We then repeat the same analysis (Steps 1 and 2) for another eigenvalue and determine the corresponding Jordan cell. Note that it is impossible that a root vector for one eigenvalue is at the same time an eigenvector or a root vector for another eigenvalue; the Jordan cells have zero intersection. During the construction, we guarantee that we are not replacing any root vectors or eigenvectors found for the previous eigenvalues. Therefore, the final result is a basis of the form

$$\{\mathbf{v}_1, ..., \mathbf{v}_r, \mathbf{x}_1, ..., \mathbf{x}_{N-r}\}, \tag{4.16}$$

where $\{\mathbf{v}_i\}$ are the various eigenvectors and $\{\mathbf{x}_i\}$ are the corresponding root vectors of various orders.

Definition: The **Jordan basis** of an operator $\hat{A}$ is a basis of the form (4.16) such that $\mathbf{v}_i$ are eigenvectors and $\mathbf{x}_i$ are root vectors. For each root vector $\mathbf{x}$ corresponding to an eigenvalue λ we have $\hat{A}\mathbf{x} = \lambda\mathbf{x} + \mathbf{y}$, where $\mathbf{y}$ is either an eigenvector or a root vector belonging to the same eigenvalue.

The construction in this section constitutes a proof of the following statement.

Theorem 1: Any linear operator $\hat{A}$ in a vector space over $\mathbb{C}$ admits a Jordan basis.

Remark: The assumption that the vector space is over *complex* numbers $\mathbb{C}$ is necessary in order to be sure that every polynomial has as many roots (counting with the algebraic multiplicity) as its degree. If we work in a vector space over $\mathbb{R}$, the construction of the Jordan basis will be complete only for operators whose characteristic polynomial has only real roots. Otherwise we will be able to construct Jordan cells only for real eigenvalues.

Example 3: An operator $\hat{A}$ defined by the matrix

$$\hat{A} = \begin{pmatrix} 0 & 1 & 0 \\ 0 & 0 & 1 \\ 0 & 0 & 0 \end{pmatrix}$$

in a basis $\{\mathbf{e}_1, \mathbf{e}_2, \mathbf{e}_3\}$ can be also written in the tensor notation as

$$\hat{A} = \mathbf{e}_1 \otimes \mathbf{e}_2^* + \mathbf{e}_2 \otimes \mathbf{e}_3^*.$$

The characteristic polynomial of $\hat{A}$ is $Q_{\hat{A}}(\lambda) = (-\lambda)^3$, so there is only one eigenvalue, $\lambda_1 = 0$. The algebraic multiplicity of λ_1 is 3. However, there is only one eigenvector, namely $\mathbf{e}_1$. The vectors $\mathbf{e}_2$ and $\mathbf{e}_3$ are root vectors since $\hat{A}\mathbf{e}_3 = \mathbf{e}_2$ and $\hat{A}\mathbf{e}_2 = \mathbf{e}_1$. Note also that the operator $\hat{A}$ is nilpotent, $\hat{A}^3 = 0$.

Example 4: An operator $\hat{A}$ defined by the matrix

$$\hat{A} = \begin{pmatrix} 6 & 1 & 0 & 0 & 0 \\ 0 & 6 & 0 & 0 & 0 \\ 0 & 0 & 6 & 0 & 0 \\ 0 & 0 & 0 & 7 & 0 \\ 0 & 0 & 0 & 0 & 7 \end{pmatrix}$$

has the characteristic polynomial $Q_{\hat{A}}(\lambda) = (6 - \lambda)^3 (7 - \lambda)^2$ and two eigenvalues, $\lambda_1 = 6$ and $\lambda_2 = 7$. The algebraic multiplicity of λ_1 is 3. However, there are only *two* eigenvectors for the eigenvalue λ_1, namely $\mathbf{e}_1$ and $\mathbf{e}_3$. The vector $\mathbf{e}_2$ is a root vector of order 1 for the eigenvalue λ_1 since

$$\hat{A}\mathbf{e}_2 = \begin{pmatrix} 6 & 1 & 0 & 0 & 0 \\ 0 & 6 & 0 & 0 & 0 \\ 0 & 0 & 6 & 0 & 0 \\ 0 & 0 & 0 & 7 & 0 \\ 0 & 0 & 0 & 0 & 7 \end{pmatrix} \begin{bmatrix} 0 \\ 1 \\ 0 \\ 0 \\ 0 \end{bmatrix} = \begin{bmatrix} 1 \\ 6 \\ 0 \\ 0 \\ 0 \end{bmatrix} = 6\mathbf{e}_2 + \mathbf{e}_1.$$

The algebraic multiplicity of λ_2 is 2, and there are two eigenvectors for λ_2, namely $\mathbf{e}_4$ and $\mathbf{e}_5$. The vectors $\{\mathbf{e}_1, \mathbf{e}_2, \mathbf{e}_3\}$ span the Jordan cell for the eigenvalue λ_1, and the vectors $\{\mathbf{e}_4, \mathbf{e}_5\}$ span the Jordan cell for the eigenvalue λ_2.

Exercise 1: Show that root vectors of order k (with $k \geq 1$) belonging to eigenvalue λ are at the same time eigenvectors of the operator $(\hat{A} - \lambda\hat{1})^{k+1}$ with eigenvalue 0. (This gives another constructive procedure for determining the root vectors.)

4.6.1 Minimal polynomial

Recalling the Cayley-Hamilton theorem, we note that the characteristic polynomial for the operator $\hat{A}$ in Example 4 in the previous subsection vanishes on $\hat{A}$:

$$(6 - \hat{A})^3(7 - \hat{A})^2 = 0.$$

However, there is a polynomial of a lower degree that also vanishes on $\hat{A}$, namely $p(x) = (6 - x)^2 (7 - x)$.

Let us consider the operator $\hat{A}$ in Example 3 in the previous subsection. Its characteristic polynomial is $(-\lambda)^3$, and it is clear that $(\hat{A})^2 \neq 0$ but $(\hat{A})^3 = 0$. Hence there is no lower-degree polynomial $p(x)$ that makes $\hat{A}$ vanish; the minimal polynomial is λ^3.

Let us also consider the operator

$$\hat{B} = \begin{pmatrix} 2 & 0 & 0 & 0 & 0 \\ 0 & 2 & 0 & 0 & 0 \\ 0 & 0 & 1 & 0 & 0 \\ 0 & 0 & 0 & 1 & 0 \\ 0 & 0 & 0 & 0 & 1 \end{pmatrix}.$$

The characteristic polynomial of this operator is $(2-\lambda)^2(1-\lambda)^3$, but it is clear that the following simpler polynomial, $p(x)=(2-x)(1-x)$, also vanishes on $\hat{B}$. If we are interested in the lowest-degree polynomial that vanishes on $\hat{B}$, we do not need to keep higher powers of the factors $(2-\lambda)$ and $(1-\lambda)$ that appear in the characteristic polynomial.

We may ask: what is the polynomial $p(x)$ of a smallest degree such that $p(\hat{A})=0$? Is this polynomial unique?

Definition: The **minimal polynomial** for an operator $\hat{A}$ is a monic polynomial $p(x)$ such that $p(\hat{A})=0$ and that no polynomial $\tilde{p}(x)$ of lower degree satisfies $\tilde{p}(\hat{A})=0$.

Exercise 1: Suppose that the characteristic polynomial of $\hat{A}$ is given as

$$Q_{\hat{A}}(\lambda)=(\lambda_1-\lambda)^{n_1}(\lambda_2-\lambda)^{n_2}...(\lambda_s-\lambda)^{n_s}.$$

Suppose that the Jordan canonical form of $\hat{A}$ includes Jordan cells for eigenvalues $\lambda_1,...,\lambda_s$ such that the largest-order root vector for λ_i has order r_i $(i=1,...,s)$. Show that the polynomial of degree $r_1+...+r_s$ defined by

$$p(x)\equiv(-1)^{r_1+...+r_s}(\lambda_1-\lambda)^{r_1}...(\lambda_s-\lambda)^{r_s}$$

is monic and satisfies $p(\hat{A})=0$. If $\tilde{p}(x)$ is another polynomial of the same degree as $p(x)$ such that $\tilde{p}(\hat{A})=0$, show that $\tilde{p}(x)$ is proportional to $p(x)$. Show that no polynomial $q(x)$ of lower degree can satisfy $q(\hat{A})=0$. Hence, $p(x)$ is the minimal polynomial for $\hat{A}$.

Hint: It suffices to prove these statements for a single Jordan cell. ■

We now formulate a criterion that shows whether a given operator $\hat{A}$ is diagonalizable.

Definition: A polynomial $p(x)$ of degree n is **square-free** if all n roots of $p(x)$ have algebraic multiplicity 1, in other words,

$$p(x)=c(x-x_1)...(x-x_n)$$

where all x_i $(i=1,...,n)$ are different. If a polynomial

$$q(x)=c(x-x_1)^{s_1}...(x-x_m)^{s_m}$$

is not square-free (i.e. some $s_i\neq1$), its **square-free reduction** is the polynomial

$$\tilde{q}(x)=c(x-x_1)...(x-x_m).$$

Remark: In order to compute the square-free reduction of a given polynomial $q(x)$, one does *not* need to obtain the roots x_i of $q(x)$. Instead, it suffices to consider the derivative $q'(x)$ and to note that $q'(x)$ and $q(x)$ have common factors only if $q(x)$ is not square-free, and moreover, the common factors are exactly the factors that we need to remove from $q(x)$ to make it square-free. Therefore, one computes the greatest common divisor of $q(x)$ and $q'(x)$ using the Euclidean algorithm and then divides $q(x)$ by $\gcd(q,q')$ to obtain the square-free reduction $\tilde{q}(x)$.

Theorem 2: An operator $\hat{A}$ is diagonalizable if and only if $p(\hat{A}) = 0$ where $p(\lambda)$ is the square-free reduction of the characteristic polynomial $Q_{\hat{A}}(\lambda)$.

Proof: The Jordan canonical form of $\hat{A}$ may contain several Jordan cells corresponding to different eigenvalues. Suppose that the set of the eigenvalues of $\hat{A}$ is $\{\lambda_i \,|\, i = 1, ..., n\}$, where λ_i are all different and have algebraic multiplicities s_i; then the characteristic polynomial of $\hat{A}$ is

$$Q_{\hat{A}}(x) = (\lambda_1 - x)^{s_1} ... (\lambda_n - x)^{s_n},$$

and its square-free reduction is the polynomial

$$p(x) = (\lambda_1 - x) ... (\lambda_n - x).$$

If the operator $\hat{A}$ is diagonalizable, its eigenvectors $\{\mathbf{v}_j \,|\, j = 1, ..., N\}$ are a basis in V. Then $p(\hat{A})\mathbf{v}_j = 0$ for all $j = 1, ..., N$. It follows that $p(\hat{A}) = \hat{0}$ as an operator. If the operator $\hat{A}$ is not diagonalizable, there exists at least one nontrivial Jordan cell with root vectors. Without loss of generality, let us assume that this Jordan cell corresponds to λ_1. Then there exists a root vector $\mathbf{x}$ such that $\hat{A}\mathbf{x} = \lambda_1\mathbf{x} + \mathbf{v}_1$ while $\hat{A}\mathbf{v}_1 = \lambda_1\mathbf{v}_1$. Then we can compute $(\lambda_1 - \hat{A})\mathbf{x} = -\mathbf{v}_1$ and

$$\begin{aligned} p(\hat{A})\mathbf{x} &= (\lambda_1 - \hat{A})...(\lambda_n - \hat{A})\mathbf{x} \\ &\overset{(1)}{=} (\lambda_n - \hat{A})...(\lambda_2 - \hat{A})(\lambda_1 - \hat{A})\mathbf{x} \\ &\overset{(2)}{=} -(\lambda_n - \lambda_1)...(\lambda_2 - \lambda_1)\,\mathbf{v}_1 \neq 0, \end{aligned}$$

where in $\overset{(1)}{=}$ we used the fact that operators $(\lambda_i - \hat{A})$ all commute with each other, and in $\overset{(2)}{=}$ we used the property of an eigenvector, $q(\hat{A})\mathbf{v}_1 = q(\lambda_1)\mathbf{v}_1$ for any polynomial $q(x)$. Thus we have shown that $p(\hat{A})$ gives a nonzero vector on $\mathbf{x}$, which means that $p(\hat{A})$ is a nonzero operator. ■

Exercise 2: a) It is given that the characteristic polynomial of an operator $\hat{A}$ (in a complex vector space) is $\lambda^3 + 1$. Prove that the operator $\hat{A}$ is invertible and diagonalizable.

b) It is given that the operator $\hat{A}$ satisfies the equation $\hat{A}^3 = \hat{A}^2$. Is $\hat{A}$ invertible? Is $\hat{A}$ diagonalizable? (If not, give explicit counterexamples, e.g., in a 2-dimensional space.)

Exercise 3: A given operator $\hat{A}$ has a Jordan cell Span $\{\mathbf{v}_1, ..., \mathbf{v}_k\}$ with eigenvalue λ. Let

$$p(x) = p_0 + p_1 x + ... + p_s x^s$$

be an arbitrary, fixed polynomial, and consider the operator $\hat{B} \equiv p(\hat{A})$. Show that Span $\{\mathbf{v}_1, ..., \mathbf{v}_k\}$ is a subspace of *some* Jordan cell of the operator $\hat{B}$ (although the eigenvalue of that cell may be different). Show that the orders of the root vectors of $\hat{B}$ are not larger than those of $\hat{A}$.

Hint: Consider for simplicity $\lambda = 0$. The vectors $\mathbf{v}_j$ belong to the eigenvalue $p_0 \equiv p(0)$ of the operator $\hat{B}$. The statement that $\{\mathbf{v}_j\}$ are within a Jordan cell for $\hat{B}$ is equivalent to

$$\mathbf{v}_1 \wedge ... \wedge (\hat{B} - p_0\hat{1})\mathbf{v}_i \wedge ... \wedge \mathbf{v}_k = 0 \quad \text{for } i = 1, ..., k.$$

If $\mathbf{v}_1$ is an eigenvector of $\hat{A}$ with eigenvalue $\lambda = 0$ then it is also an eigenvector of $\hat{B}$ with eigenvalue p_0. If $\mathbf{x}$ is a root vector of order 1 such that $\hat{A}\mathbf{x} = \mathbf{v}_1$ then $\hat{B}\mathbf{x} = p_0\mathbf{x} + p_1\mathbf{v}$, which means that $\mathbf{x}$ could be a root vector of order 1 or an eigenvector of $\hat{B}$ depending on whether $p_1 = 0$. Similarly, one can show that the root chains of $\hat{B}$ are sub-chains of the root chains $\hat{A}$ (i.e. the root chains can only get shorter).

Example 5: A nonzero nilpotent operator $\hat{A}$ such that $\hat{A}^{1000} = 0$ may have root vectors of orders up to 999. The operator $\hat{B} \equiv \hat{A}^{500}$ satisfies $\hat{B}^2 = 0$ and thus can have root vectors only up to order 1. More precisely, the root vectors of $\hat{A}$ of orders 1 through 499 are eigenvectors of $\hat{B}$, while root vectors of $\hat{A}$ of orders 500 through 999 are root vectors of $\hat{B}$ of order 1. However, the Jordan cells of these operators are the same (the entire space V is a Jordan cell with eigenvalue 0). Also, $\hat{A}$ is not expressible as a polynomial in $\hat{B}$. ■

Exercise 3 gives a *necessary* condition for being able to express an operator $\hat{B}$ as a polynomial in $\hat{A}$: It is necessary to determine whether the Jordan cells of $\hat{A}$ and $\hat{B}$ are "compatible" in the sense of Exercise 3. If $\hat{A}$'s Jordan cells cannot be embedded as subspaces within $\hat{B}$'s Jordan cells, or if $\hat{B}$ has a root chain that is not a sub-chain of some root chain of $\hat{A}$, then $\hat{B}$ cannot be a polynomial in $\hat{A}$.

Determining a *sufficient* condition for the existence of $p(x)$ for arbitrary $\hat{A}$ and $\hat{B}$ is a complicated task, and I do not consider it here. The following exercise shows how to do this in a particularly simple case.

Exercise 4: Two operators $\hat{A}$ and $\hat{B}$ are diagonalizable in the same eigenbasis $\{\mathbf{v}_1, ..., \mathbf{v}_N\}$ with eigenvalues $\lambda_1, ..., \lambda_n$ and $\mu_1, ..., \mu_n$ that all have multiplicity 1. Show that $\hat{B} = p(\hat{A})$ for some polynomial $p(x)$ of degree at most $N - 1$.

Hint: We need to map the eigenvalues $\{\lambda_j\}$ into $\{\mu_j\}$. Choose the polynomial $p(x)$ that maps $p(\lambda_j) = \mu_j$ for $j = 1, ..., N$. Such a polynomial surely exists and is unique if we restrict to polynomials of degree not more than $N - 1$. ■

4.7 * Construction of projectors onto Jordan cells

We now consider the problem of determining the Jordan cells. It turns out that we can write a general expression for a projector onto a single Jordan cell of an operator $\hat{A}$. The projector is expressed as a polynomial in $\hat{A}$ with known coefficients. (Note that $\hat{A}$ may or may not be diagonalizable.)

The required projector $\hat{P}$ can be viewed as an operator that has the same Jordan cells as $\hat{A}$ but the eigenvalues are 1 for a single chosen Jordan cell and 0 for all other Jordan cells. One way to construct the projector $\hat{P}$ is to look for a polynomial in $\hat{A}$ such that the eigenvalues and the Jordan cells are mapped as desired. Some examples of this were discussed at the end of the previous subsection; however, the construction required a complete knowledge of the Jordan canonical form of $\hat{A}$ with all eigenvectors and root vectors. We will consider a different method of computing the projector $\hat{P}$. With this method, we only need to know the characteristic polynomial of $\hat{A}$, a single eigenvalue, and the *algebraic* multiplicity of the chosen eigenvalue. We will develop this method beginning with the simplest case.

Statement 1: If the characteristic polynomial $Q(\lambda)$ of an operator $\hat{A}$ has a zero $\lambda = \lambda_0$ of multiplicity 1, i.e. if $Q(\lambda_0) = 0$ and $Q'(\lambda_0) \neq 0$, then the operator $\hat{P}_{\lambda_0}$ defined by

$$\hat{P}_{\lambda_0} \equiv -\frac{1}{Q'(\lambda_0)}\left[\wedge^{N-1}(\hat{A} - \lambda_0\hat{1}_V)^{N-1}\right]^{\wedge T}$$

is a projector onto the one-dimensional eigenspace of the eigenvalue λ_0. The prefactor can be computed also as $-Q'(\lambda_0) = \wedge^N(\hat{A} - \lambda_0\hat{1}_V)^{N-1}$.

Proof: We denote $\hat{P} \equiv \hat{P}_{\lambda_0}$ for brevity. We will first show that for any vector $\mathbf{x}$, the vector $\hat{P}\mathbf{x}$ is an eigenvector of $\hat{A}$ with eigenvalue λ_0, i.e. that the image of $\hat{P}$ is a subspace of the λ_0-eigenspace. Then it will be sufficient to show that $\hat{P}\mathbf{v}_0 = \mathbf{v}_0$ for an eigenvector $\mathbf{v}_0$; it will follow that $\hat{P}\hat{P} = \hat{P}$ and so it will be proved that $\hat{P}$ is a projector onto the eigenspace.

Without loss of generality, we may set $\lambda_0 = 0$ (or else we consider the operator $\hat{A} - \lambda_0\hat{1}_V$ instead of $\hat{A}$). Then we have $\det \hat{A} = 0$, while the number $\wedge^N \hat{A}^{N-1}$ is equal to the last-but-one coefficient in the characteristic polynomial, which is the same as $-Q'(\lambda_0)$ and is nonzero. Thus we set

$$\hat{P} = \frac{1}{\wedge^N \hat{A}^{N-1}}\left(\wedge^{N-1}\hat{A}^{N-1}\right)^{\wedge T} = \frac{1}{\wedge^N \hat{A}^{N-1}}\tilde{\hat{A}}$$

and note that by Lemma 1 in Sec. 4.2.1

$$\hat{P}\hat{A} = \frac{1}{\wedge^N \hat{A}^{N-1}}(\det \hat{A})\hat{1}_V = \hat{0}_V.$$

Since $\hat{P}$ is a polynomial in $\hat{A}$, we have $\hat{P}\hat{A} = \hat{A}\hat{P} = 0$. Therefore $\hat{A}(\hat{P}\mathbf{x}) = 0$ for all $\mathbf{x} \in V$, so $\mathrm{im}\hat{P}$ is indeed a subspace of the eigenspace of the eigenvalue $\lambda_0 = 0$.

It remains to show that $\hat{P}\mathbf{v}_0 = \mathbf{v}_0$ for an eigenvector $\mathbf{v}_0$ such that $\hat{A}\mathbf{v}_0 = 0$. This is verified by a calculation: We use Lemma 1 in Sec. 4.2.1, which is the identity

$$\left(\wedge^{N-1}\hat{A}^{N-n}\right)^{\wedge T}\hat{A} + \left(\wedge^{N-1}\hat{A}^{N-n+1}\right)^{\wedge T} = \left(\wedge^N \hat{A}^{N-n+1}\right)\hat{1}_V$$

valid for all $n = 1, ..., N$, and apply both sides to the vector $\mathbf{v}_0$ with $n = 2$:

$$(\wedge^{N-1}\hat{A}^{N-2})^{\wedge T}\hat{A}\mathbf{v}_0 + (\wedge^{N-1}\hat{A}^{N-1})^{\wedge T}\mathbf{v}_0 = (\wedge^N\hat{A}^{N-1})\mathbf{v}_0,$$

which yields the required formula,

$$\frac{(\wedge^{N-1}\hat{A}^{N-1})^{\wedge T}\mathbf{v}_0}{\wedge^N\hat{A}^{N-1}} = \mathbf{v}_0,$$

since $\hat{A}\mathbf{v}_0 = 0$. Therefore, $\hat{P}\mathbf{v}_0 = \mathbf{v}_0$ as required. ■

Remark: The projector $\hat{P}_{\lambda_0}$ is a polynomial in $\hat{A}$ with coefficients that are known if the characteristic polynomial $Q(\lambda)$ is known. The quantity $Q'(\lambda_0)$ is also an algebraically constructed object that can be calculated without taking derivatives. More precisely, the following formula holds.

Exercise 1: If $\hat{A}$ is any operator in V, prove that

$$\begin{aligned}(-1)^k\frac{\partial^k}{\partial\lambda^k}Q_{\hat{A}}(\lambda) &\equiv (-1)^k\frac{\partial^k}{\partial\lambda^k}\wedge^N(\hat{A}-\lambda\hat{1}_V)^N \\ &= k!\wedge^N(\hat{A}-\lambda\hat{1}_V)^{N-k}.\end{aligned} \qquad (4.17)$$

Solution: An easy calculation. For example, with $k = 2$ and $N = 2$,

$$\begin{aligned}\frac{\partial^2}{\partial\lambda^2}\wedge^2(\hat{A}-\lambda\hat{1}_V)^2\mathbf{u}\wedge\mathbf{v} &= \frac{\partial^2}{\partial\lambda^2}\left[(\hat{A}-\lambda\hat{1}_V)\mathbf{u}\wedge(\hat{A}-\lambda\hat{1}_V)\mathbf{v}\right] \\ &= 2\mathbf{u}\wedge\mathbf{v}.\end{aligned}$$

The formula (4.17) shows that the derivatives of the characteristic polynomial are algebraically defined quantities with a polynomial dependence on the operator $\hat{A}$. ■

Example 1: We illustrate this construction of the projector in a two-dimensional space for simplicity. Let V be a space of polynomials in x of degree at most 1, i.e. polynomials of the form $\alpha + \beta x$ with $\alpha, \beta \in \mathbb{C}$, and consider the linear operator $\hat{A} = x\frac{d}{dx}$ in this space. The basis in V is $\{\underline{1}, \underline{x}\}$, where we use an underbar to distinguish the *polynomials* $\underline{1}$ and $\underline{x}$ from *numbers* such as 1. We first determine the characteristic polynomial,

$$Q_{\hat{A}}(\lambda) = \det(\hat{A}-\lambda\hat{1}) = \frac{(\hat{A}-\lambda)\underline{1}\wedge(\hat{A}-\lambda)\underline{x}}{\underline{1}\wedge\underline{x}} = -\lambda(1-\lambda).$$

Let us determine the projector onto the eigenspace of $\lambda = 0$. We have $\wedge^2\hat{A}^1 = -Q'(0) = 1$ and

$$\hat{P}_0 = -\frac{1}{Q'(0)}(\wedge^1\hat{A}^1)^{\wedge T} = (\wedge^2\hat{A}^1)\hat{1} - \hat{A} = \hat{1} - x\frac{d}{dx}.$$

Since $\hat{P}_0\underline{1} = \underline{1}$ while $\hat{P}_0\underline{x} = 0$, the image of $\hat{P}$ is the subspace spanned by $\underline{1}$. Hence, the eigenspace of $\lambda = 0$ is Span$\{\underline{1}\}$. ■

What if the eigenvalue λ_0 has an algebraic multiplicity larger than 1? Let us first consider the easier case when the geometric multiplicity is equal to the algebraic multiplicity.

Statement 2: If λ_0 is an eigenvalue of both geometric and algebraic multiplicity n then the operator $\hat{P}^{(n)}_{\lambda_0}$ defined by

$$\hat{P}^{(n)}_{\lambda_0} \equiv \left[\wedge^N \hat{A}^{N-n}\right]^{-1} \left[\wedge^{N-1}(\hat{A} - \lambda_0 \hat{1}_V)^{N-n}\right]^{\wedge T} \tag{4.18}$$

is a projector onto the subspace of eigenvectors with eigenvalue λ_0.

Proof: As in the proof of Statement 1, we first show that the image (im $\hat{P}^{(n)}_{\lambda_0}$) is a subspace of the λ_0-eigenspace of $\hat{A}$, and then show that any eigenvector $\mathbf{v}_0$ of $\hat{A}$ with eigenvalue λ_0 satisfies $\hat{P}^{(n)}_{\lambda_0}\mathbf{v}_0 = \mathbf{v}_0$. Let us write $\hat{P} \equiv \hat{P}^{(n)}_{\lambda_0}$ for brevity.

We first need to show that $(\hat{A} - \lambda_0\hat{1})\hat{P} = 0$. Since by assumption λ_0 has algebraic multiplicity n, the characteristic polynomial is of the form $Q_{\hat{A}}(\lambda) = (\lambda_0 - \lambda)^n p(\lambda)$, where $p(\lambda)$ is another polynomial such that $p(\lambda_0) \neq 0$. Without loss of generality we set $\lambda_0 = 0$. With $\lambda_0 = 0$, the factor $(-\lambda^n)$ in the characteristic polynomial means that many of its coefficients $q_k \equiv \wedge^N \hat{A}^{N-k}$ are equal to zero: $q_k = 0$ for $k = 0, ..., n-1$ but $q_n \neq 0$. (Thus the denominator in Eq. (4.18) is nonzero.)

By Lemma 1 in Sec. 4.2.1, for every $k = 1, ..., N$ we have the identity

$$\left(\wedge^{N-1}\hat{A}^{N-k}\right)^{\wedge T}\hat{A} + \left(\wedge^{N-1}\hat{A}^{N-k+1}\right)^{\wedge T} = (\wedge^N \hat{A}^{N-k+1})\hat{1}_V.$$

We can rewrite this as

$$\hat{A}_{(k)}\hat{A} + \hat{A}_{(k-1)} = q_{k-1}\hat{1}, \tag{4.19}$$

where we denoted, as before,

$$\hat{A}_{(k)} \equiv \left(\wedge^{N-1}\hat{A}^{N-k}\right)^{\wedge T}.$$

Setting $k = n$, we find

$$\hat{A}_{(n)}\hat{A} = q_n\hat{P}^{(n)}\hat{A} = 0.$$

Since $q_n \neq 0$, we find $\hat{P}\hat{A} = 0$. Since $\hat{P}$ is a polynomial in $\hat{A}$, it commutes with $\hat{A}$, so $\hat{P}\hat{A} = \hat{A}\hat{P} = 0$. Hence the image of $\hat{P}$ is a subspace of the eigenspace of $\hat{A}$ with $\lambda_0 = 0$.

Now it remains to show that all $\mathbf{v}_i$'s are eigenvectors of $\hat{P}$ with eigenvalue 1. We set $k = n + 1$ in Eq. (4.19) and obtain

$$\hat{A}_{(n+1)}\hat{A}\mathbf{v}_i + \hat{A}_{(n)}\mathbf{v}_i = q_n\mathbf{v}_i.$$

Since $\hat{A}\mathbf{v}_i = 0$, it follows that $\hat{A}_{(n)}\mathbf{v}_i = q_n\mathbf{v}_i$. Therefore $\hat{P}\mathbf{v}_1 = \mathbf{v}_1$. ■

It remains to consider the case when the geometric multiplicity of λ_0 is less than the algebraic multiplicity, i.e. if there exist some root vectors.

Statement 3: We work with an operator $\hat{A}$ whose characteristic polynomial is known,

$$Q_{\hat{A}}(\lambda) = q_0 + (-\lambda)\, q_1 + ... + (-\lambda)^{N-1} q_{N-1} + (-\lambda)^N .$$

Without loss of generality, we assume that $\hat{A}$ has an eigenvalue $\lambda_0 = 0$ of algebraic multiplicity $n \geq 1$. The geometric multiplicity of λ_0 may be less than or equal to n. (For nonzero eigenvalues λ_0, we consider the operator $\hat{A} - \lambda_0 \hat{1}$ instead of $\hat{A}$.)

(1) A projector onto the Jordan cell of dimension n belonging to eigenvalue λ_0 is given by the operator

$$\hat{P}_{\lambda_0} \equiv \sum_{k=1}^{n} c_k \hat{A}_{(k)} = \hat{1} + \sum_{k=1}^{n} \sum_{i=n}^{N-k} c_k q_{i+k} (-\hat{A})^i, \tag{4.20}$$

where

$$\hat{A}_{(k)} \equiv (\wedge^{N-1} \hat{A}^{N-k})^{\wedge T}, \quad 1 \leq k \leq N-1,$$

and $c_1, ..., c_n$ are the numbers that solve the system of equations

$$\begin{pmatrix} q_n & q_{n+1} & q_{n+2} & \cdots & q_{2n-1} \\ 0 & q_n & q_{n+1} & \cdots & q_{2n-2} \\ \vdots & 0 & \ddots & \ddots & \vdots \\ 0 & \vdots & \ddots & q_n & q_{n+1} \\ 0 & 0 & \cdots & 0 & q_n \end{pmatrix} \begin{bmatrix} c_1 \\ c_2 \\ \vdots \\ c_{n-1} \\ c_n \end{bmatrix} = \begin{bmatrix} 0 \\ 0 \\ \vdots \\ 0 \\ 1 \end{bmatrix}.$$

For convenience, we have set $q_N \equiv 1$ and $q_i \equiv 0$ for $i > N$.

(2) No polynomial in $\hat{A}$ can be a projector onto the subspace of *eigenvectors* within the Jordan cell (rather than a projector onto the entire Jordan cell) when the geometric multiplicity is strictly less than the algebraic.

Proof: **(1)** The Jordan cell consists of all vectors $\mathbf{x}$ such that $\hat{A}^n \mathbf{x} = 0$. We proceed as in the proof of Statement 2, starting from Eq. (4.19). By induction in k, starting from $k = 1$ until $k = n$, we obtain

$$\begin{aligned} \hat{A}\hat{A}_{(1)} &= q_0 \hat{1} = 0, \\ \hat{A}^2 \hat{A}_{(2)} + \hat{A}\hat{A}_{(1)} &= \hat{A} q_1 \hat{1} = 0 \Rightarrow \hat{A}^2 \hat{A}_{(2)} = 0, \\ ..., &\quad \Rightarrow \hat{A}^n \hat{A}_{(n)} = 0. \end{aligned}$$

So we find $\hat{A}^n \hat{A}_{(k)} = 0$ for all k ($1 \leq k \leq n$). Since $\hat{P}_{\lambda_0}$ is by construction equal to a linear combination of these $\hat{A}_{(k)}$, we have $\hat{A}^n \hat{P}_{\lambda_0} = 0$, i.e. the image of $\hat{P}_{\lambda_0}$ is contained in the Jordan cell.

It remains to prove that the Jordan cell is also *contained* in the image of $\hat{P}_{\lambda_0}$, that is, to show that $\hat{A}^n \mathbf{x} = 0$ implies $\hat{P}_{\lambda_0} \mathbf{x} = \mathbf{x}$. We use the explicit formulas for $\hat{A}_{(k)}$ that can be obtained by induction from Eq. (4.19) starting with $k = N$:

we have $\hat{A}_{(N)} = 0$, $\hat{A}_{(N-1)} = q_{N-1}\hat{1} - \hat{A}$, and finally

$$\hat{A}_{(k)} = q_k\hat{1} - q_{k+1}\hat{A} + ... + q_N(-\hat{A})^{N-k} = \sum_{i=0}^{N-k} q_{k+i}(-\hat{A})^i, \quad k \geq 1. \quad (4.21)$$

The operator $\hat{P}_{\lambda_0}$ is a linear combination of $\hat{A}_{(k)}$ with $1 \leq k \leq n$. The Jordan cell of dimension n consists of all $\mathbf{x} \in V$ such that $\hat{A}^n\mathbf{x} = 0$. Therefore, while computing $\hat{P}_{\lambda_0}\mathbf{x}$ for any $\mathbf{x}$ such that $\hat{A}^n\mathbf{x} = 0$, we can restrict the summation over i to $0 \leq i \leq n-1$,

$$\hat{P}_{\lambda_0}\mathbf{x} = \sum_{k=1}^{n} c_k \sum_{i=0}^{N-k} q_{k+i}(-\hat{A})^i\mathbf{x} = \sum_{k=1}^{n}\sum_{i=0}^{n-1} c_k q_{k+i}(-\hat{A})^i\mathbf{x}.$$

We would like to choose the coefficients c_k such that the sum above contains only the term $(-\hat{A})^0\mathbf{x} = \mathbf{x}$ with coefficient 1, while all other powers of $\hat{A}$ will enter with zero coefficient. In other words, we require that

$$\sum_{k=1}^{n}\sum_{i=0}^{n-1} c_k q_{k+i}(-\hat{A})^i = \hat{1} \quad (4.22)$$

identically as polynomial in $\hat{A}$. This will happen if the coefficients c_k satisfy

$$\sum_{k=1}^{n} c_k q_k = 1,$$
$$\sum_{k=1}^{n} c_k q_{k+i} = 0, \quad i = 1, ..., n-1.$$

This system of equations for the unknown coefficients c_k can be rewritten in matrix form as

$$\begin{pmatrix} q_n & q_{n+1} & q_{n+2} & \cdots & q_{2n-1} \\ q_{n-1} & q_n & q_{n+1} & \cdots & q_{2n-2} \\ \vdots & q_{n-1} & \ddots & \ddots & \vdots \\ q_2 & \vdots & \ddots & q_n & q_{n+1} \\ q_1 & q_2 & \cdots & q_{n-1} & q_n \end{pmatrix} \begin{bmatrix} c_1 \\ c_2 \\ \vdots \\ c_{n-1} \\ c_n \end{bmatrix} = \begin{bmatrix} 0 \\ 0 \\ \vdots \\ 0 \\ 1 \end{bmatrix}.$$

However, it is given that $\lambda_0 = 0$ is a root of multiplicity n, therefore $q_0 = ... = q_{n-1} = 0$ while $q_n \neq 0$. Therefore, the system of equations has the triangular form as given in Statement 3. Its solution is unique since $q_n \neq 0$. Thus, we are able to choose c_k such that $\hat{P}_{\lambda_0}\mathbf{x} = \mathbf{x}$ for any $\mathbf{x}$ within the Jordan cell.

The formula for $\hat{P}_{\lambda_0}$ can be simplified by writing

$$\hat{P}_{\lambda_0} = \sum_{k=1}^{n}\left[\sum_{i=0}^{n-1} c_k q_{k+i}(-\hat{A})^i + \sum_{i=n}^{N-k} c_k q_{k+i}(-\hat{A})^i\right].$$

The first sum yields $\hat{1}$ by Eq. (4.22), and so we obtain Eq. (4.20).

(2) A simple counterexample is the (non-diagonalizable) operator

$$\hat{A} = \begin{pmatrix} 0 & 1 \\ 0 & 0 \end{pmatrix} = \mathbf{e}_1 \otimes \mathbf{e}_2^*.$$

This operator has a Jordan cell with eigenvalue 0 spanned by the basis vectors $\mathbf{e}_1$ and $\mathbf{e}_2$. The eigenvector with eigenvalue 0 is $\mathbf{e}_1$, and a possible projector onto this eigenvector is $\hat{P} = \mathbf{e}_1 \otimes \mathbf{e}_1^*$. However, no polynomial in $\hat{A}$ can yield $\hat{P}$ or any other projector only onto $\mathbf{e}_1$. This can be seen as follows. We note that $\hat{A}\hat{A} = 0$, and thus any polynomial in $\hat{A}$ can be rewritten as $a_0\hat{1}_V + a_1\hat{A}$. However, if an operator of the form $a_0\hat{1}_V + a_1\hat{A}$ is a projector, and $\hat{A}\hat{A} = 0$, then we can derive that $a_0^2 = a_0$ and $a_1 = 2a_0a_1$, which forces $a_0 = 1$ and $a_1 = 0$. Therefore the only result of a polynomial formula can be the projector $\mathbf{e}_1 \otimes \mathbf{e}_1^* + \mathbf{e}_2 \otimes \mathbf{e}_2^*$ onto the entire Jordan cell. ■

Example 2: Consider the space of polynomials in x and y of degree at most 1, i.e. the space spanned by $\{\underline{1}, \underline{x}, \underline{y}\}$, and the operator

$$\hat{A} = x\frac{\partial}{\partial x} + \frac{\partial}{\partial y}.$$

The characteristic polynomial of $\hat{A}$ is found as

$$\begin{aligned} Q_{\hat{A}}(\lambda) &= \frac{(\hat{A}-\lambda)\underline{1} \wedge (\hat{A}-\lambda)\underline{x} \wedge (\hat{A}-\lambda)\underline{y}}{\underline{1} \wedge \underline{x} \wedge \underline{y}} \\ &= \lambda^2 - \lambda^3 \equiv q_0 - q_1\lambda + q_2\lambda^2 - q_3\lambda^3. \end{aligned}$$

Hence $\lambda = 0$ is an eigenvalue of algebraic multiplicity 2. It is easy to guess the eigenvectors, $\mathbf{v}_1 = \underline{1}$ $(\lambda = 0)$ and $\mathbf{v}_2 = \underline{x}$ $(\lambda = 1)$, as well as the root vector $\mathbf{v}_3 = \underline{y}$ $(\lambda = 0)$. However, let us pretend that we do not know the Jordan basis, and instead determine the projector $\hat{P}_0$ onto the Jordan cell belonging to the eigenvalue $\lambda_0 = 0$ using Statement 3 with $n = 2$ and $N = 3$.

We have $q_0 = q_1 = 0$, $q_2 = q_3 = 1$. The system of equations for the coefficients c_k is

$$\begin{aligned} q_2c_1 + q_3c_2 &= 0, \\ q_2c_2 &= 1, \end{aligned}$$

and the solution is $c_1 = -1$ and $c_2 = 1$. We note that in our example,

$$\hat{A}^2 = x\frac{\partial}{\partial x}.$$

So we can compute the projector $\hat{P}_0$ by using Eq. (4.20):

$$\begin{aligned} \hat{P}_0 &= \hat{1} + \sum_{k=1}^{2}\sum_{i=2}^{3-k} c_k q_{i+k}(-\hat{A})^i \\ &= \hat{1} + c_1q_3\hat{A}^2 = \hat{1} - x\frac{\partial}{\partial x}. \end{aligned}$$

(The summation over k and i collapses to a single term $k = 1$, $i = 2$.) The image of $\hat{P}_0$ is Span $\{\underline{1}, \underline{y}\}$, and we have $\hat{P}_0\hat{P}_0 = \hat{P}_0$. Hence $\hat{P}_0$ is indeed a projector onto the Jordan cell Span $\{\underline{1}, \underline{y}\}$ that belongs to the eigenvalue $\lambda = 0$.

Exercise 2: Suppose the operator $\hat{A}$ has eigenvalue λ_0 with algebraic multiplicity n. Show that one can choose a basis $\{\mathbf{v}_1, ..., \mathbf{v}_n, \mathbf{e}_{n+1}, ..., \mathbf{e}_N\}$ such that $\mathbf{v}_i$ are eigenvalues or root vectors belonging to the eigenvalue λ_0, and $\mathbf{e}_j$ are such that the vectors $(\hat{A} - \lambda_0\hat{1})\mathbf{e}_j$ (with $j = n+1,...,N$) belong to the subspace Span $\{\mathbf{e}_{n+1}, ..., \mathbf{e}_N\}$. Deduce that the subspace Span $\{\mathbf{e}_{n+1}, ..., \mathbf{e}_N\}$ is mapped one-to-one onto itself by the operator $\hat{A} - \lambda_0\hat{1}$.

Hint: Assume that the Jordan canonical form of $\hat{A}$ is known. Show that

$$\wedge^{N-n}(\hat{A} - \lambda_0\hat{1})^{N-n}(\mathbf{e}_{n+1} \wedge ... \wedge \mathbf{e}_N) \neq 0.$$

(Otherwise, a linear combination of $\mathbf{e}_j$ is an eigenvector with eigenvalue λ_0.)

Remark: Operators of the form

$$\hat{R}_k \equiv \left[\wedge^{N-1}(\hat{A} - \lambda_0\hat{1}_V)^{N-k}\right]^{\wedge T} \qquad (4.23)$$

with $k \leq n$ are used in the construction of projectors onto the Jordan cell. What if we use Eq. (4.23) with other values of k? It turns out that the resulting operators are not projectors. If $k \geq n$, the operator $\hat{R}_k$ does not map into the Jordan cell. If $k < n$, the operator $\hat{R}_k$ does not map onto the *entire* Jordan cell but rather onto a subspace of the Jordan cell; the image of $\hat{R}_k$ contains eigenvectors or root vectors of a certain order. An example of this property will be shown in Exercise 3.

Exercise 3: Suppose an operator $\hat{A}$ has an eigenvalue λ_0 with algebraic multiplicity n and geometric multiplicity $n-1$. This means (according to the theory of the Jordan canonical form) that there exist $n-1$ eigenvectors and *one* root vector of order 1. Let us denote that root vector by $\mathbf{x}_1$ and let $\mathbf{v}_2, ..., \mathbf{v}_n$ be the $(n-1)$ eigenvectors with eigenvalue λ_0. Moreover, let us choose $\mathbf{v}_2$ such that $\hat{A}\mathbf{v}_1 = \lambda_0\mathbf{x}_1 + \mathbf{v}_2$ (i.e. the vectors $\mathbf{x}_1, \mathbf{v}_2$ are a root chain). Show that the operator $\hat{R}_k$ given by the formula (4.23), with $k = n - 1$, satisfies

$$\hat{R}_{n-1}\mathbf{x}_1 = \text{const} \cdot \mathbf{v}_2; \quad \hat{R}_{n-1}\mathbf{v}_j = 0, \quad j = 2, ..., n;$$
$$\hat{R}_{n-1}\mathbf{e}_j = 0, \quad j = n+1, ..., N.$$

In other words, the image of the operator $\hat{R}_{n-1}$ contains only the eigenvector $\mathbf{v}_2$; that is, the image contains the eigenvector related to a root vector of order 1.

Hint: Use a basis of the form $\{\mathbf{x}_1, \mathbf{v}_2, ..., \mathbf{v}_n, \mathbf{e}_{n+1}, ..., \mathbf{e}_N\}$ as in Exercise 2.

5 Scalar product

Until now we did not use any scalar product in our vector spaces. In this chapter we explore the properties of spaces with a scalar product. The exterior product techniques are especially powerful when used together with a scalar product.

5.1 Vector spaces with scalar product

As you already know, the scalar product of vectors is related to the geometric notions of angle and length. These notions are most useful in vector spaces over *real* numbers, so in most of this chapter I will assume that $\mathbb{K}$ is a field where it makes sense to compare numbers (i.e. the comparison $x > y$ is defined and has the usual properties) and where statements such as $\lambda^2 \geq 0$ ($\forall \lambda \in \mathbb{K}$) hold. (Scalar products in complex spaces are defined in a different way and will be considered in Sec. 5.6.)

In order to understand the properties of spaces with a scalar product, it is helpful to define the scalar product in a purely algebraic way, without any geometric constructions. The geometric interpretation will be developed subsequently.

The scalar product of two vectors is a *number*, i.e. the scalar product maps a pair of vectors into a number. We will denote the scalar product by $\langle \mathbf{u}, \mathbf{v} \rangle$, or sometimes by writing it in a functional form, $S(\mathbf{u}, \mathbf{v})$.

A scalar product must be compatible with the linear structure of the vector space, so it cannot be an arbitrary map. The precise definition is the following.

Definition: A map $B : V \times V \to \mathbb{K}$ is a **bilinear form** in a vector space V if for any vectors $\mathbf{u}, \mathbf{v}, \mathbf{w} \in V$ and for any $\lambda \in \mathbb{K}$,

$$B(\mathbf{u}, \mathbf{v} + \lambda \mathbf{w}) = B(\mathbf{u}, \mathbf{v}) + \lambda B(\mathbf{u}, \mathbf{w}),$$
$$B(\mathbf{v} + \lambda \mathbf{w}, \mathbf{u}) = B(\mathbf{v}, \mathbf{u}) + \lambda B(\mathbf{w}, \mathbf{u}).$$

A bilinear form B is **symmetric** if $B(\mathbf{v}, \mathbf{w}) = B(\mathbf{w}, \mathbf{v})$ for any $\mathbf{v}$, $\mathbf{w}$. A bilinear form is **nondegenerate** if for any nonzero vector $\mathbf{v} \neq 0$ there exists another vector $\mathbf{w}$ such that $B(\mathbf{v}, \mathbf{w}) \neq 0$. A bilinear form is **positive-definite** if $B(\mathbf{v}, \mathbf{v}) > 0$ for all nonzero vectors $\mathbf{v} \neq 0$.

A **scalar product** in V is a nondegenerate, positive-definite, symmetric bilinear form $S : V \times V \to \mathbb{K}$. The action of the scalar product on pairs of vectors is also denoted by $\langle \mathbf{v}, \mathbf{w} \rangle \equiv S(\mathbf{v}, \mathbf{w})$. A finite-dimensional vector space over $\mathbb{R}$ with a scalar product is called a **Euclidean space**. The **length** of a vector $\mathbf{v}$ is the non-negative number $\sqrt{\langle \mathbf{v}, \mathbf{v} \rangle}$. (This number is also called the **norm** of $\mathbf{v}$.) ■

Verifying that a map $S : V \times V \to \mathbb{K}$ is a scalar product in V requires proving that S is a bilinear form satisfying certain properties. For instance, the zero function $B(\mathbf{v}, \mathbf{w}) = 0$ is symmetric but is not a scalar product because it is degenerate.

Remark: The above definition of the scalar product is an "abstract definition" because it does not specify any particular scalar product in a given vector space. To specify a scalar product, one usually gives an explicit formula for computing $\langle \mathbf{a}, \mathbf{b} \rangle$. In the same space V, one could consider different scalar products.

Example 1: In the space $\mathbb{R}^n$, the standard scalar product is

$$\langle (x_1, ..., x_N), (y_1, ..., y_N) \rangle \equiv \sum_{j=1}^{N} x_j y_j. \tag{5.1}$$

Let us verify that this defines a symmetric, nondegenerate, and positive-definite bilinear form. This is a bilinear form because it depends linearly on each x_j and on each y_j. This form is symmetric because it is invariant under the interchange of x_j with y_j. This form is nondegenerate because for any $\mathbf{x} \neq 0$ at least one of x_j, say x_1, is nonzero; then the scalar product of $\mathbf{x}$ with the vector $\mathbf{w} \equiv (1, 0, 0, ..., 0)$ is nonzero. So for any $\mathbf{x} \neq 0$ there exists $\mathbf{w}$ such that $\langle \mathbf{x}, \mathbf{w} \rangle \neq 0$, which is the nondegeneracy property. Finally, the scalar product is positive-definite because for any nonzero $\mathbf{x}$ there is at least one nonzero x_j and thus

$$\langle \mathbf{x}, \mathbf{x} \rangle = \langle (x_1, ..., x_N), (x_1, ..., x_N) \rangle \equiv \sum_{j=1}^{N} x_j^2 > 0.$$

Remark: The fact that a bilinear form is nondegenerate does not mean that it must always be nonzero on any two vectors. It is perfectly possible that $\langle \mathbf{a}, \mathbf{b} \rangle = 0$ while $\mathbf{a} \neq 0$ and $\mathbf{b} \neq 0$. In the usual Euclidean space, this would mean that $\mathbf{a}$ and $\mathbf{b}$ are orthogonal to each other. Nondegeneracy means that no vector is orthogonal to *every* other vector. It is also *impossible* that $\langle \mathbf{a}, \mathbf{a} \rangle = 0$ while $\mathbf{a} \neq 0$ (this contradicts the positive-definiteness).

Example 2: Consider the space $\text{End}\, V$ of linear operators in V. We can define a bilinear form in the space $\text{End}\, V$ as follows: For any two operators $\hat{A}, \hat{B} \in \text{End}\, V$ we set $\langle \hat{A}, \hat{B} \rangle \equiv \text{Tr}(\hat{A}\hat{B})$. This bilinear form is *not* positive-definite. For example, if there is an operator $\hat{J}$ such that $\hat{J}^2 = -\hat{1}_V$ then $\text{Tr}(\hat{J}\hat{J}) = -N < 0$ while $\text{Tr}(\hat{1}\hat{1}) = N > 0$, so neither $\text{Tr}(\hat{A}\hat{B})$ nor $-\text{Tr}(\hat{A}\hat{B})$ can be positive-definite. (See Exercise 4 in Sec. 5.1.2 below for more information.)

Remark: Bilinear forms that are not positive-definite (or even degenerate) are sometimes useful as "pseudo-scalar products." We will not discuss these cases here.

Exercise 1: Prove that two vectors are equal, $\mathbf{u} = \mathbf{v}$, if and only if $\langle \mathbf{u}, \mathbf{x} \rangle = \langle \mathbf{v}, \mathbf{x} \rangle$ for all vectors $\mathbf{x} \in V$.

Hint: Consider the vector $\mathbf{u} - \mathbf{v}$ and the definition of nondegeneracy of the scalar product.

Solution: If $\mathbf{u}-\mathbf{v}=0$ then by the linearity of the scalar product $\langle \mathbf{u}-\mathbf{v}, \mathbf{x}\rangle = 0 = \langle \mathbf{u}, \mathbf{x}\rangle - \langle \mathbf{v}, \mathbf{x}\rangle$. Conversely, suppose that $\mathbf{u} \neq \mathbf{v}$; then $\mathbf{u}-\mathbf{v} \neq 0$, and (by definition of nondegeneracy of the scalar product) there exists a vector $\mathbf{x}$ such that $\langle \mathbf{u}-\mathbf{v}, \mathbf{x}\rangle \neq 0$. ■

Exercise 2: Prove that two linear operators $\hat{A}$ and $\hat{B}$ are equal as operators, $\hat{A}=\hat{B}$, if and only if $\langle \hat{A}\mathbf{x}, \mathbf{y}\rangle = \langle \hat{B}\mathbf{x}, \mathbf{y}\rangle$ for all vectors $\mathbf{x}, \mathbf{y} \in V$.

Hint: Consider the vector $\hat{A}\mathbf{x} - \hat{B}\mathbf{x}$. ■

5.1.1 Orthonormal bases

A scalar product defines an important property of a basis in V.

Definition: A set of vectors $\{\mathbf{e}_1, ..., \mathbf{e}_k\}$ in a space V is **orthonormal** with respect to the scalar product if

$$\langle \mathbf{e}_i, \mathbf{e}_j\rangle = \delta_{ij}, \quad 1 \leq i, j \leq k.$$

If an orthonormal set $\{\mathbf{e}_j\}$ is a basis in V, it is called an **orthonormal basis**.

Example 2: In the space $\mathbb{R}^N$ of N-tuples of real numbers $(x_1, ..., x_N)$, the natural scalar product is defined by the formula (5.1). Then the standard basis in $\mathbb{R}^N$, i.e. the basis consisting of vectors $(1, 0, ..., 0)$, $(0, 1, 0, ..., 0)$, ..., $(0, ..., 0, 1)$, is orthonormal with respect to this scalar product. ■

The standard properties of orthonormal bases are summarized in the following theorems.

Statement: Any orthonormal set of vectors is linearly independent.

Proof: If an orthonormal set $\{\mathbf{e}_1, ..., \mathbf{e}_k\}$ is linearly dependent, there exist numbers λ_j, not all equal to zero, such that

$$\sum_{j=1}^{k} \lambda_j \mathbf{e}_j = 0.$$

By assumption, there exists an index s such that $\lambda_s \neq 0$; then the scalar product of the above sum with $\mathbf{e}_s$ yields a contradiction,

$$0 = \langle 0, \mathbf{e}_s\rangle = \left\langle \sum_{j=1}^{k} \lambda_j \mathbf{e}_j \,, \mathbf{e}_s \right\rangle = \sum_{j=1}^{k} \delta_{js}\lambda_j = \lambda_s \neq 0.$$

Hence, any orthonormal set is linearly independent (although it is not necessarily a basis). ■

Theorem 1: Assume that V is a finite-dimensional vector space with a scalar product and $\mathbb{K}$ is a field where one can compute square roots (i.e. for any $\lambda \in \mathbb{K}$, $\lambda > 0$ there exists another number $\mu \equiv \sqrt{\lambda} \in \mathbb{K}$ such that $\lambda = \mu^2$). Then there exists an orthonormal basis in V.

Proof: We can build a basis by the standard orthogonalization procedure (the **Gram-Schmidt procedure**). This procedure uses induction to determine a sequence of orthonormal sets $\{\mathbf{e}_1, ..., \mathbf{e}_k\}$ for $k = 1, ..., N$.

Basis of induction: Choose any nonzero vector $\mathbf{v} \in V$ and compute $\langle \mathbf{v}, \mathbf{v} \rangle$; since $\mathbf{v} \neq 0$, we have $\langle \mathbf{v}, \mathbf{v} \rangle > 0$, so $\sqrt{\langle \mathbf{v}, \mathbf{v} \rangle}$ exists, and we can define $\mathbf{e}_1$ by

$$\mathbf{e}_1 \equiv \frac{\mathbf{v}}{\sqrt{\langle \mathbf{v}, \mathbf{v} \rangle}}.$$

It follows that $\langle \mathbf{e}_1, \mathbf{e}_1 \rangle = 1$.

Induction step: If $\{\mathbf{e}_1, ..., \mathbf{e}_k\}$ is an orthonormal set, we need to find a vector $\mathbf{e}_{k+1}$ such that $\{\mathbf{e}_1, ..., \mathbf{e}_k, \mathbf{e}_{k+1}\}$ is again an orthonormal set. To find a suitable vector $\mathbf{e}_{k+1}$, we first take any vector $\mathbf{v}$ such that the set $\{\mathbf{e}_1, ..., \mathbf{e}_k, \mathbf{v}\}$ is linearly independent; such $\mathbf{v}$ exists if $k < N$, while for $k = N$ there is nothing left to prove. Then we define a new vector

$$\mathbf{w} \equiv \mathbf{v} - \sum_{j=1}^{k} \langle \mathbf{e}_j, \mathbf{v} \rangle \mathbf{e}_j.$$

This vector has the property $\langle \mathbf{e}_j, \mathbf{w} \rangle = 0$ for $1 \leq j \leq k$. We have $\mathbf{w} \neq 0$ because (by construction) $\mathbf{v}$ is not a linear combination of $\mathbf{e}_1$, ..., $\mathbf{e}_k$; therefore $\langle \mathbf{w}, \mathbf{w} \rangle > 0$. Finally, we define

$$\mathbf{e}_{k+1} \equiv \frac{\mathbf{w}}{\sqrt{\langle \mathbf{w}, \mathbf{w} \rangle}},$$

so that $\langle \mathbf{e}_{k+1}, \mathbf{e}_{k+1} \rangle = 1$; then the set $\{\mathbf{e}_1, ..., \mathbf{e}_k, \mathbf{e}_{k+1}\}$ is orthonormal. So the required set $\{\mathbf{e}_1, ..., \mathbf{e}_{k+1}\}$ is now constructed. ■

Question: What about number fields $\mathbb{K}$ where the square root does not exist, for example the field of rational numbers $\mathbb{Q}$?

Answer: In that case, an orthonormal basis may or may not exist. For example, suppose that we consider vectors in $\mathbb{Q}^2$ and the scalar product

$$\langle (x_1, x_2), (y_1, y_2) \rangle = x_1 y_1 + 5 x_2 y_2.$$

Then we cannot normalize the vectors: there exists no vector $\mathbf{x} \equiv (x_1, x_2) \in \mathbb{Q}^2$ such that $\langle \mathbf{x}, \mathbf{x} \rangle = x_1^2 + 5x_2^2 = 1$. The proof of this is similar to the ancient proof of the irrationality of $\sqrt{2}$. Thus, there exists no orthonormal basis in this space with this scalar product.

Theorem 2: If $\{\mathbf{e}_j\}$ is an orthonormal basis then any vector $\mathbf{v} \in V$ is expanded according to the formula

$$\mathbf{v} = \sum_{j=1}^{N} v_j \mathbf{e}_j, \quad v_j \equiv \langle \mathbf{e}_j, \mathbf{v} \rangle.$$

In other words, the j-th component of the vector $\mathbf{v}$ in the basis $\{\mathbf{e}_1, ..., \mathbf{e}_N\}$ is equal to the scalar product $\langle \mathbf{e}_j, \mathbf{v} \rangle$.

Proof: Compute the scalar product $\langle \mathbf{e}_j, \mathbf{v} \rangle$ and obtain $v_j \equiv \langle \mathbf{e}_j, \mathbf{v} \rangle$. ■

Remark: Theorem 2 shows that the components of a vector in an orthonormal basis can be computed quickly. As we have seen before, the component v_j of a vector $\mathbf{v}$ in the basis $\{\mathbf{e}_j\}$ is given by the covector $\mathbf{e}_j^*$ from the dual basis, $v_j = \mathbf{e}_j^*(\mathbf{v})$. Hence, the dual basis $\{\mathbf{e}_j^*\}$ consists of linear functions

$$\mathbf{e}_j^* : \mathbf{x} \mapsto \langle \mathbf{e}_j, \mathbf{x} \rangle . \tag{5.2}$$

In contrast, determining the dual basis for a general (non-orthonormal) basis requires a complicated construction, such as that given in Sec. 2.3.3.

Corollary: If $\{\mathbf{e}_1, ..., \mathbf{e}_N\}$ is an arbitrary basis in V, there exists a scalar product with respect to which $\{\mathbf{e}_j\}$ is an orthonormal basis.

Proof: Let $\{\mathbf{e}_1^*, ..., \mathbf{e}_N^*\}$ be the dual basis in V^*. The required scalar product is defined by the bilinear form

$$S(\mathbf{u}, \mathbf{v}) = \sum_{j=1}^{N} \mathbf{e}_j^*(\mathbf{u})\, \mathbf{e}_j^*(\mathbf{v}) .$$

It is easy to show that the basis $\{\mathbf{e}_j\}$ is orthonormal with respect to the bilinear form S, namely $S(\mathbf{e}_i, \mathbf{e}_j) = \delta_{ij}$ (where δ_{ij} is the Kronecker symbol). It remains to prove that S is nondegenerate and positive-definite. To prove the nondegeneracy: Suppose that $\mathbf{u} \neq 0$; then we can decompose $\mathbf{u}$ in the basis $\{\mathbf{e}_j\}$,

$$\mathbf{u} = \sum_{j=1}^{N} u_j \mathbf{e}_j .$$

There will be at least one nonzero coefficient u_s, thus $S(\mathbf{e}_s, \mathbf{u}) = u_s \neq 0$. To prove that S is positive-definite, compute

$$S(\mathbf{u}, \mathbf{u}) = \sum_{j=1}^{N} u_j^2 > 0$$

as long as at least one coefficient u_j is nonzero. ■

Exercise 1: Let $\{\mathbf{v}_1, ..., \mathbf{v}_N\}$ be a basis in V, and let $\{\mathbf{e}_1, ..., \mathbf{e}_N\}$ be an orthonormal basis. Show that the linear operator

$$\hat{A}\mathbf{x} \equiv \sum_{i=1}^{N} \langle \mathbf{e}_i, \mathbf{x} \rangle\, \mathbf{v}_i$$

maps the basis $\{\mathbf{e}_i\}$ into the basis $\{\mathbf{v}_i\}$.

Exercise 2: Let $\{\mathbf{v}_1, ..., \mathbf{v}_n\}$ with $n < N$ be a linearly independent set (not necessarily orthonormal). Show that this set can be completed to a basis $\{\mathbf{v}_1, ..., \mathbf{v}_n, \mathbf{e}_{n+1}, ..., \mathbf{e}_N\}$ in V, such that every vector $\mathbf{e}_j$ $(j = n+1, ..., N)$ is orthogonal to every vector $\mathbf{v}_i$ $(i = 1, ..., n)$.

Hint: Follow the proof of Theorem 1 but begin the Gram-Schmidt procedure at step n, without orthogonalizing the vectors $\mathbf{v}_i$.

Exercise 3: Let $\{\mathbf{e}_1, ..., \mathbf{e}_N\}$ be an orthonormal basis, and let $v_i \equiv \langle \mathbf{v}, \mathbf{e}_i \rangle$. Show that

$$\langle \mathbf{v}, \mathbf{v} \rangle = \sum_{i=1}^{N} |v_i|^2 .$$

Exercise 4: Consider the space of polynomials of degree at most 2 in the variable x. Let us define the scalar product of two polynomials $p_1(x)$ and $p_2(x)$ by the formula

$$\langle p_1, p_2 \rangle = \frac{1}{2} \int_{-1}^{1} p_1(x) p_2(x) dx.$$

Find a linear polynomial $q_1(x)$ and a quadratic polynomial $q_2(x)$ such that $\{\underline{1}, q_1, q_2\}$ is an orthonormal basis in this space.

Remark: Some of the properties of the scalar product are related in an essential way to the assumption that we are working with real numbers. As an example of what could go wrong if we naively extended the same results to complex vector spaces, let us consider a vector $\mathbf{x} = (1, \mathrm{i}) \in \mathbb{C}^2$ and compute its scalar product with itself by the formula

$$\langle \mathbf{x}, \mathbf{x} \rangle = x_1^2 + x_2^2 = 1^2 + \mathrm{i}^2 = 0.$$

Hence we have a nonzero vector whose "length" is zero. To correct this problem when working with complex numbers, one usually considers a different kind of scalar product designed for complex vector spaces. For instance, the scalar product in $\mathbb{C}^n$ is defined by the formula

$$\langle (x_1, ..., x_n), (y_1, ..., y_n) \rangle = \sum_{j=1}^{n} x_j^* y_j,$$

where x_j^* is the complex conjugate of the component x_j. This scalar product is called **Hermitian** and has the property

$$\langle \mathbf{x}, \mathbf{y} \rangle = \langle \mathbf{y}, \mathbf{x} \rangle^* ,$$

that is, it is not symmetric but becomes complex-conjugated when the order of vectors is interchanged. According to this scalar product, we have for the vector $\mathbf{x} = (1, \mathrm{i}) \in \mathbb{C}^2$ a sensible result,

$$\langle \mathbf{x}, \mathbf{x} \rangle = x_1^* x_1 + x_2^* x_2 = |1|^2 + |\mathrm{i}|^2 = 2.$$

More generally, for $\mathbf{x} \neq 0$

$$\langle \mathbf{x}, \mathbf{x} \rangle = \sum_{i=1}^{N} |x_i|^2 > 0.$$

In this text, I will use this kind of scalar product only once (Sec. 5.6).

5.1.2 Correspondence between vectors and covectors

Let us temporarily consider the scalar product $\langle \mathbf{v}, \mathbf{x} \rangle$ as a function of $\mathbf{x}$ for a *fixed* $\mathbf{v}$. We may denote this function by $\mathbf{f}^*$. So $\mathbf{f}^* : \mathbf{x} \mapsto \langle \mathbf{v}, \mathbf{x} \rangle$ is a linear map $V \to \mathbb{K}$, i.e. (by definition) an element of V^*. Thus, a covector $\mathbf{f}^* \in V^*$ is determined for every $\mathbf{v}$. Therefore we have defined a map $V \to V^*$ whereby a vector $\mathbf{v}$ is mapped to the covector $\mathbf{f}^*$, which is defined by its action on vectors $\mathbf{x}$ as follows,

$$\mathbf{v} \mapsto \mathbf{f}^*; \quad \mathbf{f}^*(\mathbf{x}) \equiv \langle \mathbf{v}, \mathbf{x} \rangle, \quad \forall \mathbf{x} \in V. \tag{5.3}$$

This map is an isomorphism between V and V^* (not a canonical one, since it depends on the choice of the scalar product), as the following statement shows.

Statement 1: A nondegenerate bilinear form $B : V \otimes V \to \mathbb{K}$ defines an isomorphism $V \to V^*$ by the formula $\mathbf{v} \mapsto \mathbf{f}^*$, $\mathbf{f}^*(\mathbf{x}) \equiv B(\mathbf{v}, \mathbf{x})$.

Proof: We need to show that the map $\hat{B} : V \to V^*$ is a linear one-to-one (bijective) map. Linearity easily follows from the bilinearity of B. Bijectivity requires that no two different vectors are mapped into one and the same covector, and that any covector is an image of some vector. If two vectors $\mathbf{u} \neq \mathbf{v}$ are mapped into one covector $\mathbf{f}^*$ then $\hat{B}(\mathbf{u} - \mathbf{v}) = \mathbf{f}^* - \mathbf{f}^* = 0 \in V^*$, in other words, $B(\mathbf{u} - \mathbf{v}, \mathbf{x}) = 0$ for all $\mathbf{x}$. However, from the nondegeneracy of B it follows that there exists $\mathbf{x} \in V$ such that $B(\mathbf{u} - \mathbf{v}, \mathbf{x}) \neq 0$, which gives a contradiction. Finally, consider a basis $\{\mathbf{v}_j\}$ in V. Its image $\{\hat{B}\mathbf{v}_1, ..., \hat{B}\mathbf{v}_N\}$ must be a linearly independent set in V^* because a vanishing linear combination

$$\sum_k \lambda_k \hat{B}\mathbf{v}_k = 0 = \hat{B}\big(\sum_k \lambda_k \mathbf{v}_k\big)$$

entails $\sum_k \lambda_k \mathbf{v}_k = 0$ (we just proved that a nonzero vector cannot be mapped into the zero covector). Therefore $\{\hat{B}\mathbf{v}_1, ..., \hat{B}\mathbf{v}_N\}$ is a basis in V^*, and any covector $\mathbf{f}^*$ is a linear combination

$$\mathbf{f}^* = \sum_k f_k^* \hat{B}\mathbf{v}_k = \hat{B}\big(\sum_k f_k^* \mathbf{v}_k\big).$$

It follows that any vector $\mathbf{f}^*$ is an image of some vector from V. Thus $\hat{B}$ is a one-to-one map. ■

Let us show explicitly how to use the scalar product in order to map vectors to covectors and vice versa.

Example: We use the scalar product as the bilinear form B, so $B(\mathbf{x}, \mathbf{y}) \equiv \langle \mathbf{x}, \mathbf{y} \rangle$. Suppose $\{\mathbf{e}_j\}$ is an orthonormal basis. What is the covector $\hat{B}\mathbf{e}_1$? By Eq. (5.3), this covector acts on an arbitrary vector $\mathbf{x}$ as

$$\hat{B}\mathbf{e}_1(\mathbf{x}) = \langle \mathbf{e}_1, \mathbf{x} \rangle \equiv x_1,$$

where x_1 is the first component of the vector $\mathbf{x}$ in the basis $\{\mathbf{e}_j\}$, i.e. $\mathbf{x} = \sum_{i=1}^N x_i \mathbf{e}_i$. We find that $\hat{B}\mathbf{e}_1$ is the same as the covector $\mathbf{e}_1^*$ from the basis $\{\mathbf{e}_j^*\}$ dual to $\{\mathbf{e}_j\}$.

Suppose $\mathbf{f}^* \in V^*$ is a given covector. What is its pre-image $\hat{B}^{-1}\mathbf{f}^* \in V$? It is a vector $\mathbf{v}$ such that $\mathbf{f}^*(\mathbf{x}) = \langle \mathbf{v}, \mathbf{x} \rangle$ for any $\mathbf{x} \in V$. In order to determine $\mathbf{v}$, let us substitute the basis vectors $\mathbf{e}_j$ instead of $\mathbf{x}$; we then obtain

$$\mathbf{f}^*(\mathbf{e}_j) = \langle \mathbf{v}, \mathbf{e}_j \rangle .$$

Since the covector $\mathbf{f}^*$ is given, the numbers $\mathbf{f}^*(\mathbf{e}_j)$ are known, and hence

$$\mathbf{v} = \sum_{i=1}^{n} \mathbf{e}_j \langle \mathbf{v}, \mathbf{e}_j \rangle = \sum_{i=1}^{N} \mathbf{e}_j \, \mathbf{f}^*(\mathbf{e}_j).$$

■

Bilinear forms can be viewed as elements of the space $V^* \otimes V^*$.

Statement 2: All bilinear forms in V constitute a vector space canonically isomorphic to $V^* \otimes V^*$. A basis $\{\mathbf{e}_j\}$ is orthonormal with respect to the bilinear form

$$B \equiv \sum_{j=1}^{N} \mathbf{e}_j^* \otimes \mathbf{e}_j^* .$$

Proof: Left as exercise. ■

Exercise 1: Let $\{\mathbf{v}_1, ..., \mathbf{v}_N\}$ be a basis in V (not necessarily orthonormal), and denote by $\{\mathbf{v}_i^*\}$ the dual basis to $\{\mathbf{v}_i\}$. The dual basis is a basis in V^*. Now, we can map $\{\mathbf{v}_i^*\}$ into a basis $\{\mathbf{u}_i\}$ in V using the covector-vector correspondence. Show that $\langle \mathbf{v}_i, \mathbf{u}_j \rangle = \delta_{ij}$. Use this formula to show that this construction, applied to an orthonormal basis $\{\mathbf{e}_i\}$, yields again the same basis $\{\mathbf{e}_i\}$.

Hint: If vectors $\mathbf{x}$ and $\mathbf{y}$ have the same scalar products $\langle \mathbf{v}_i, \mathbf{x} \rangle = \langle \mathbf{v}_i, \mathbf{y} \rangle$ (for $i = 1, ..., N$) then $\mathbf{x} = \mathbf{y}$.

Exercise 2: Let $\{\mathbf{v}_1, ..., \mathbf{v}_N\}$ be a given (not necessarily orthonormal) basis in V, and denote by $\{\mathbf{v}_i^*\}$ the dual basis to $\{\mathbf{v}_i\}$. Due to the vector-covector correspondence, $\{\mathbf{v}_i^*\}$ is mapped into a basis $\{\mathbf{u}_j\}$ in V, so the tensor

$$\hat{1}_V \equiv \sum_{i=1}^{N} \mathbf{v}_i \otimes \mathbf{v}_i^*$$

is mapped into a bilinear form B acting as

$$B(\mathbf{x}, \mathbf{y}) = \sum_{i=1}^{N} \langle \mathbf{v}_i, \mathbf{x} \rangle \langle \mathbf{u}_i, \mathbf{y} \rangle .$$

Show that this bilinear form coincides with the scalar product, i.e.

$$B(\mathbf{x}, \mathbf{y}) = \langle \mathbf{x}, \mathbf{y} \rangle , \quad \forall \mathbf{x}, \mathbf{y} \in V.$$

Hint: Since $\sum_{i=1}^{N} \mathbf{v}_i \otimes \mathbf{v}_i^* = \hat{1}_V$, we have $\sum_{i=1}^{N} \mathbf{v}_i \langle \mathbf{u}_i, \mathbf{y} \rangle = \mathbf{y}$.

Exercise 3: If a scalar product $\langle \cdot, \cdot \rangle$ is given in V, a scalar product $\langle \cdot, \cdot \rangle_*$ can be constructed also in V^* as follows: Given any two covectors $\mathbf{f}^*, \mathbf{g}^* \in V^*$, we map them into vectors $\mathbf{u}, \mathbf{v} \in V$ and then define

$$\langle \mathbf{f}^*, \mathbf{g}^* \rangle_* \equiv \langle \mathbf{u}, \mathbf{v} \rangle .$$

Show that this scalar product is bilinear and positive-definite if $\langle \cdot, \cdot \rangle$ is. For an orthonormal basis $\{\mathbf{e}_j\}$, show that the dual basis $\{\mathbf{e}_j^*\}$ in V^* is also orthonormal with respect to this scalar product.

Exercise 4:* Consider the space End V of linear operators in a vector space V with $\dim V \geq 2$. A bilinear form in the space End V is defined as follows: for any two operators $\hat{A}, \hat{B} \in \text{End}\, V$ we set $\langle \hat{A}, \hat{B} \rangle \equiv \text{Tr}(\hat{A}\hat{B})$. Show that $\langle \hat{A}, \hat{B} \rangle$ is bilinear, symmetric, and nondegenerate, but *not* positive-definite.

Hint: To show nondegeneracy, consider a nonzero operator $\hat{A}$; there exists $\mathbf{v} \in V$ such that $\hat{A}\mathbf{v} \neq 0$, and then one can choose $\mathbf{f}^* \in V^*$ such that $\mathbf{f}^*(\hat{A}\mathbf{v}) \neq 0$; then define $\hat{B} \equiv \mathbf{v} \otimes \mathbf{f}^*$ and verify that $\langle \hat{A}, \hat{B} \rangle$ is nonzero. To show that the scalar product is not positive-definite, consider $\hat{C} = \mathbf{v} \otimes \mathbf{f}^* + \mathbf{w} \otimes \mathbf{g}^*$ and choose the vectors and the covectors appropriately so that $\text{Tr}(\hat{C}^2) < 0$.

5.1.3 * Example: bilinear forms on $V \oplus V^*$

If V is a vector space then the space $V \oplus V^*$ has *two* canonically defined bilinear forms that could be useful under certain circumstances (when positive-definiteness is not required). This construction is used in abstract algebra, and I mention it here as an example of a purely algebraic and basis-free definition of a bilinear form.

If $(\mathbf{u}, \mathbf{f}^*)$ and $(\mathbf{v}, \mathbf{g}^*)$ are two elements of $V \oplus V^*$, a canonical bilinear form is defined by the formula

$$\langle (\mathbf{u}, \mathbf{f}^*), (\mathbf{v}, \mathbf{g}^*) \rangle = \mathbf{f}^*(\mathbf{v}) + \mathbf{g}^*(\mathbf{u}) . \qquad (5.4)$$

This formula does *not* define a positive-definite bilinear form because

$$\langle (\mathbf{u}, \mathbf{f}^*), (\mathbf{u}, \mathbf{f}^*) \rangle = 2\mathbf{f}^*(\mathbf{u}) ,$$

which can be positive, negative, or zero for some $(\mathbf{u}, \mathbf{f}^*) \in V \oplus V^*$.

Statement: The bilinear form defined by Eq. (5.4) is symmetric and nondegenerate.

Proof: The symmetry is obvious from Eq. (5.4). Then for any nonzero vector $(\mathbf{u}, \mathbf{f}^*)$ we need to find a vector $(\mathbf{v}, \mathbf{g}^*)$ such that $\langle (\mathbf{u}, \mathbf{f}^*), (\mathbf{v}, \mathbf{g}^*) \rangle \neq 0$. By assumption, either $\mathbf{u} \neq 0$ or $\mathbf{f}^* \neq 0$ or both. If $\mathbf{u} \neq 0$, there exists a covector $\mathbf{g}^*$ such that $\mathbf{g}^*(\mathbf{u}) \neq 0$; then we choose $\mathbf{v} = 0$. If $\mathbf{f}^* \neq 0$, there exists a vector $\mathbf{v}$ such that $\mathbf{f}^*(\mathbf{v}) \neq 0$, and then we choose $\mathbf{g}^* = 0$. Thus the nondegeneracy is proved. ■

Alternatively, there is a canonically defined *antisymmetric* bilinear form (or 2-form),

$$\langle (\mathbf{u}, \mathbf{f}^*), (\mathbf{v}, \mathbf{g}^*) \rangle = \mathbf{f}^*(\mathbf{v}) - \mathbf{g}^*(\mathbf{u}) .$$

This bilinear form is also nondegenerate (the same proof goes through as for the symmetric bilinear form above). Nevertheless, none of the two bilinear forms can serve as a scalar product: the former lacks positive-definiteness, the latter is antisymmetric rather than symmetric.

5.1.4 Scalar product in index notation

In the index notation, the scalar product tensor $S \in V^* \otimes V^*$ is represented by a matrix S_{ij} (with lower indices), and so the scalar product of two vectors is written as

$$\langle \mathbf{u}, \mathbf{v} \rangle = u^i v^j S_{ij}.$$

Alternatively, one uses the vector-to-covector map $\hat{S} : V \to V^*$ and writes

$$\langle \mathbf{u}, \mathbf{v} \rangle = \mathbf{u}^* (\mathbf{v}) = u_i v^i,$$

where the covector $\mathbf{u}^*$ is defined by

$$\mathbf{u}^* \equiv \hat{S}\mathbf{u} \;\Rightarrow\; u_i \equiv S_{ij} u^j.$$

Typically, in the index notation one uses the same symbol to denote a vector, u^i, and the corresponding covector, u_i. This is unambiguous as long as the scalar product is fixed.

5.2 Orthogonal subspaces

From now on, we work in a real, N-dimensional vector space V equipped with a scalar product.

We call two subspaces $V_1 \subset V$ and $V_2 \subset V$ **orthogonal** if every vector from V_1 is orthogonal to every vector from V_2. An important example of orthogonal subspaces is given by the construction of the orthogonal complement.

Definition: The set of vectors orthogonal to a given vector $\mathbf{v}$ is denoted by $\mathbf{v}^\perp$ and is called the **orthogonal complement** of the vector $\mathbf{v}$. Written as a formula:

$$\mathbf{v}^\perp = \{ \mathbf{x} \,|\, \mathbf{x} \in V,\, \langle \mathbf{x}, \mathbf{v} \rangle = 0 \}.$$

Similarly, the set of vectors orthogonal to each of the vectors $\{\mathbf{v}_1, ..., \mathbf{v}_n\}$ is denoted by $\{\mathbf{v}_1, ..., \mathbf{v}_n\}^\perp$.

Examples: If $\{\mathbf{e}_1, \mathbf{e}_2, \mathbf{e}_3, \mathbf{e}_4\}$ is an orthonormal basis in V then the subspace Span $\{\mathbf{e}_1, \mathbf{e}_3\}$ is orthogonal to the subspace Span $\{\mathbf{e}_2, \mathbf{e}_4\}$ because any linear combination of $\mathbf{e}_1$ and $\mathbf{e}_3$ is orthogonal to any linear combination of $\mathbf{e}_2$ and $\mathbf{e}_4$. The orthogonal complement of $\mathbf{e}_1$ is

$$\mathbf{e}_1^\perp = \text{Span}\,\{\mathbf{e}_2, \mathbf{e}_3, \mathbf{e}_4\}.$$

Statement 1: **(1)** The orthogonal complement $\{\mathbf{v}_1, ..., \mathbf{v}_n\}^\perp$ is a subspace of V.

(2) Every vector from the subspace Span $\{\mathbf{v}_1, ..., \mathbf{v}_n\}$ is orthogonal to every vector from $\{\mathbf{v}_1, ..., \mathbf{v}_n\}^\perp$.

Proof: **(1)** If two vectors $\mathbf{x}, \mathbf{y}$ belong to $\{\mathbf{v}_1, ..., \mathbf{v}_n\}^\perp$, it means that $\langle \mathbf{v}_i, \mathbf{x} \rangle = 0$ and $\langle \mathbf{v}_i, \mathbf{y} \rangle = 0$ for $i = 1, ..., n$. Since the scalar product is linear, it follows that

$$\langle \mathbf{v}_i, \mathbf{x} + \lambda \mathbf{y} \rangle = 0, \quad i = 1, ..., n.$$

Therefore, any linear combination of $\mathbf{x}$ and $\mathbf{y}$ also belongs to $\{\mathbf{v}_1, ..., \mathbf{v}_n\}^\perp$. This is the same as to say that $\{\mathbf{v}_1, ..., \mathbf{v}_n\}^\perp$ is a subspace of V.

(2) Suppose $\mathbf{x} \in \text{Span}\{\mathbf{v}_1, ..., \mathbf{v}_n\}$ and $\mathbf{y} \in \{\mathbf{v}_1, ..., \mathbf{v}_n\}^\perp$; then we may express $\mathbf{x} = \sum_{i=1}^n \lambda_i \mathbf{v}_i$ with some coefficients λ_i, while $\langle \mathbf{v}_i, \mathbf{y} \rangle = 0$ for $i = 1, ..., n$. It follows from the linearity of the scalar product that

$$\langle \mathbf{x}, \mathbf{y} \rangle = \sum_{i=1}^{n} \langle \lambda_i \mathbf{v}_i, \mathbf{y} \rangle = 0.$$

Hence, every such $\mathbf{x}$ is orthogonal to every such $\mathbf{y}$. ■

Definition: If $U \subset V$ is a given subspace, the **orthogonal complement** $U^\perp$ is defined as the subspace of vectors that are orthogonal to every vector from U. (It is easy to see that all these vectors form a subspace.)

Exercise 1: Given a subspace $U \subset V$, we may choose a basis $\{\mathbf{u}_1, ..., \mathbf{u}_n\}$ in U and then construct the orthogonal complement $\{\mathbf{u}_1, ..., \mathbf{u}_n\}^\perp$ as defined above. Show that the subspace $\{\mathbf{u}_1, ..., \mathbf{u}_n\}^\perp$ is the same as $U^\perp$ independently of the choice of the basis $\{\mathbf{u}_j\}$ in U. ■

The space V can be decomposed into a direct sum of orthogonal subspaces.

Statement 2: Given a subspace $U \subset V$, we can construct its orthogonal complement $U^\perp \subset V$. Then $V = U \oplus U^\perp$; in other words, every vector $\mathbf{x} \in V$ can be uniquely decomposed as $\mathbf{x} = \mathbf{u} + \mathbf{w}$ where $\mathbf{u} \in U$ and $\mathbf{w} \in U^\perp$.

Proof: Choose a basis $\{\mathbf{u}_1, ..., \mathbf{u}_n\}$ of U. If $n = N$, the orthogonal complement $U^\perp$ is the zero-dimensional subspace, so there is nothing left to prove. If $n < N$, we may choose some additional vectors $\mathbf{e}_{n+1}, ..., \mathbf{e}_N$ such that the set $\{\mathbf{u}_1, ..., \mathbf{u}_n, \mathbf{e}_{n+1}, ..., \mathbf{e}_N\}$ is a basis in V and *every* vector $\mathbf{e}_j$ is orthogonal to *every* vector $\mathbf{u}_i$. Such a basis exists (see Exercise 2 in Sec. 5.1.1). Then every vector $\mathbf{x} \in V$ can be decomposed as

$$\mathbf{x} = \sum_{i=1}^{n} \lambda_i \mathbf{u}_i + \sum_{i=n+1}^{N} \mu_i \mathbf{e}_i \equiv \mathbf{u} + \mathbf{w}.$$

This decomposition provides the required decomposition of $\mathbf{x}$ into two vectors.

It remains to show that this decomposition is unique (in particular, independent of the choice of bases). If there were two different such decompositions, say $\mathbf{x} = \mathbf{u} + \mathbf{w} = \mathbf{u}' + \mathbf{w}'$, we would have

$$0 \stackrel{!}{=} \langle \mathbf{u} - \mathbf{u}' + \mathbf{w} - \mathbf{w}', \mathbf{y} \rangle, \quad \forall \mathbf{y} \in V.$$

Let us now show that $\mathbf{u} = \mathbf{u}'$ and $\mathbf{w} = \mathbf{w}'$: Taking an arbitrary $\mathbf{y} \in U$, we have $\langle \mathbf{w} - \mathbf{w}', \mathbf{y} = 0 \rangle$ and hence find that $\mathbf{u} - \mathbf{u}'$ is orthogonal to $\mathbf{y}$. It means that the vector $\mathbf{u}-\mathbf{u}' \in U$ is orthogonal to *every* vector $\mathbf{y} \in U$, e.g. to $\mathbf{y} \equiv \mathbf{u}-\mathbf{u}'$; since the scalar product of a nonzero vector with itself cannot be equal to zero, we must have $\mathbf{u}-\mathbf{u}'=0$. Similarly, by taking an arbitrary $\mathbf{z} \in U^{\perp}$, we find that $\mathbf{w}-\mathbf{w}'$ is orthogonal to $\mathbf{z}$, hence we must have $\mathbf{w}-\mathbf{w}'=0$. ■

An important operation is the orthogonal projection onto a subspace.

Statement 3: There are many projectors onto a given subspace $U \subset V$, but only one projector $\hat{P}_U$ that preserves the scalar product with vectors from U. Namely, there exists a unique linear operator $\hat{P}_U$, called the **orthogonal projector** onto the subspace U, such that

$$\begin{aligned} \hat{P}_U \hat{P}_U = \hat{P}_U; \quad (\hat{P}_U \mathbf{x}) \in U \text{ for } \forall \mathbf{x} \in V \quad &\text{— projection property;} \\ \langle \hat{P}_U \mathbf{x}, \mathbf{a} \rangle = \langle \mathbf{x}, \mathbf{a} \rangle, \ \forall \mathbf{x} \in V, \ \mathbf{a} \in U \quad &\text{— preserves } \langle \cdot, \cdot \rangle. \end{aligned}$$

Remark: The name "orthogonal projections" (this is quite different from "orthogonal transformations" defined in the next section!) comes from a geometric analogy: Projecting a three-dimensional vector orthogonally onto a plane means that the projection does not add to the vector any components parallel to the plane. The vector is "cast down" in the direction normal to the plane. The projection modifies a vector $\mathbf{x}$ by adding to it some vector orthogonal to the plane; this modification preserves the scalar products of $\mathbf{x}$ with vectors in the plane. Perhaps a better word would be "normal projection."

Proof: Suppose $\{\mathbf{u}_1, ..., \mathbf{u}_n\}$ is a basis in the subspace U, and assume that $n < N$ (or else $U = V$ and there exists only one projector onto U, namely the identity operator, which preserves the scalar product, so there is nothing left to prove). We may complete the basis $\{\mathbf{u}_1, ..., \mathbf{u}_n\}$ of U to a basis $\{\mathbf{u}_1, ..., \mathbf{u}_n, \mathbf{e}_{n+1}, ..., \mathbf{e}_N\}$ in the entire space V. Let $\{\mathbf{u}_1^*, ..., \mathbf{u}_n^*, \mathbf{e}_{n+1}^*, ..., \mathbf{e}_N^*\}$ be the corresponding dual basis. Then a projector onto U can be defined by

$$\hat{P} = \sum_{i=1}^{n} \mathbf{u}_i \otimes \mathbf{u}_i^*,$$

that is, $\hat{P}\mathbf{x}$ simply omits the components of the vector $\mathbf{x}$ parallel to any $\mathbf{e}_j$ ($j = n+1, ..., N$). For example, the operator $\hat{P}$ maps the linear combination $\lambda \mathbf{u}_1 + \mu \mathbf{e}_{n+1}$ to $\lambda \mathbf{u}_1$, omitting the component parallel to $\mathbf{e}_{n+1}$. There are infinitely many ways of choosing $\{\mathbf{e}_j \,|\, j = n+1, ..., N\}$; for instance, one can add to $\mathbf{e}_{n+1}$ an arbitrary linear combination of $\{\mathbf{u}_j\}$ and obtain another possible choice of $\mathbf{e}_{n+1}$. Hence there are infinitely many possible projectors onto U.

While all these projectors satisfy the projection property, not all of them preserve the scalar product. The orthogonal projector is the one obtained from a particular completion of the basis, namely such that every vector $\mathbf{e}_j$ is orthogonal to every vector $\mathbf{u}_i$. Such a basis exists (see Exercise 2 in Sec. 5.1.1). Using the construction shown above, we obtain a projector that we will denote $\hat{P}_U$. We will now show that this projector is unique and satisfies the scalar product preservation property.

The scalar product is preserved for the following reason. For any $\mathbf{x} \in V$, we have a unique decomposition $\mathbf{x} = \mathbf{u} + \mathbf{w}$, where $\mathbf{u} \in U$ and $\mathbf{w} \in U^{\perp}$. The definition of $\hat{P}_U$ guarantees that $\hat{P}_U \mathbf{x} = \mathbf{u}$. Hence

$$\langle \mathbf{x}, \mathbf{a} \rangle = \langle \mathbf{u} + \mathbf{w}, \mathbf{a} \rangle = \langle \mathbf{u}, \mathbf{a} \rangle = \langle \hat{P}_U \mathbf{x}, \mathbf{a} \rangle, \quad \forall \mathbf{x} \in V,\ \mathbf{a} \in U.$$

Now the uniqueness: If there were two projectors $\hat{P}_U$ and $\hat{P}'_U$, both satisfying the scalar product preservation property, then

$$\langle (\hat{P}_U - \hat{P}'_U)\mathbf{x}, \mathbf{u} \rangle = 0 \quad \forall \mathbf{x} \in V,\ \mathbf{u} \in U.$$

For a given $\mathbf{x} \in V$, the vector $\mathbf{y} \equiv (\hat{P}_U - \hat{P}'_U)\mathbf{x}$ belongs to U and is orthogonal to every vector in U. Therefore $\mathbf{y} = 0$. It follows that $(\hat{P}_U - \hat{P}'_U)\mathbf{x} = 0$ for any $\mathbf{x} \in V$, i.e. the operator $(\hat{P}_U - \hat{P}'_U)$ is equal to zero. ■

Example: Given a nonzero vector $\mathbf{v} \in V$, let us construct the orthogonal projector onto the subspace $\mathbf{v}^{\perp}$. It seems (judging from the proof of Statement 3) that we need to chose a basis in $\mathbf{v}^{\perp}$. However, the projector (as we know) is in fact independent of the choice of the basis and can be constructed as follows:

$$\hat{P}_{\mathbf{v}^{\perp}} \mathbf{x} \equiv \mathbf{x} - \mathbf{v} \frac{\langle \mathbf{v}, \mathbf{x} \rangle}{\langle \mathbf{v}, \mathbf{v} \rangle}.$$

It is easy to check that this is indeed a projector onto $\mathbf{v}^{\perp}$, namely we can check that $\langle \hat{P}_{\mathbf{v}^{\perp}} \mathbf{x}, \mathbf{v} \rangle = 0$ for all $\mathbf{x} \in V$, and that $\mathbf{v}^{\perp}$ is an invariant subspace under $\hat{P}_{\mathbf{v}^{\perp}}$.

Exercise 2: Construct an orthogonal projector $\hat{P}_{\mathbf{v}}$ onto the space spanned by the vector $\mathbf{v}$.

Answer: $\hat{P}_{\mathbf{v}} \mathbf{x} = \mathbf{v} \frac{\langle \mathbf{v}, \mathbf{x} \rangle}{\langle \mathbf{v}, \mathbf{v} \rangle}$.

5.2.1 Affine hyperplanes

Suppose $\mathbf{n} \in V$ is a given vector and α a given number. The set of vectors $\mathbf{x}$ satisfying the equation

$$\langle \mathbf{n}, \mathbf{x} \rangle = \alpha$$

is called an **affine hyperplane**. Note that an affine hyperplane is not necessarily a subspace of V because $\mathbf{x} = 0$ does not belong to the hyperplane when $\alpha \neq 0$.

The geometric interpretation of a hyperplane follows from the fact that the difference of any two vectors $\mathbf{x}_1$ and $\mathbf{x}_2$, both belonging to the hyperplane, satisfies

$$\langle \mathbf{n}, \mathbf{x}_1 - \mathbf{x}_2 \rangle = 0.$$

Hence, all vectors in the hyperplane can be represented as a sum of one such vector, say $\mathbf{x}_0$, and an arbitrary vector orthogonal to $\mathbf{n}$. Geometrically, this means that the hyperplane is orthogonal to the vector $\mathbf{n}$ and may be shifted from the origin.

Example: Let us consider an affine hyperplane given by the equation $\langle \mathbf{n}, \mathbf{x} \rangle = 1$, and let us compute the shortest vector belonging to the hyperplane. Any vector $\mathbf{x} \in V$ can be written as

$$\mathbf{x} = \lambda \mathbf{n} + \mathbf{b},$$

where $\mathbf{b}$ is some vector such that $\langle \mathbf{n}, \mathbf{b} \rangle = 0$. If $\mathbf{x}$ belongs to the hyperplane, we have

$$1 = \langle \mathbf{n}, \mathbf{x} \rangle = \langle \mathbf{n}, \lambda \mathbf{n} + \mathbf{b} \rangle = \lambda \langle \mathbf{n}, \mathbf{n} \rangle .$$

Hence, we must have

$$\lambda = \frac{1}{\langle \mathbf{n}, \mathbf{n} \rangle}.$$

The squared length of $\mathbf{x}$ is then computed as

$$\begin{aligned} \langle \mathbf{x}, \mathbf{x} \rangle &= \lambda^2 \langle \mathbf{n}, \mathbf{n} \rangle + \langle \mathbf{b}, \mathbf{b} \rangle \\ &= \frac{1}{\langle \mathbf{n}, \mathbf{n} \rangle} + \langle \mathbf{b}, \mathbf{b} \rangle \geq \frac{1}{\langle \mathbf{n}, \mathbf{n} \rangle}. \end{aligned}$$

The inequality becomes an equality when $\mathbf{b} = 0$, i.e. when $\mathbf{x} = \lambda \mathbf{n}$. Therefore, the smallest possible length of $\mathbf{x}$ is equal to $\sqrt{\lambda}$, which is equal to the inverse length of $\mathbf{n}$.

Exercise: Compute the shortest distance between two parallel hyperplanes defined by equations $\langle \mathbf{n}, \mathbf{x} \rangle = \alpha$ and $\langle \mathbf{n}, \mathbf{x} \rangle = \beta$.

Answer:

$$\frac{|\alpha - \beta|}{\sqrt{\langle \mathbf{n}, \mathbf{n} \rangle}}.$$

5.3 Orthogonal transformations

Definition: An operator $\hat{A}$ is called an **orthogonal transformation** with respect to the scalar product $\langle , \rangle$ if

$$\langle \hat{A}\mathbf{v}, \hat{A}\mathbf{w} \rangle = \langle \mathbf{v}, \mathbf{w} \rangle , \quad \forall \mathbf{v}, \mathbf{w} \in V.$$

(We use the words "transformation" and "operator" interchangeably since we are always working within the same vector space V.)

5.3.1 Examples and properties

Example 1: Rotation by a fixed angle is an orthogonal transformation in a Euclidean plane. It is easy to see that such a rotation preserves scalar products (angles and lengths are preserved by a rotation). Let us define this transformation by a formula. If $\{\mathbf{e}_1, \mathbf{e}_2\}$ is a positively oriented orthonormal basis in the Euclidean plane, then we define the **rotation** $\hat{R}_\alpha$ of the plane by angle α in the counter-clockwise direction by

$$\begin{aligned} \hat{R}_\alpha \mathbf{e}_1 &\equiv \mathbf{e}_1 \cos \alpha - \mathbf{e}_2 \sin \alpha, \\ \hat{R}_\alpha \mathbf{e}_2 &\equiv \mathbf{e}_1 \sin \alpha + \mathbf{e}_2 \cos \alpha. \end{aligned}$$

One can quickly verify that the transformed basis $\{\hat{R}_\alpha \mathbf{e}_1, \hat{R}_\alpha \mathbf{e}_2\}$ is also an orthonormal basis; for example,

$$\langle \hat{R}_\alpha \mathbf{e}_1, \hat{R}_\alpha \mathbf{e}_1 \rangle = \langle \mathbf{e}_1, \mathbf{e}_1 \rangle \cos^2 \alpha + \langle \mathbf{e}_2, \mathbf{e}_2 \rangle \sin^2 \alpha = 1.$$

Example 2: Mirror reflections are also orthogonal transformations. A mirror reflection with respect to the basis vector $\mathbf{e}_1$ maps a vector $\mathbf{x} = \lambda_1 \mathbf{e}_1 + \lambda_2 \mathbf{e}_2 + ... + \lambda_N \mathbf{e}_N$ into $\hat{M}_{\mathbf{e}_1} \mathbf{x} = -\lambda_1 \mathbf{e}_1 + \lambda_2 \mathbf{e}_2 + ... + \lambda_N \mathbf{e}_N$, i.e. only the first coefficient changes sign. A mirror reflection with respect to an arbitrary axis $\mathbf{n}$ (where $\mathbf{n}$ is a **unit** vector, i.e. $\langle \mathbf{n}, \mathbf{n} \rangle = 1$) can be defined as the transformation

$$\hat{M}_{\mathbf{n}} \mathbf{x} \equiv \mathbf{x} - 2 \langle \mathbf{n}, \mathbf{x} \rangle \mathbf{n}.$$

This transformation is interpreted geometrically as mirror reflection with respect to the hyperplane $\mathbf{n}^\perp$. ■

An interesting fact is that orthogonality *entails* linearity.

Statement 1: If a map $\hat{A} : V \to V$ is orthogonal then it is a linear map, $\hat{A}(\mathbf{u} + \lambda \mathbf{v}) = \hat{A}\mathbf{u} + \lambda \hat{A}\mathbf{v}$.

Proof: Consider an orthonormal basis $\{\mathbf{e}_1, ..., \mathbf{e}_N\}$. The set $\{\hat{A}\mathbf{e}_1, ..., \hat{A}\mathbf{e}_N\}$ is orthonormal because

$$\langle \hat{A}\mathbf{e}_i, \hat{A}\mathbf{e}_j \rangle = \langle \mathbf{e}_i, \mathbf{e}_j \rangle = \delta_{ij}.$$

By Theorem 1 of Sec. 5.1 the set $\{\hat{A}\mathbf{e}_1, ..., \hat{A}\mathbf{e}_N\}$ is linearly independent and is therefore an *orthonormal basis* in V. Consider an arbitrary vector $\mathbf{v} \in V$ and its image $\hat{A}\mathbf{v}$ after the transformation $\hat{A}$. By Theorem 2 of Sec. 5.1.1, we can decompose $\mathbf{v}$ in the basis $\{\mathbf{e}_j\}$ and $\hat{A}\mathbf{v}$ in the basis $\{\hat{A}\mathbf{e}_j\}$ as follows,

$$\mathbf{v} = \sum_{j=1}^{N} \langle \mathbf{e}_j, \mathbf{v} \rangle \mathbf{e}_j,$$
$$\hat{A}\mathbf{v} = \sum_{j=1}^{N} \langle \hat{A}\mathbf{e}_j, \hat{A}\mathbf{v} \rangle \hat{A}\mathbf{e}_j = \sum_{j=1}^{N} \langle \mathbf{e}_j, \mathbf{v} \rangle \hat{A}\mathbf{e}_j.$$

Any other vector $\mathbf{u} \in V$ can be similarly decomposed, and so we obtain

$$\begin{aligned} \hat{A}(\mathbf{u} + \lambda \mathbf{v}) &= \sum_{j=1}^{N} \langle \mathbf{e}_j, \mathbf{u} + \lambda \mathbf{v} \rangle \hat{A}\mathbf{e}_j \\ &= \sum_{j=1}^{N} \langle \mathbf{e}_j, \mathbf{u} \rangle \hat{A}\mathbf{e}_j + \lambda \sum_{j=1}^{N} \langle \mathbf{e}_j, \mathbf{v} \rangle \hat{A}\mathbf{e}_j \\ &= \hat{A}\mathbf{u} + \lambda \hat{A}\mathbf{v}, \quad \forall \mathbf{u}, \mathbf{v} \in V, \ \lambda \in \mathbb{K}, \end{aligned}$$

showing that the map $\hat{A}$ is linear. ■

An orthogonal operator always maps an orthonormal basis into another orthonormal basis (this was shown in the proof of Statement 1). The following exercise shows that the converse is also true.

Exercise 1: Prove that a transformation is orthogonal if and only if it maps *some* orthonormal basis into another orthonormal basis. Deduce that any orthogonal transformation is invertible.

Exercise 2: If a linear transformation $\hat{A}$ satisfies $\langle \hat{A}\mathbf{x}, \hat{A}\mathbf{x}\rangle = \langle \mathbf{x}, \mathbf{x}\rangle$ for all $\mathbf{x} \in V$, show that $\hat{A}$ is an orthogonal transformation. (This shows how to check more easily whether a given linear transformation is orthogonal.)

Hint: Substitute $\mathbf{x} = \mathbf{y} + \mathbf{z}$.

Exercise 3: Show that for any two orthonormal bases $\{\mathbf{e}_j \,|\, j = 1, ..., N\}$ and $\{\mathbf{f}_j \,|\, j = 1, ..., N\}$, there exists an orthogonal operator $\hat{R}$ that maps the basis $\{\mathbf{e}_j\}$ into the basis $\{\mathbf{f}_j\}$, i.e. $\hat{R}\mathbf{e}_j = \mathbf{f}_j$ for $j = 1, ..., N$.

Hint: A linear operator mapping $\{\mathbf{e}_j\}$ into $\{\mathbf{f}_j\}$ exists; show that this operator is orthogonal.

Exercise 4: Prove that $\hat{M}_{\mathbf{n}}$ (as defined in Example 2) is an orthogonal transformation by showing that $\langle \hat{M}_{\mathbf{n}}\mathbf{x}, \hat{M}_{\mathbf{n}}\mathbf{x}\rangle = \langle \mathbf{x}, \mathbf{x}\rangle$ for any $\mathbf{x}$.

Exercise 5: Consider the orthogonal transformations $\hat{R}_\alpha$ and $\hat{M}_{\mathbf{n}}$ and an orthonormal basis $\{\mathbf{e}_1, \mathbf{e}_2\}$ as defined in Examples 1 and 2. Show by a direct calculation that

$$(\hat{R}_\alpha \mathbf{e}_1) \wedge (\hat{R}_\alpha \mathbf{e}_2) = \mathbf{e}_1 \wedge \mathbf{e}_2$$

and that

$$(\hat{M}_{\mathbf{n}} \mathbf{e}_1) \wedge (\hat{M}_{\mathbf{n}} \mathbf{e}_2) = -\mathbf{e}_1 \wedge \mathbf{e}_2.$$

This is the same as to say that $\det \hat{R}_\alpha = 1$ and $\det \hat{M}_{\mathbf{n}} = -1$. This indicates that rotations preserve orientation while mirror reflections reverse orientation. ■

5.3.2 Transposition

Another way to characterize orthogonal transformations is by using transposed operators. Recall that the canonically defined transpose to $\hat{A}$ is $\hat{A}^T : V^* \to V^*$ (see Sec. 1.8.4, p. 59 for a definition). In a (finite-dimensional) space with a scalar product, the one-to-one correspondence between V and V^* means that $\hat{A}^T$ can be identified with some operator acting in V (rather than in V^*). Let us also denote that operator by $\hat{A}^T$ and call it the **transposed** to $\hat{A}$. (This transposition is not canonical but depends on the scalar product.) We can formulate the definition of $\hat{A}^T$ as follows.

Definition 1: In a finite-dimensional space with a scalar product, the **transposed** operator $\hat{A}^T : V \to V$ is defined by

$$\langle \hat{A}^T \mathbf{x}, \mathbf{y}\rangle \equiv \langle \mathbf{x}, \hat{A}\mathbf{y}\rangle, \quad \forall \mathbf{x}, \mathbf{y} \in V.$$

Exercise 1: Show that $(\hat{A}\hat{B})^T = \hat{B}^T \hat{A}^T$.

Statement 1: If $\hat{A}$ is orthogonal then $\hat{A}^T \hat{A} = \hat{1}_V$.

Proof: By definition of orthogonal transformation, $\langle \hat{A}\mathbf{x}, \hat{A}\mathbf{y}\rangle = \langle \mathbf{x}, \mathbf{y}\rangle$ for all $\mathbf{x}, \mathbf{y} \in V$. Then we use the definition of $\hat{A}^T$ and obtain

$$\langle \mathbf{x}, \mathbf{y}\rangle = \langle \hat{A}\mathbf{x}, \hat{A}\mathbf{y}\rangle = \langle \hat{A}^T \hat{A}\mathbf{x}, \mathbf{y}\rangle.$$

Since this holds for all $\mathbf{x}, \mathbf{y} \in V$, we conclude that $\hat{A}^T \hat{A} = \hat{1}_V$ (see Exercise 2 in Sec. 5.1). ■

Let us now see how transposed operators appear in matrix form. Suppose $\{\mathbf{e}_j\}$ is an orthonormal basis in V; then the operator $\hat{A}$ can be represented by some matrix A_{ij} in this basis. Then the operator $\hat{A}^T$ is represented by the matrix A_{ji} in the same basis (i.e. by the matrix transpose of A_{ij}), as shown in the following exercise. (Note that the operator $\hat{A}^T$ is *not* represented by the transposed matrix when the basis is not orthonormal.)

Exercise 2: Show that the operator $\hat{A}^T$ is represented by the transposed matrix A_{ji} in the same (orthonormal) basis in which the operator $\hat{A}$ has the matrix A_{ij}. Deduce that $\det \hat{A} = \det(\hat{A}^T)$.

Solution: The matrix element A_{ij} with respect to an orthonormal basis $\{\mathbf{e}_j\}$ is the coefficient in the tensor decomposition $\hat{A} = \sum_{i,j=1}^N A_{ij}\mathbf{e}_i \otimes \mathbf{e}_j^*$ and can be computed using the scalar product as

$$A_{ij} = \langle \mathbf{e}_i, \hat{A}\mathbf{e}_j\rangle.$$

The transposed operator satisfies

$$\langle \mathbf{e}_i, \hat{A}^T\mathbf{e}_j\rangle = \langle \hat{A}\mathbf{e}_i, \mathbf{e}_j\rangle = A_{ji}.$$

Hence, the matrix elements of $\hat{A}^T$ are A_{ji}, i.e. the matrix elements of the transposed matrix. We know that $\det(A_{ji}) = \det(A_{ij})$. If the basis $\{\mathbf{e}_j\}$ is not orthonormal, the property $A_{ij} = \langle \mathbf{e}_i, \hat{A}\mathbf{e}_j\rangle$ does not hold and the argument fails. ■

We have seen in Exercise 5 (Sec. 5.3.1) that the determinants of some orthogonal transformations were equal to $+1$ or -1. This is, in fact, a general property.

Statement 2: The determinant of an orthogonal transformation is equal to 1 or to -1.

Proof: An orthogonal transformation $\hat{A}$ satisfies $\hat{A}^T \hat{A} = \hat{1}_V$. Compute the determinant of both sides; since the determinant of the transposed operator is equal to that of the original operator, we have $(\det \hat{A})^2 = 1$. ■

5.4 Applications of exterior product

We will now apply the exterior product techniques to spaces with a scalar product and obtain several important results.

5.4.1 Orthonormal bases, volume, and $\wedge^N V$

If an orthonormal basis $\{\mathbf{e}_j\}$ is chosen, we can consider a special tensor in $\wedge^N V$, namely

$$\omega \equiv \mathbf{e}_1 \wedge ... \wedge \mathbf{e}_N.$$

Since $\omega \neq 0$, the tensor ω can be considered a basis tensor in the one-dimensional space $\wedge^N V$. This choice allows one to identify the space $\wedge^N V$ with scalars (the one-dimensional space of numbers, $\mathbb{K}$). Namely, any tensor $\tau \in \wedge^N V$ must be proportional to ω (since $\wedge^N V$ is one-dimensional), so $\tau = t\omega$ where $t \in \mathbb{K}$ is some number. The number t corresponds uniquely to each $\tau \in \wedge^N V$.

As we have seen before, tensors from $\wedge^N V$ have the interpretation of oriented volumes. In this interpretation, ω represents the volume of a parallelepiped spanned by the unit basis vectors $\{\mathbf{e}_j\}$. Since the vectors $\{\mathbf{e}_j\}$ are orthonormal and have unit length, it is reasonable to assume that they span a *unit* volume. Hence, the oriented volume represented by ω is equal to ± 1 depending on the orientation of the basis $\{\mathbf{e}_j\}$. The tensor ω is called the **unit volume tensor**.

Once ω is fixed, the (oriented) volume of a parallelepiped spanned by arbitrary vectors $\{\mathbf{v}_1, ..., \mathbf{v}_N\}$ is equal to the constant C in the equality

$$\mathbf{v}_1 \wedge ... \wedge \mathbf{v}_N = C\omega. \tag{5.5}$$

In our notation of "tensor division," we can also write

$$\mathrm{Vol}\,\{\mathbf{v}_1, ..., \mathbf{v}_N\} \equiv C = \frac{\mathbf{v}_1 \wedge ... \wedge \mathbf{v}_N}{\omega}.$$

It might appear that ω is arbitrarily chosen and will change when we select another orthonormal basis. However, it turns out that the basis tensor ω does not actually depend on the choice of the orthonormal basis, *up to a sign*. (The sign of ω is necessarily ambiguous because one can always interchange, say, $\mathbf{e}_1$ and $\mathbf{e}_2$ in the orthonormal basis, and then the sign of ω will be flipped.) We will now prove that a different orthonormal basis yields again either ω or $-\omega$, depending on the order of vectors. In other words, ω depends on the choice of the scalar product but not on the choice of an orthonormal basis, *up to a sign*.

Statement: Given two orthonormal bases $\{\mathbf{e}_j\}$ and $\{\mathbf{f}_j\}$, let us define two tensors $\omega \equiv \mathbf{e}_1 \wedge ... \wedge \mathbf{e}_N$ and $\omega' \equiv \mathbf{f}_1 \wedge ... \wedge \mathbf{f}_N$. Then $\omega' = \pm\omega$.

Proof: There exists an orthogonal transformation $\hat{R}$ that maps the basis $\{\mathbf{e}_j\}$ into the basis $\{\mathbf{f}_j\}$, i.e. $\hat{R}\mathbf{e}_j = \mathbf{f}_j$ for $j = 1, ..., N$. Then $\det \hat{R} = \pm 1$ and thus

$$\omega' = \hat{R}\mathbf{e}_1 \wedge ... \wedge \hat{R}\mathbf{e}_N = (\det \hat{R})\omega = \pm\omega.$$

■

The sign factor ± 1 in the definition of the unit-volume tensor ω is an essential ambiguity that cannot be avoided; instead, one simply chooses some orthonormal basis $\{\mathbf{e}_j\}$, computes $\omega \equiv \mathbf{e}_1 \wedge ... \wedge \mathbf{e}_N$, and declares this ω to be "positively oriented." Any other nonzero N-vector $\psi \in \wedge^N V$ can then be compared with ω as $\psi = C\omega$, yielding a constant $C \neq 0$. If $C > 0$ then ψ is also "positively oriented," otherwise ψ is "negatively oriented." Similarly, any given basis $\{\mathbf{v}_j\}$ is then deemed to be "positively oriented" if Eq. (5.5) holds with $C > 0$. Choosing ω is therefore called "fixing the **orientation of space**."

Remark: right-hand rule. To fix the orientation of the basis in the 3-dimensional space, frequently the "right-hand rule" is used: The thumb, the index finger, and the middle finger of a relaxed *right hand* are considered the "positively oriented" basis vectors $\{\mathbf{e}_1, \mathbf{e}_2, \mathbf{e}_3\}$. However, this is not really a definition in the mathematical sense because the concept of "fingers of a right hand" is undefined and actually *cannot* be defined in geometric terms. In other words, it is impossible to give a purely algebraic or geometric definition of a "positively oriented" basis in terms of any properties of the vectors $\{\mathbf{e}_j\}$ alone! (Not to mention that there is no human hand in N dimensions.) However, once an *arbitrary* basis $\{\mathbf{e}_j\}$ is selected and declared to be "positively oriented," we may look at any other basis $\{\mathbf{v}_j\}$, compute

$$C \equiv \frac{\mathbf{v}_1 \wedge ... \wedge \mathbf{v}_N}{\mathbf{e}_1 \wedge ... \wedge \mathbf{e}_N} = \frac{\mathbf{v}_1 \wedge ... \wedge \mathbf{v}_N}{\omega},$$

and examine the sign of C. We will have $C \neq 0$ since $\{\mathbf{v}_j\}$ is a basis. If $C > 0$, the basis $\{\mathbf{v}_j\}$ is positively oriented. If $C < 0$, we need to change the ordering of vectors in $\{\mathbf{v}_j\}$; for instance, we may swap the first two vectors and use $\{\mathbf{v}_2, \mathbf{v}_1, \mathbf{v}_3, ..., \mathbf{v}_N\}$ as the positively oriented basis. In other words, "a positive orientation of space" simply means choosing a certain ordering of vectors in each basis. As we have seen, it suffices to choose the unit volume tensor ω (rather than a basis) to fix the orientation of space. The choice of sign of ω is quite arbitrary and does not influence the results of any calculations because the tensor ω always appears on both sides of equations or in a quadratic combination. ■

5.4.2 Vector product in $\mathbb{R}^3$ and Levi-Civita symbol ε

In the familiar three-dimensional Euclidean space, $V = \mathbb{R}^3$, there is a vector product $\mathbf{a} \times \mathbf{b}$ and a scalar product $\mathbf{a} \cdot \mathbf{b}$. We will now show how the vector product can be expressed through the exterior product.

A positively oriented orthonormal basis $\{\mathbf{e}_1, \mathbf{e}_2, \mathbf{e}_3\}$ defines the unit volume tensor $\omega \equiv \mathbf{e}_1 \wedge \mathbf{e}_2 \wedge \mathbf{e}_3$ in $\wedge^3 V$. Due to the presence of the scalar product, V can be identified with V^*, as we have seen.

Further, the space $\wedge^2 V$ can be identified with V by the following construction. A 2-vector $A \in \wedge^2 V$ generates a covector $\mathbf{f}^*$ by the formula

$$\mathbf{f}^*(\mathbf{x}) \equiv \frac{\mathbf{x} \wedge A}{\omega}, \quad \forall \mathbf{x} \in V.$$

Now the identification of vectors and covectors shows that $\mathbf{f}^*$ corresponds to a certain vector $\mathbf{c}$. Thus, a 2-vector $A \in \wedge^2 V$ is mapped to a vector $\mathbf{c} \in V$. Let us denote this map by the "star" symbol and write $\mathbf{c} = *A$. This map is called the **Hodge star**; it is a linear map $\wedge^2 V \to V$.

Example 1: Let us compute $*(\mathbf{e}_2 \wedge \mathbf{e}_3)$. The 2-vector $\mathbf{e}_2 \wedge \mathbf{e}_3$ is mapped to the covector $\mathbf{f}^*$ defined by

$$\mathbf{f}^*(\mathbf{x})\mathbf{e}_1 \wedge \mathbf{e}_2 \wedge \mathbf{e}_3 \equiv \mathbf{x} \wedge \mathbf{e}_2 \wedge \mathbf{e}_3 = x_1 \mathbf{e}_1 \wedge \mathbf{e}_2 \wedge \mathbf{e}_3,$$

where $\mathbf{x}$ is an arbitrary vector and $x_1 \equiv \mathbf{e}_1^*(\mathbf{x})$ is the first component of $\mathbf{x}$ in the basis. Therefore $\mathbf{f}^* = \mathbf{e}_1^*$. By the vector-covector correspondence, $\mathbf{f}^*$ is mapped to the vector $\mathbf{e}_1$ since

$$x_1 = \mathbf{e}_1^*(\mathbf{x}) = \langle \mathbf{e}_1, \mathbf{x} \rangle .$$

Therefore $*(\mathbf{e}_2 \wedge \mathbf{e}_3) = \mathbf{e}_1$.

Similarly we compute $*(\mathbf{e}_1 \wedge \mathbf{e}_3) = -\mathbf{e}_2$ and $*(\mathbf{e}_1 \wedge \mathbf{e}_2) = \mathbf{e}_3$. ■

Generalizing Example 1 to a single-term product $\mathbf{a} \wedge \mathbf{b}$, where $\mathbf{a}$ and $\mathbf{b}$ are vectors from V, we find that the vector $\mathbf{c} = *(\mathbf{a} \wedge \mathbf{b})$ is equal to the usually defined **vector product** or "cross product" $\mathbf{c} = \mathbf{a} \times \mathbf{b}$. We note that the vector product depends on the choice of the *orientation* of the basis; exchanging the order of any two basis vectors will change the sign of the tensor ω and hence will change the sign of the vector product.

Exercise 1: The vector product in $\mathbb{R}^3$ is usually defined through the components of vectors in an orthogonal basis, as in Eq. (1.2). Show that the definition

$$\mathbf{a} \times \mathbf{b} \equiv *(\mathbf{a} \wedge \mathbf{b})$$

is equivalent to that.

Hint: Since the vector product is bilinear, it is sufficient to show that $*(\mathbf{a} \wedge \mathbf{b})$ is linear in both $\mathbf{a}$ and $\mathbf{b}$, and then to consider the pairwise vector products $\mathbf{e}_1 \times \mathbf{e}_2$, $\mathbf{e}_2 \times \mathbf{e}_3$, $\mathbf{e}_3 \times \mathbf{e}_1$ for an orthonormal basis $\{\mathbf{e}_1, \mathbf{e}_2, \mathbf{e}_3\}$. Some of these calculations were performed in Example 1. ■

The Hodge star is a one-to-one map because $*(\mathbf{a} \wedge \mathbf{b}) = 0$ if and only if $\mathbf{a} \wedge \mathbf{b} = 0$. Hence, the inverse map $V \to \wedge^2 V$ exists. It is convenient to denote the inverse map also by the same "star" symbol, so that we have the map $* : V \to \wedge^2 V$. For example,

$$*(\mathbf{e}_1) = \mathbf{e}_2 \wedge \mathbf{e}_3, \quad *(\mathbf{e}_2) = -\mathbf{e}_1 \wedge \mathbf{e}_3,$$
$$* * (\mathbf{e}_1) = *(\mathbf{e}_2 \wedge \mathbf{e}_3) = \mathbf{e}_1.$$

We may then write symbolically $** = \hat{1}$; here one of the stars stands for the map $V \to \wedge^2 V$, and the other star is the map $\wedge^2 V \to V$.

The **triple product** is defined by the formula

$$(\mathbf{a}, \mathbf{b}, \mathbf{c}) \equiv \langle \mathbf{a}, \mathbf{b} \times \mathbf{c} \rangle .$$

The triple product is fully antisymmetric,

$$(\mathbf{a}, \mathbf{b}, \mathbf{c}) = -(\mathbf{b}, \mathbf{a}, \mathbf{c}) = -(\mathbf{a}, \mathbf{c}, \mathbf{b}) = +(\mathbf{c}, \mathbf{a}, \mathbf{b}) = ...$$

The geometric interpretation of the triple product is that of the oriented volume of the parallelepiped spanned by the vectors $\mathbf{a}$, $\mathbf{b}$, $\mathbf{c}$. This suggests a connection with the exterior power $\wedge^3(\mathbb{R}^3)$.

Indeed, the triple product can be expressed through the exterior product. We again use the tensor $\omega = \mathbf{e}_1 \wedge \mathbf{e}_2 \wedge \mathbf{e}_3$. Since $\{\mathbf{e}_j\}$ is an orthonormal basis, the volume of the parallelepiped spanned by $\mathbf{e}_1, \mathbf{e}_2, \mathbf{e}_3$ is equal to 1. Then we can express $\mathbf{a} \wedge \mathbf{b} \wedge \mathbf{c}$ as

$$\mathbf{a} \wedge \mathbf{b} \wedge \mathbf{c} = \langle \mathbf{a}, *(\mathbf{b} \wedge \mathbf{c}) \rangle \, \omega = \langle \mathbf{a}, \mathbf{b} \times \mathbf{c} \rangle \, \omega = (\mathbf{a}, \mathbf{b}, \mathbf{c}) \, \omega.$$

Therefore we may write

$$(\mathbf{a}, \mathbf{b}, \mathbf{c}) = \frac{\mathbf{a} \wedge \mathbf{b} \wedge \mathbf{c}}{\omega}.$$

In the index notation, the triple product is written as

$$(\mathbf{a}, \mathbf{b}, \mathbf{c}) \equiv \varepsilon_{jkl} a^j b^k c^l.$$

Here the symbol ε_{jkl} (the **Levi-Civita symbol**) is by definition $\varepsilon_{123} = 1$ and $\varepsilon_{ijk} = -\varepsilon_{jik} = -\varepsilon_{ikj}$. This antisymmetric array of numbers, ε_{ijk}, can be also thought of as the index representation of the unit volume tensor $\omega = \mathbf{e}_1 \wedge \mathbf{e}_2 \wedge \mathbf{e}_3$ because

$$\omega = \mathbf{e}_1 \wedge \mathbf{e}_2 \wedge \mathbf{e}_3 = \frac{1}{3!} \sum_{i,j,k=1}^{3} \varepsilon_{ijk} \mathbf{e}_i \wedge \mathbf{e}_j \wedge \mathbf{e}_k.$$

Remark: Geometric interpretation. The Hodge star is useful in conjunction with the interpretation of bivectors as oriented areas. If a bivector $\mathbf{a} \wedge \mathbf{b}$ represents the oriented area of a parallelogram spanned by the vectors $\mathbf{a}$ and $\mathbf{b}$, then $*(\mathbf{a} \wedge \mathbf{b})$ is the vector $\mathbf{a} \times \mathbf{b}$, i.e. the vector orthogonal to the plane of the parallelogram whose length is numerically equal to the area of the parallelogram. Conversely, if $\mathbf{n}$ is a vector then $*(\mathbf{n})$ is a bivector that may represent some parallelogram orthogonal to $\mathbf{n}$ with the appropriate area.

Another geometric example is the computation of the intersection of two planes: If $\mathbf{a} \wedge \mathbf{b}$ and $\mathbf{c} \wedge \mathbf{d}$ represent two parallelograms in space then

$$*\big([*(\mathbf{a} \wedge \mathbf{b})] \wedge [*(\mathbf{c} \wedge \mathbf{d})]\big) = (\mathbf{a} \times \mathbf{b}) \times (\mathbf{c} \times \mathbf{d})$$

is a vector parallel to the line of intersection of the two planes containing the two parallelograms. While in three dimensions the Hodge star yields the same results as the cross product, the advantage of the Hodge star is that it is defined in any dimensions, as the next section shows. ■

5.4.3 Hodge star and Levi-Civita symbol in N dimensions

We would like to generalize our results to an N-dimensional space. We begin by defining the unit volume tensor $\omega = \mathbf{e}_1 \wedge ... \wedge \mathbf{e}_N$, where $\{\mathbf{e}_j\}$ is a positively oriented orthonormal basis. As we have seen, the tensor ω is independent of the choice of the orthonormal basis $\{\mathbf{e}_j\}$ and depends only on the scalar product and on the choice of the orientation of space. (Alternatively, the choice of ω rather than $-\omega$ as the unit volume tensor defines the fact that the basis $\{\mathbf{e}_j\}$

is positively oriented.) Below we will always assume that the orthonormal basis $\{\mathbf{e}_j\}$ is chosen to be positively oriented.

The **Hodge star** is now defined as a linear map $V \to \wedge^{N-1}V$ through its action on the basis vectors,

$$*(\mathbf{e}_j) \equiv (-1)^{j-1}\mathbf{e}_1 \wedge ... \wedge \mathbf{e}_{j-1} \wedge \mathbf{e}_{j+1} \wedge ... \wedge \mathbf{e}_N,$$

where we write the exterior product of all the basis vectors except $\mathbf{e}_j$. To check the sign, we note the identity

$$\mathbf{e}_j \wedge *(\mathbf{e}_j) = \omega, \quad 1 \leq j \leq N.$$

Remark: The Hodge star map depends on the scalar product and on the choice of the orientation of the space V, i.e. on the choice of the *sign* in the basis tensor $\omega \equiv \mathbf{e}_1 \wedge ... \wedge \mathbf{e}_N$, but not on the choice of the vectors $\{\mathbf{e}_j\}$ in a positively oriented orthonormal basis. This is in contrast with the "complement" operation defined in Sec. 2.3.3, where the scalar product was not available: the "complement" operation depends on the choice of *every* vector in the basis. The "complement" operation is equivalent to the Hodge star only if we use an orthonormal basis.

Alternatively, given some basis $\{\mathbf{v}_j\}$, we may temporarily introduce a new scalar product such that $\{\mathbf{v}_j\}$ is orthonormal. The "complement" operation is then the same as the Hodge star defined with respect to the new scalar product. The "complement" operation was introduced by H. Grassmann (1844) long before the now standard definitions of vector space and scalar product were developed. ■

The Hodge star can be also defined more generally as a map of $\wedge^k V$ to $\wedge^{N-k}V$. The construction of the Hodge star map is as follows. We require that it be a linear map. So it suffices to define the Hodge star on single-term products of the form $\mathbf{a}_1 \wedge ... \wedge \mathbf{a}_k$. The vectors $\{\mathbf{a}_i \,|\, i = 1, ..., k\}$ define a subspace of V, which we temporarily denote by $U \equiv \text{Span}\,\{\mathbf{a}_i\}$. Through the scalar product, we can construct the orthogonal complement subspace $U^\perp$; this subspace consists of all vectors that are orthogonal to every $\mathbf{a}_i$. Thus, U is an $(N-k)$-dimensional subspace of V. We can find a basis $\{\mathbf{b}_i \,|\, i = k+1, ..., N\}$ in $U^\perp$ such that

$$\mathbf{a}_1 \wedge ... \wedge \mathbf{a}_k \wedge \mathbf{b}_{k+1} \wedge ... \wedge \mathbf{b}_N = \omega. \tag{5.6}$$

Then we define

$$*(\mathbf{a}_1 \wedge ... \wedge \mathbf{a}_k) \equiv \mathbf{b}_{k+1} \wedge ... \wedge \mathbf{b}_N \in \wedge^{N-k}V.$$

Examples:

$$*(\mathbf{e}_1 \wedge \mathbf{e}_3) = -\mathbf{e}_2 \wedge \mathbf{e}_4 \wedge ... \wedge \mathbf{e}_N;$$
$$*(1) = \mathbf{e}_1 \wedge ... \wedge \mathbf{e}_N; \quad *(\mathbf{e}_1 \wedge ... \wedge \mathbf{e}_N) = 1.$$

The fact that we denote different maps by the same star symbol will not cause confusion because in each case we will write the tensor to which the Hodge star is applied. ■

Even though (by definition) $\mathbf{e}_j \wedge *(\mathbf{e}_j) = \omega$ for the basis vectors $\mathbf{e}_j$, it is *not* true that $\mathbf{x} \wedge *(\mathbf{x}) = \omega$ for any $\mathbf{x} \in V$.

Exercise 1: Show that $\mathbf{x} \wedge (*\mathbf{x}) = \langle \mathbf{x}, \mathbf{x} \rangle \, \omega$ for any $\mathbf{x} \in V$. Then set $\mathbf{x} = \mathbf{a} + \mathbf{b}$ and show (using $*\omega = 1$) that

$$\langle \mathbf{a}, \mathbf{b} \rangle = *(\mathbf{a} \wedge *\mathbf{b}) = *(\mathbf{b} \wedge *\mathbf{a}), \quad \forall \mathbf{a}, \mathbf{b} \in V.$$

Statement: The Hodge star map $* : \wedge^k V \to \wedge^{N-k} V$, as defined above, is independent of the choice of the basis in $U^\perp$.

Proof: A different choice of basis in $U^\perp$, say $\{\mathbf{b}'_i\}$ instead of $\{\mathbf{b}_i\}$, will yield a tensor $\mathbf{b}'_{k+1} \wedge ... \wedge \mathbf{b}'_N$ that is proportional to $\mathbf{b}_{k+1} \wedge ... \wedge \mathbf{b}_N$. The coefficient of proportionality is fixed by Eq. (5.6). Therefore, no ambiguity remains. ■

The insertion map $\iota_{\mathbf{a}^*}$ was defined in Sec. 2.3.1 for covectors $\mathbf{a}^*$. Due to the correspondence between vectors and covectors, we may now use the insertion map with vectors. Namely, we define

$$\iota_{\mathbf{x}} \psi \equiv \iota_{\mathbf{x}^*} \psi,$$

where the covector $\mathbf{x}^*$ is defined by

$$\mathbf{x}^*(\mathbf{v}) \equiv \langle \mathbf{x}, \mathbf{v} \rangle, \quad \forall \mathbf{v} \in V.$$

For example, we then have

$$\iota_{\mathbf{x}}(\mathbf{a} \wedge \mathbf{b}) = \langle \mathbf{x}, \mathbf{a} \rangle \, \mathbf{b} - \langle \mathbf{x}, \mathbf{b} \rangle \, \mathbf{a}.$$

Exercise 2: Show that $*(\mathbf{e}_i) = \iota_{\mathbf{e}_i} \omega$ for basis vectors $\mathbf{e}_i$. Deduce that $*\mathbf{x} = \iota_{\mathbf{x}} \omega$ for any $\mathbf{x} \in V$.

Exercise 3: Show that

$$*\mathbf{x} = \sum_{i=1}^{N} \langle \mathbf{x}, \mathbf{e}_i \rangle \, \iota_{\mathbf{e}_i} \omega = \sum_{i=1}^{N} (\iota_{\mathbf{e}_i} \mathbf{x})(\iota_{\mathbf{e}_i} \omega).$$

Here $\iota_{\mathbf{a}} \mathbf{b} \equiv \langle \mathbf{a}, \mathbf{b} \rangle$. ■

In the previous section, we saw that $**\mathbf{e}_1 = \mathbf{e}_1$ (in three dimensions). The following exercise shows what happens in N dimensions: we may get a minus sign.

Exercise 4: a) Given a vector $\mathbf{x} \in V$, define $\psi \in \wedge^{N-1} V$ as $\psi \equiv *\mathbf{x}$. Then show that

$$*\psi \equiv *(*\mathbf{x}) = (-1)^{N-1} \mathbf{x}.$$

b) Show that $** = (-1)^{k(N-k)} \hat{1}$ when applied to the space $\wedge^k V$ or $\wedge^{N-k} V$.

Hint: Since $*$ is a linear map, it is sufficient to consider its action on a basis vector, say $\mathbf{e}_1$, or a basis tensor $\mathbf{e}_1 \wedge ... \wedge \mathbf{e}_k \in \wedge^k V$, where $\{\mathbf{e}_j\}$ is an orthonormal basis.

Exercise 5: Suppose that $\mathbf{a}_1$, ..., $\mathbf{a}_k$, $\mathbf{x} \in V$ are such that $\langle \mathbf{x}, \mathbf{a}_i \rangle = 0$ for all $i = 1, ..., k$ while $\langle \mathbf{x}, \mathbf{x} \rangle = 1$. The k-vector $\psi \in \wedge^k V$ is then defined as a function of t by

$$\psi(t) \equiv (\mathbf{a}_1 + t\mathbf{x}) \wedge ... \wedge (\mathbf{a}_k + t\mathbf{x}).$$

Show that $t\partial_t \psi = \mathbf{x} \wedge \iota_{\mathbf{x}} \psi$.

Exercise 6: For $\mathbf{x} \in V$ and $\psi \in \wedge^k V$ $(1 \leq k \leq N)$, the tensor $\iota_{\mathbf{x}}\psi \in \wedge^{k-1}V$ is called the **interior product** of $\mathbf{x}$ and ψ. Show that

$$\iota_{\mathbf{x}}\psi = *(\mathbf{x} \wedge *\psi).$$

(Note however that $\psi \wedge *\mathbf{x} = 0$ for $k \geq 2$.)
Exercise 7: a) Suppose $\mathbf{x} \in V$ and $\psi \in \wedge^k V$ are such that $\mathbf{x} \wedge \psi = 0$ while $\langle \mathbf{x}, \mathbf{x} \rangle = 1$. Show that

$$\psi = \mathbf{x} \wedge \iota_{\mathbf{x}}\psi.$$

Hint: Use Exercise 2 in Sec. 2.3.2 with a suitable $\mathbf{f}^*$.
b) For any $\psi \in \wedge^k V$, show that

$$\psi = \frac{1}{k} \sum_{j=1}^{N} \mathbf{e}_j \wedge \iota_{\mathbf{e}_j}\psi,$$

where $\{\mathbf{e}_j\}$ is an orthonormal basis.
Hint: It suffices to consider $\psi = \mathbf{e}_{i_1} \wedge ... \wedge \mathbf{e}_{i_k}$. ∎

The Levi-Civita symbol $\varepsilon_{i_1 ... i_N}$ is defined in an N-dimensional space as the coordinate representation of the unit volume tensor $\omega \equiv \mathbf{e}_1 \wedge ... \wedge \mathbf{e}_N \in \wedge^N V$ (see also Sections 2.3.6 and 3.4.1). When a scalar product is fixed, the tensor ω is unique up to a sign; if we assume that ω corresponds to a positively oriented basis, the Levi-Civita symbol is the index representation of ω in *any* positively oriented orthonormal basis. It is instructive to see how one writes the Hodge star in the index notation using the Levi-Civita symbol. (I will write the summations explicitly here, but keep in mind that in the physics literature the summations are implicit.)

Given an orthonormal basis $\{\mathbf{e}_j\}$, the natural basis in $\wedge^k V$ is the set of tensors $\{\mathbf{e}_{i_1} \wedge ... \wedge \mathbf{e}_{i_k}\}$ where all indices $i_1, ..., i_k$ are different (or else the exterior product vanishes). Therefore, an arbitrary tensor $\psi \in \wedge^k V$ can be expanded in this basis as

$$\psi = \frac{1}{k!} \sum_{i_1,...,i_k=1}^{N} A^{i_1...i_k} \mathbf{e}_{i_1} \wedge ... \wedge \mathbf{e}_{i_k},$$

where $A^{i_1...i_k}$ are some scalar coefficients. I have included the prefactor $1/k!$ in order to cancel the combinatorial factor $k!$ that appears due to the summation over all the indices $i_1, ..., i_k$.

Let us write the tensor $\psi \equiv *(\mathbf{e}_1)$ in this way. The corresponding coefficients $A^{i_1...i_{N-1}}$ are zero unless the set of indices $(i_1, ..., i_{N-1})$ is a permutation of the set $(2, 3, ..., N)$. This statement can be written more concisely as

$$(*\mathbf{e}_1)^{i_1...i_{N-1}} \equiv A^{i_1...i_{N-1}} = \varepsilon^{1 i_1...i_{N-1}}.$$

Generalizing to an arbitrary vector $\mathbf{x} = \sum_{j=1}^{N} x_j \mathbf{e}_j$, we find

$$(*\mathbf{x})^{i_1...i_{N-1}} \equiv \sum_{j=1}^{N} x^j (*\mathbf{e}_j)^{i_1...i_{N-1}} = \sum_{i,j=1}^{N} x^j \delta_{ji} \varepsilon^{i i_1...i_{N-1}}.$$

Remark: The extra Kronecker symbol above is introduced for consistency of the notation (summing only over a pair of opposite indices). However, this Kronecker symbol can be interpreted as the coordinate representation of the scalar product in the orthonormal basis. This formula then shows how to write the Hodge star in another basis: replace δ_{ji} with the matrix representation of the scalar product. ■

Similarly, we can write the Hodge star of an arbitrary k-vector in the index notation through the ε symbol. For example, in a four-dimensional space one maps a 2-vector $\sum_{i,j} A^{ij}\mathbf{e}_i \wedge \mathbf{e}_j$ into

$$*\big(\sum_{i,j} A^{ij}\mathbf{e}_i \wedge \mathbf{e}_j\big) = \sum_{k,l} B^{kl}\mathbf{e}_k \wedge \mathbf{e}_l,$$

where

$$B^{kl} \equiv \frac{1}{2!} \sum_{i,j,m,n} \delta^{km}\delta^{ln}\varepsilon_{ijmn}A^{ij}.$$

A vector $\mathbf{v} = \sum_i v^i \mathbf{e}_i$ is mapped into

$$*(\mathbf{v}) = *\big(\sum_i v^i\mathbf{e}_i\big) = \frac{1}{3!}\sum_{i,j,k,l} \varepsilon_{ijkl}v^i\mathbf{e}_j \wedge \mathbf{e}_k \wedge \mathbf{e}_l.$$

Note the combinatorial factors 2! and 3! appearing in these formulas, according to the number of indices in ε that are being summed over.

5.4.4 Reciprocal basis

Suppose $\{\mathbf{v}_1, ..., \mathbf{v}_N\}$ is a basis in V, not necessarily orthonormal. For any $\mathbf{x} \in V$, we can compute the components of $\mathbf{x}$ in the basis $\{\mathbf{v}_j\}$ by first computing the dual basis, $\{\mathbf{v}_j^*\}$, as in Sec. 2.3.3, and then writing

$$\mathbf{x} = \sum_{i=1}^{N} x_i\mathbf{v}_i, \quad x_i \equiv \mathbf{v}_i^*(\mathbf{x}).$$

The scalar product in V provides a vector-covector correspondence. Hence, each $\mathbf{v}_i^*$ has a corresponding vector; let us denote that vector temporarily by $\mathbf{u}_i$. We then obtain a set of N vectors, $\{\mathbf{u}_1, ..., \mathbf{u}_N\}$. By definition of the vector-covector correspondence, the vector $\mathbf{u}_i$ is such that

$$\langle \mathbf{u}_i, \mathbf{x}\rangle = \mathbf{v}_i^*(\mathbf{x}) \equiv x_i, \quad \forall \mathbf{x} \in V.$$

We will now show that the set $\{\mathbf{u}_1, ..., \mathbf{u}_N\}$ is a basis in V. It is called the **reciprocal basis** for the basis $\{\mathbf{v}_j\}$. The reciprocal basis is useful, in particular, because the components of a vector $\mathbf{x}$ in the basis $\{\mathbf{v}_j\}$ are computed conveniently through scalar products with the vectors $\{\mathbf{u}_j\}$, as shown by the formula above.

Statement 1: The set $\{\mathbf{u}_1, ..., \mathbf{u}_N\}$ is a basis in V.

Proof: We first note that

$$\langle \mathbf{u}_i, \mathbf{v}_j \rangle \equiv \mathbf{v}_i^*(\mathbf{v}_j) = \delta_{ij}.$$

We need to show that the set $\{\mathbf{u}_1, ..., \mathbf{u}_N\}$ is linearly independent. Suppose a vanishing linear combination exists,

$$\sum_{i=1}^{N} \lambda_i \mathbf{u}_i = 0,$$

and take its scalar product with the vector $\mathbf{v}_1$,

$$0 = \big\langle \mathbf{v}_1, \sum_{i=1}^{N} \lambda_i \mathbf{u}_i \big\rangle = \sum_{i=1}^{N} \lambda_i \delta_{1i} = \lambda_1.$$

In the same way we show that all λ_i are zero. A linearly independent set of N vectors in an N-dimensional space is always a basis, hence $\{\mathbf{u}_j\}$ is a basis. ■

Exercise 1: Show that computing the reciprocal basis to an *orthonormal* basis $\{\mathbf{e}_j\}$ gives again the same basis $\{\mathbf{e}_j\}$. ■

The following statement shows that, in some sense, the reciprocal basis is the "inverse" of the basis $\{\mathbf{v}_j\}$.

Statement 2: The oriented volume of the parallelepiped spanned by $\{\mathbf{u}_j\}$ is the inverse of that spanned by $\{\mathbf{v}_j\}$.

Proof: The volume of the parallelepiped spanned by $\{\mathbf{u}_j\}$ is found as

$$\mathrm{Vol}\,\{\mathbf{u}_j\} = \frac{\mathbf{u}_1 \wedge ... \wedge \mathbf{u}_N}{\mathbf{e}_1 \wedge ... \wedge \mathbf{e}_N},$$

where $\{\mathbf{e}_j\}$ is a positively oriented orthonormal basis. Let us introduce an auxiliary transformation $\hat{M}$ that maps $\{\mathbf{e}_j\}$ into $\{\mathbf{v}_j\}$; such a transformation surely exists and is invertible. Since $\hat{M}\mathbf{e}_j = \mathbf{v}_j$ $(j = 1, ..., N)$, we have

$$\det \hat{M} = \frac{\hat{M}\mathbf{e}_1 \wedge ... \wedge \hat{M}\mathbf{e}_N}{\mathbf{e}_1 \wedge ... \wedge \mathbf{e}_N} = \frac{\mathbf{v}_1 \wedge ... \wedge \mathbf{v}_N}{\mathbf{e}_1 \wedge ... \wedge \mathbf{e}_N} = \mathrm{Vol}\,\{\mathbf{v}_j\}.$$

Consider the transposed operator $\hat{M}^T$ (the transposition is performed using the scalar product, see Definition 1 in Sec. 5.3.1). We can now show that $\hat{M}^T$ maps the dual basis $\{\mathbf{u}_j\}$ into $\{\mathbf{e}_j\}$. To show this, we consider the scalar products

$$\langle \mathbf{e}_i, \hat{M}^T \mathbf{u}_j \rangle = \langle \hat{M}\mathbf{e}_i, \mathbf{u}_j \rangle = \langle \mathbf{v}_i, \mathbf{u}_j \rangle = \delta_{ij}.$$

Since the above is true for any $i, j = 1, ..., N$, it follows that $\hat{M}^T \mathbf{u}_j = \mathbf{e}_j$ as desired.

Since $\det \hat{M}^T = \det \hat{M}$, we have

$$\mathbf{e}_1 \wedge ... \wedge \mathbf{e}_N = \hat{M}^T \mathbf{u}_1 \wedge ... \wedge \hat{M}^T \mathbf{u}_N = (\det \hat{M}) \mathbf{u}_1 \wedge ... \wedge \mathbf{u}_N.$$

It follows that

$$\mathrm{Vol}\,\{\mathbf{u}_j\} = \frac{\mathbf{u}_1 \wedge ... \wedge \mathbf{u}_N}{\mathbf{e}_1 \wedge ... \wedge \mathbf{e}_N} = \frac{1}{\det \hat{M}} = \frac{1}{\mathrm{Vol}\,\{\mathbf{v}_j\}}.$$

■

The vectors of the reciprocal basis can be also computed using the Hodge star, as follows.

Exercise 2: Suppose that $\{\mathbf{v}_j\}$ is a basis (not necessarily orthonormal) and $\{\mathbf{u}_j\}$ is its reciprocal basis. Show that

$$\mathbf{u}_1 = *(\mathbf{v}_2 \wedge ... \wedge \mathbf{v}_N)\frac{\omega}{\mathbf{v}_1 \wedge ... \wedge \mathbf{v}_N},$$

where $\omega \equiv \mathbf{e}_1 \wedge ... \wedge \mathbf{e}_N$, $\{\mathbf{e}_j\}$ is a positively oriented orthonormal basis, and we use the Hodge star as a map from $\wedge^{N-1}V$ to V.

Hint: Use the formula for the dual basis (Sec. 2.3.3),

$$\mathbf{v}_1^*(\mathbf{x}) = \frac{\mathbf{x} \wedge \mathbf{v}_2 \wedge ... \wedge \mathbf{v}_N}{\mathbf{v}_1 \wedge \mathbf{v}_2 \wedge ... \wedge \mathbf{v}_N},$$

and the property

$$\langle \mathbf{x}, \mathbf{u} \rangle \,\omega = \mathbf{x} \wedge *\mathbf{u}.$$

5.5 Scalar product in $\wedge^k V$

In this section we will apply the techniques developed until now to the problem of computing k-dimensional volumes.

If a scalar product is given in V, one can naturally define a scalar product also in each of the spaces $\wedge^k V$ ($k = 2, ..., N$). We will show that this scalar product allows one to compute the ordinary (number-valued) volumes represented by tensors from $\wedge^k V$. This is fully analogous to computing the lengths of vectors through the scalar product in V. A vector $\mathbf{v}$ in a Euclidean space represents at once the orientation and the length of a straight line segment between two points; the length is found as $\sqrt{\langle \mathbf{v}, \mathbf{v}\rangle}$ using the scalar product in V. Similarly, a tensor $\psi = \mathbf{v}_1 \wedge ... \wedge \mathbf{v}_k \in \wedge^k V$ represents at once the orientation and the volume of a parallelepiped spanned by the vectors $\{\mathbf{v}_j\}$; the unoriented volume of the parallelepiped will be found as $\sqrt{\langle \psi, \psi \rangle}$ using the scalar product in $\wedge^k V$.

We begin by considering the space $\wedge^N V$.

5.5.1 Scalar product in $\wedge^N V$

Suppose $\{\mathbf{u}_j\}$ and $\{\mathbf{v}_j\}$ are two bases in V, not necessarily orthonormal, and consider the pairwise scalar products

$$G_{jk} \equiv \langle \mathbf{u}_j, \mathbf{v}_k \rangle\,, \quad j, k = 1, ..., N.$$

The coefficients G_{jk} can be arranged into a square-shaped table, i.e. into a **matrix**. The determinant of this matrix, $\det(G_{jk})$, can be computed using Eq. (3.1). Now consider two tensors $\omega_1, \omega_2 \in \wedge^N V$ defined as

$$\omega_1 \equiv \mathbf{u}_1 \wedge ... \wedge \mathbf{u}_N, \quad \omega_2 \equiv \mathbf{v}_1 \wedge ... \wedge \mathbf{v}_N.$$

Then $\det(G_{jk})$, understood as a *function* of the tensors ω_1 and ω_2, is bilinear and symmetric, and thus can be interpreted as the **scalar product** of ω_1 and ω_2. After some work proving the necessary properties, we obtain a scalar product in the space $\wedge^N V$, given a scalar product in V.

Exercise 1: We try to define the scalar product in the space $\wedge^N V$ as follows: Given a scalar product $\langle \cdot, \cdot \rangle$ in V and given two tensors $\omega_1, \omega_2 \in \wedge^N V$, we first represent these tensors *in some way* as products

$$\omega_1 \equiv \mathbf{u}_1 \wedge ... \wedge \mathbf{u}_N, \quad \omega_2 \equiv \mathbf{v}_1 \wedge ... \wedge \mathbf{v}_N,$$

where $\{\mathbf{u}_i\}$ and $\{\mathbf{v}_i\}$ are *some* suitable sets of vectors, then consider the matrix of pairwise scalar products $\langle \mathbf{u}_i, \mathbf{v}_j \rangle$, and finally define the scalar product $\langle \omega_1, \omega_2 \rangle$ as the determinant of that matrix:

$$\langle \omega_1, \omega_2 \rangle \equiv \det \langle \mathbf{u}_i, \mathbf{v}_j \rangle .$$

Prove that this definition really yields a symmetric bilinear form in $\wedge^N V$, independently of the particular representation of ω_1, ω_2 through vectors.

Hint: The known properties of the determinant show that $\langle \omega_1, \omega_2 \rangle$ is an antisymmetric and multilinear function of every $\mathbf{u}_i$ and $\mathbf{v}_j$. A linear transformation of the vectors $\{\mathbf{u}_i\}$ that leaves ω_1 constant will also leave $\langle \omega_1, \omega_2 \rangle$ constant. Therefore, it can be considered as a linear function of the tensors ω_1 and ω_2. Symmetry follows from $\det(G_{ij}) = \det(G_{ji})$.

Exercise 2: Given an orthonormal basis $\{\mathbf{e}_j \,|\, j = 1, ..., N\}$, let us consider the unit volume tensor $\omega \equiv \mathbf{e}_1 \wedge ... \wedge \mathbf{e}_N \in \wedge^N V$.

a) Show that $\langle \omega, \omega \rangle = 1$, where the scalar product in $\wedge^N V$ is chosen according to the definition in Exercise 1.

b) Given a linear operator $\hat{A}$, show that $\det \hat{A} = \langle \omega, \wedge^N \hat{A}^N \omega \rangle$.

Exercise 3: For any $\phi, \psi \in \wedge^N V$, show that

$$\langle \phi, \psi \rangle = \frac{\phi}{\omega} \frac{\psi}{\omega},$$

where ω is the unit volume tensor. Deduce that $\langle \phi, \psi \rangle$ is a positive-definite bilinear form.

Statement: The volume of a parallelepiped spanned by vectors $\mathbf{v}_1, ..., \mathbf{v}_N$ is equal to $\sqrt{\det(G_{ij})}$, where $G_{ij} \equiv \langle \mathbf{v}_i, \mathbf{v}_j \rangle$ is the matrix of the pairwise scalar products.

Proof: If $\mathbf{v}_1 \wedge ... \wedge \mathbf{v}_N \neq 0$, the set of vectors $\{\mathbf{v}_j \,|\, j = 1, ..., N\}$ is a basis in V. Let us also choose some orthonormal basis $\{\mathbf{e}_j \,|\, j = 1, ..., N\}$. There exists a linear transformation $\hat{A}$ that maps the basis $\{\mathbf{e}_j\}$ into the basis $\{\mathbf{v}_j\}$. Then we have $\hat{A}\mathbf{e}_j = \mathbf{v}_j$ and hence

$$G_{ij} = \langle \mathbf{v}_i, \mathbf{v}_j \rangle = \langle \hat{A}\mathbf{e}_i, \hat{A}\mathbf{e}_j \rangle = \langle \hat{A}^T \hat{A}\mathbf{e}_i, \mathbf{e}_j \rangle.$$

It follows that the matrix G_{ij} is equal to the matrix representation of the operator $\hat{A}^T \hat{A}$ in the basis $\{\mathbf{e}_j\}$. Therefore,

$$\det(G_{ij}) = \det(\hat{A}^T \hat{A}) = (\det \hat{A})^2.$$

Finally, we note that the volume v of the parallelepiped spanned by $\{\mathbf{v}_j\}$ is the coefficient in the tensor equality

$$v \mathbf{e}_1 \wedge ... \wedge \mathbf{e}_N = \mathbf{v}_1 \wedge ... \wedge \mathbf{v}_N = (\det \hat{A}) \mathbf{e}_1 \wedge ... \wedge \mathbf{e}_N.$$

Hence $v^2 = (\det \hat{A})^2 = \det(G_{ij})$. ■

We have found that the (unoriented, i.e. number-valued) N-dimensional volume of a parallelepiped spanned by a set of N vectors $\{\mathbf{v}_j\}$ is expressed as $v = \sqrt{\langle \psi, \psi \rangle}$, where $\psi \equiv \mathbf{v}_1 \wedge ... \wedge \mathbf{v}_N$ is the tensor representing the oriented volume of the parallelepiped, and $\langle \psi, \psi \rangle$ is the scalar product in the space $\wedge^N V$. The expression $|\psi| \equiv \sqrt{\langle \psi, \psi \rangle}$ is naturally interpreted as the "length" of the tensor ψ. In this way, we obtain a geometric interpretation of tensors $\psi \in \wedge^N V$ as oriented volumes of parallelepipeds: The tensor ψ represents at once the orientation of the parallelepiped and the magnitude of the volume.

5.5.2 Volumes of k-dimensional parallelepipeds

In a similar way we treat k-dimensional volumes.

We begin by defining a scalar product in the spaces $\wedge^k V$ for $2 \leq k \leq N$. Let us choose an orthonormal basis $\{\mathbf{e}_j\}$ in V and consider the set of $\binom{N}{k}$ tensors

$$\omega_{i_1 ... i_k} \equiv \mathbf{e}_{i_1} \wedge ... \wedge \mathbf{e}_{i_k} \in \wedge^k V.$$

Since the set of these tensors (for all admissible sets of indices) is a basis in $\wedge^k V$, it is sufficient to define the scalar product of any two tensors $\omega_{i_1 ... i_k}$. It is natural to define the scalar product such that $\omega_{i_1 ... i_k}$ are *orthonormal*:

$$\langle \omega_{i_1 ... i_k}, \omega_{i_1 ... i_k} \rangle = 1,$$
$$\langle \omega_{i_1 ... i_k}, \omega_{j_1 ... j_k} \rangle = 0 \quad \text{if} \quad \omega_{i_1 ... i_k} \neq \pm \omega_{j_1 ... j_k}.$$

For any two tensors $\psi_1, \psi_2 \in \wedge^k V$, we then define $\langle \psi_1, \psi_2 \rangle$ by expressing ψ_1, ψ_2 through the basis tensors $\omega_{i_1 ... i_k}$ and requiring the bilinearity of the scalar product.

In the following exercise, we derive an explicit formula for the scalar product $\langle \psi_1, \psi_2 \rangle$ through scalar products of the constituent vectors.

Exercise 1: Use the definition above to prove that

$$\langle \mathbf{u}_1 \wedge ... \wedge \mathbf{u}_k, \mathbf{v}_1 \wedge ... \wedge \mathbf{v}_k \rangle = \det \langle \mathbf{u}_i, \mathbf{v}_j \rangle . \quad (5.7)$$

Hints: The right side of Eq. (5.7) is a totally antisymmetric, linear function of every $\mathbf{u}_i$ due to the known properties of the determinant. Also, the function is invariant under the interchange of $\mathbf{u}_j$ with $\mathbf{v}_j$. The left side of Eq. (5.7) has the same symmetry and linearity properties. Therefore, it is sufficient to verify Eq. (5.7) when vectors $\mathbf{u}_i$ and $\mathbf{v}_j$ are chosen from the set of orthonormal basis vectors $\{\mathbf{e}_j\}$. Then $\mathbf{u}_1 \wedge ... \wedge \mathbf{u}_k$ and $\mathbf{v}_1 \wedge ... \wedge \mathbf{v}_k$ are among the basis tensors $\omega_{i_1...i_k}$. Show that the matrix $\langle \mathbf{u}_i, \mathbf{v}_j \rangle$ has at least one row or one column of zeros unless the sets $\{\mathbf{u}_i\}$ and $\{\mathbf{v}_j\}$ *coincide* as unordered sets of vectors, i.e. unless

$$\mathbf{u}_1 \wedge ... \wedge \mathbf{u}_k = \pm \mathbf{v}_1 \wedge ... \wedge \mathbf{v}_k .$$

If the above does not hold, both sides of Eq. (5.7) are zero. It remains to verify that both sides of Eq. (5.7) are equal to 1 when we choose identical vectors $\mathbf{u}_i = \mathbf{v}_i$ from the orthonormal basis, for instance if $\mathbf{u}_j = \mathbf{v}_j = \mathbf{e}_j$ for $j = 1, ..., k$. ■

We now come back to the problem of computing the volume of a k-dimensional parallelepiped spanned by vectors $\{\mathbf{v}_1, ..., \mathbf{v}_k\}$ in an n-dimensional Euclidean space $\mathbb{R}^n$. In Sec. 2.1.2 we considered a parallelogram (i.e. we had $k = 2$), and we projected the parallelogram onto the $\binom{N}{2}$ coordinate planes to define a "vector-valued" area. We now generalize that construction to k-dimensional parallelepipeds. We project the given parallelepiped onto each of the k-dimensional coordinate hyperplanes in the space, which are the subspaces $\text{Span}\{\mathbf{e}_{i_1}, ..., \mathbf{e}_{i_k}\}$ (with $1 \le i_1 < ... < i_k \le n$). There will be $\binom{N}{k}$ such coordinate hyperplanes and, accordingly, we may determine the $\binom{N}{k}$ oriented k-dimensional volumes of these projections. It is natural to view these numbers as the components of the *oriented volume* of the k-dimensional parallelepiped in some basis in the $\binom{N}{k}$-dimensional "space of oriented volumes." As we have shown before, oriented volumes are antisymmetric in the vectors $\mathbf{v}_j$. The space of all antisymmetric combinations of k vectors is, in our present notation, $\wedge^k V$. Thus the oriented volume of the k-dimensional parallelepiped is represented by the tensor $\mathbf{v}_1 \wedge ... \wedge \mathbf{v}_k \in \wedge^k V$. The unoriented volume is computed as the "length" of the oriented volume, defined via the scalar product in $\wedge^k V$.

Statement: The unoriented k-dimensional volume v of a parallelepiped spanned by k vectors $\{\mathbf{v}_1, ..., \mathbf{v}_k\}$ is equal to $\sqrt{\langle \psi, \psi \rangle}$, where $\psi \equiv \mathbf{v}_1 \wedge ... \wedge \mathbf{v}_k$ and $\langle \psi, \psi \rangle$ is the scalar product defined above.

Proof: Consider the orthogonal projection of the given k-dimensional parallelepiped onto some k-dimensional coordinate hyperplane, e.g. onto the hyperplane $\text{Span}\{\mathbf{e}_1, ..., \mathbf{e}_k\}$. Each vector $\mathbf{v}_i$ is projected orthogonally, i.e. by omitting the components of $\mathbf{v}_i$ at $\mathbf{e}_{k+1}$, ..., $\mathbf{e}_N$. Let us denote the projected vectors by $\tilde{\mathbf{v}}_i$ ($i = 1, ..., k$). The projection is a k-dimensional parallelepiped spanned by $\{\tilde{\mathbf{v}}_i\}$ in the coordinate hyperplane. Let us now restrict attention to the subspace $\text{Span}\{\mathbf{e}_1, ..., \mathbf{e}_k\}$. In this subspace, the oriented k-

dimensional volume of the projected parallelepiped is represented by the tensor $\tilde{\psi} \equiv \tilde{\mathbf{v}}_1 \wedge ... \wedge \tilde{\mathbf{v}}_k$. By construction, $\tilde{\psi}$ is proportional to the unit volume tensor in the subspace, $\tilde{\psi} = \lambda \mathbf{e}_1 \wedge ... \wedge \mathbf{e}_k$ for some λ. Therefore, the oriented k-dimensional volume of the projected parallelepiped is equal to λ.

Let us now decompose the tensor ψ into the basis tensors in $\wedge^k V$,

$$\psi = \sum_{1 \leq i_1 < ... < i_k \leq N} c_{i_1...i_k} \omega_{i_1...i_k}$$
$$= c_{1...k} \mathbf{e}_1 \wedge ... \wedge \mathbf{e}_k + c_{13...(k+1)} \mathbf{e}_1 \wedge \mathbf{e}_3 \wedge ... \wedge \mathbf{e}_{k+1} + ...,$$

where we have only written down the first two of the $\binom{N}{k}$ possible terms of the expansion. The projection of $\{\mathbf{v}_i\}$ onto the hyperplane Span $\{\mathbf{e}_1, ..., \mathbf{e}_k\}$ removes the components proportional to $\mathbf{e}_{k+1}, ..., \mathbf{e}_N$, hence $\tilde{\psi}$ is equal to the first term $c_{1...k} \mathbf{e}_1 \wedge ... \wedge \mathbf{e}_k$. Therefore, the oriented volume of the projection onto the hyperplane Span $\{\mathbf{e}_1, ..., \mathbf{e}_k\}$ is equal to $c_{1...k}$.

By definition of the scalar product in $\wedge^k V$, all the basis tensors $\omega_{i_1...i_k}$ are orthonormal. Hence, the coefficients $c_{i_1...i_k}$ can be computed as

$$c_{i_1...i_k} = \langle \psi, \mathbf{e}_{i_1} \wedge ... \wedge \mathbf{e}_{i_k} \rangle \equiv \langle \psi, \omega_{i_1...i_k} \rangle .$$

For brevity, we may introduce the **multi-index** $I \equiv \{i_1, ..., i_k\}$ and rewrite the above as

$$c_I = \langle \psi, \omega_I \rangle .$$

Then the value $\langle \psi, \psi \rangle$ can be computed as

$$\langle \psi, \psi \rangle = \Big\langle \sum_I c_I \omega_I, \sum_J c_J \omega_J \Big\rangle = \sum_{I,J} c_I c_J \langle \omega_I, \omega_J \rangle$$
$$= \sum_{I,J} c_I c_J \delta_{IJ} = \sum_I |c_I|^2 .$$

In other words, we have shown that $\langle \psi, \psi \rangle$ is equal to the sum of all $\binom{N}{k}$ squared projected volumes,

$$\langle \psi, \psi \rangle = \sum_{1 \leq i_1 < ... < i_k \leq N} |c_{i_1...i_k}|^2 .$$

It remains to show that $\sqrt{\langle \psi, \psi \rangle}$ is actually equal to the unoriented volume v of the parallelepiped. To this end, let us choose a new orthonormal basis $\{\tilde{\mathbf{e}}_j\}$ ($j = 1, ..., N$) such that every vector $\mathbf{v}_i$ ($i = 1, ..., k$) lies entirely within the hyperplane spanned by the first k basis vectors. (This choice of basis is certainly possible, for instance, by choosing an orthonormal basis in Span $\{\mathbf{v}_i\}$ and then completing it to an orthonormal basis in V.) Then we will have $\psi = \tilde{\lambda} \tilde{\mathbf{e}}_1 \wedge ... \wedge \tilde{\mathbf{e}}_k$, i.e. with zero coefficients for all other basis tensors. Restricting attention to the subspace Span $\{\tilde{\mathbf{e}}_1, ..., \tilde{\mathbf{e}}_k\}$, we can use the results of Sec. 5.5.1 to find that the volume v is equal to $|\tilde{\lambda}|$. It remains to show that $\sqrt{\langle \psi, \psi \rangle} = |\tilde{\lambda}|$.

The transformation from the old basis $\{\mathbf{e}_j\}$ to $\{\tilde{\mathbf{e}}_j\}$ can be performed using a certain orthogonal transformation $\hat{R}$ such that $\hat{R}\mathbf{e}_j = \tilde{\mathbf{e}}_j$ $(j = 1, ..., N)$. Since the scalar product in $\wedge^k V$ is defined directly through scalar products of vectors in V (Exercise 1) and since $\hat{R}$ is orthogonal, we have for any $\{\mathbf{a}_i\}$ and $\{\mathbf{b}_i\}$ that

$$\begin{aligned}&\langle \hat{R}\mathbf{a}_1 \wedge ... \wedge \hat{R}\mathbf{a}_k, \hat{R}\mathbf{b}_1 \wedge ... \wedge \hat{R}\mathbf{b}_k \rangle = \det \langle \hat{R}\mathbf{a}_i, \hat{R}\mathbf{b}_j \rangle \\ &\quad = \det \langle \mathbf{a}_i, \mathbf{b}_j \rangle = \langle \mathbf{a}_1 \wedge ... \wedge \mathbf{a}_k, \mathbf{b}_1 \wedge ... \wedge \mathbf{b}_k \rangle .\end{aligned}$$

In other words, the operator $\wedge^k \hat{R}^k$ is an *orthogonal transformation* in $\wedge^k V$. Therefore,

$$\begin{aligned}\psi &= \tilde{\lambda}\tilde{\mathbf{e}}_1 \wedge ... \wedge \tilde{\mathbf{e}}_k = \tilde{\lambda}\hat{R}\mathbf{e}_1 \wedge ... \wedge \hat{R}\mathbf{e}_k = \tilde{\lambda}(\wedge^k \hat{R}^k \omega_{1...k});\\ \langle \psi, \psi \rangle &= \tilde{\lambda}^2 \langle \wedge^k \hat{R}^k \omega_{1...k}, \wedge^k \hat{R}^k \omega_{1...k} \rangle = \tilde{\lambda}^2 \langle \omega_{1...k}, \omega_{1...k} \rangle = \tilde{\lambda}^2.\end{aligned}$$

Therefore, $\sqrt{\langle \psi, \psi \rangle} = |\tilde{\lambda}| = v$ as required. ■

Remark: The scalar product in the space $\wedge^k V$ is related the k-dimensional volume of a body embedded in the space V, in the same way as the scalar product in V is related to the length of a straight line segment embedded in V. The tensor $\psi = \mathbf{v}_1 \wedge ... \wedge \mathbf{v}_k$ fully represents the orientation of the k-dimensional parallelepiped spanned by the vectors $\{\mathbf{v}_1, ..., \mathbf{v}_k\}$, while the "length" $\sqrt{\langle \psi, \psi \rangle}$ of this tensor gives the numerical value of the volume of the parallelepiped. This is a multidimensional generalization of the Pythagoras theorem that is not easy to visualize! The techniques of exterior algebra enables us to calculate these quantities without visualizing them.

Example 1: In a Euclidean space $\mathbb{R}^4$ with a standard orthonormal basis $\{\mathbf{e}_j\}$, a three-dimensional parallelepiped is spanned by the given vectors

$$\mathbf{a} = \mathbf{e}_1 + 2\mathbf{e}_2, \quad \mathbf{b} = \mathbf{e}_3 - \mathbf{e}_1, \quad \mathbf{c} = \mathbf{e}_2 + \mathbf{e}_3 + \mathbf{e}_4.$$

We would like to determine the volume of the parallelepiped. We compute the wedge product $\psi \equiv \mathbf{a} \wedge \mathbf{b} \wedge \mathbf{c}$ using Gaussian elimination,

$$\begin{aligned}\psi &= (\mathbf{e}_1 + 2\mathbf{e}_2) \wedge (\mathbf{e}_3 - \mathbf{e}_1) \wedge (\mathbf{e}_2 + \mathbf{e}_3 + \mathbf{e}_4)\\ &= (\mathbf{e}_1 + 2\mathbf{e}_2) \wedge (\mathbf{e}_3 + 2\mathbf{e}_2) \wedge (\mathbf{e}_2 + \mathbf{e}_3 + \mathbf{e}_4)\\ &= [(\mathbf{e}_1 + 2\mathbf{e}_2) \wedge \mathbf{e}_3 + 2\mathbf{e}_1 \wedge \mathbf{e}_2] \wedge \left(\tfrac{1}{2}\mathbf{e}_3 + \mathbf{e}_4\right)\\ &= \mathbf{e}_1 \wedge \mathbf{e}_2 \wedge \mathbf{e}_3 + \mathbf{e}_1 \wedge \mathbf{e}_3 \wedge \mathbf{e}_4\\ &\quad + 2\mathbf{e}_2 \wedge \mathbf{e}_3 \wedge \mathbf{e}_4 + 2\mathbf{e}_1 \wedge \mathbf{e}_2 \wedge \mathbf{e}_4.\end{aligned}$$

We see that the volumes of the projections onto the four coordinate hyperplanes are 1, 1, 2, 2. Therefore the numerical value of the volume is

$$v = \sqrt{\langle \psi, \psi \rangle} = \sqrt{1 + 1 + 4 + 4} = \sqrt{10}.$$

Exercise 2: Show that the scalar product of two tensors $\psi_1, \psi_2 \in \wedge^k V$ can be expressed through the Hodge star as

$$\langle \psi_1, \psi_2 \rangle = *\big(\psi_1 \wedge *\psi_2\big) \quad \text{or as} \quad \langle \psi_1, \psi_2 \rangle = *\big(\psi_2 \wedge *\psi_1\big),$$

depending on whether $2k \leq N$ or $2k \geq N$.

Hint: Since both sides are linear in ψ_1 and ψ_2, it is sufficient to show that the relationship holds for basis tensors $\omega_{i_1...i_k} \equiv \mathbf{e}_{i_1} \wedge ... \wedge \mathbf{e}_{i_k}$.

Exercise 3: Intersection of hyperplanes. Suppose $U_1, ..., U_{N-1} \subset V$ are some $(N-1)$-dimensional subspaces (hyperplanes) in V. Each U_i can be represented by a tensor $\psi_i \in \wedge^{N-1} V$, e.g. by choosing ψ_i as the exterior product of all vectors in a basis in U. Define the vector

$$\mathbf{v} \equiv *\big[(*\psi_1) \wedge ... \wedge (*\psi_{N-1})\big].$$

If $\mathbf{v} \neq 0$, show that $\mathbf{v}$ belongs to the intersection of all the $(N-1)$-dimensional hyperplanes.

Hint: Show that $\mathbf{v} \wedge \psi_i = 0$ for each $i = 1, ..., N-1$. Use Exercise 2.

Exercise 4: Show that $\langle \mathbf{v}, \mathbf{v} \rangle = \langle *\mathbf{v}, *\mathbf{v} \rangle$ for $\mathbf{v} \in V$ (noting that $*\mathbf{v} \in \wedge^{N-1} V$ and using the scalar product in that space). Show more generally that

$$\langle \psi_1, \psi_2 \rangle = \langle *\psi_1, *\psi_2 \rangle,$$

where $\psi_1, \psi_2 \in \wedge^k V$ and thus $*\psi_1$ and $*\psi_2$ belong to $\wedge^{N-k} V$. Deduce that the Hodge star is an orthogonal transformation in $\wedge^{N/2} V$ (if N is even).

Hint: Use Exercise 2.

5.6 Scalar product for complex spaces

In complex spaces, one can get useful results if one defines the scalar product in a different way. In this section we work in a complex vector space V.

A **Hermitian scalar product** is a complex function of two vectors $\mathbf{a}, \mathbf{b} \in V$ with the properties

$$\langle \mathbf{a}, \lambda \mathbf{b} \rangle = \lambda \langle \mathbf{a}, \mathbf{b} \rangle, \quad \langle \lambda \mathbf{a}, \mathbf{b} \rangle = \lambda^* \langle \mathbf{a}, \mathbf{b} \rangle,$$
$$\langle \mathbf{a} + \mathbf{b}, \mathbf{c} \rangle = \langle \mathbf{a}, \mathbf{c} \rangle + \langle \mathbf{b}, \mathbf{c} \rangle, \quad \langle \mathbf{b}, \mathbf{a} \rangle = \langle \mathbf{a}, \mathbf{b} \rangle^*,$$

and nondegeneracy ($\forall \mathbf{a} \in V$, $\exists \mathbf{b} \in V$ such that $\langle \mathbf{a}, \mathbf{b} \neq 0 \rangle$). (Note that λ^* in the formula above means the complex conjugate to λ.) It follows that $\langle \mathbf{x}, \mathbf{x} \rangle$ is real-valued. One usually also imposes the property $\langle \mathbf{x}, \mathbf{x} \rangle > 0$ for $\mathbf{x} \neq 0$, which is positive-definiteness.

Remark: Note that the scalar product is not linear in the first argument because we have the factor λ^* instead of λ; one says that it is **antilinear**. One can also define a Hermitian scalar product that is linear in the *first* argument but antilinear in the second argument, i.e. $\langle \mathbf{a}, \lambda \mathbf{b} \rangle = \lambda^* \langle \mathbf{a}, \mathbf{b} \rangle$ and $\langle \lambda \mathbf{a}, \mathbf{b} \rangle = \lambda \langle \mathbf{a}, \mathbf{b} \rangle$. Here we follow the definition used in the physics literature. This definition is designed to be compatible with the Dirac notation for complex spaces (see Example 3 below).

Example 1: In the vector space $\mathbb{C}^n$, vectors are n-tuples of complex numbers, $\mathbf{x} = (x_1, ..., x_n)$. A Hermitian scalar product is defined by the formula

$$\langle \mathbf{x}, \mathbf{y} \rangle = \sum_{i=1}^{n} x_i^* y_i.$$

This scalar product is nondegenerate and positive-definite.

Example 2: Suppose we have a real, N-dimensional vector space V with an ordinary (real) scalar product $\langle \cdot, \cdot \rangle$. We can construct a *complex* vector space out of V by the following construction (called the **complexification** of V). First we consider the space $\mathbb{C}$ as a real, two-dimensional vector space over $\mathbb{R}$. Then we consider the tensor product $V \otimes \mathbb{C}$, still a vector space over $\mathbb{R}$. Elements of $V \otimes \mathbb{C}$ are linear combinations of terms of the form $\mathbf{v} \otimes \lambda$, where $\mathbf{v} \in V$ and $\lambda \in \mathbb{C}$. However, the ($2N$-dimensional, real) vector space $V \otimes \mathbb{C}$ can be also viewed as a vector space over $\mathbb{C}$: the multiplication of $\mathbf{v} \otimes \lambda$ by a complex number z yields $\mathbf{v} \otimes (\lambda z)$. Then $V \otimes \mathbb{C}$ is interpreted as an N-dimensional, complex vector space. A Hermitian scalar product in this space is defined by

$$\langle \mathbf{a} \otimes \lambda, \mathbf{b} \otimes \mu \rangle \equiv \langle \mathbf{a}, \mathbf{b} \rangle \lambda^* \mu.$$

Here $\langle \mathbf{a}, \mathbf{b} \rangle$ is the ordinary (real) scalar product in V. It is easy to verify that the properties of a Hermitian scalar product are satisfied by the above definition. ■

Using the Hermitian scalar product, one defines an orthonormal basis and other constructions analogous to those defined using the ordinary (real) scalar product. For instance, the Hermitian scalar product allows one to identify vectors and covectors.

Example 3: The vector-covector correspondence in complex spaces is slightly different from that in real spaces. Consider a vector $\mathbf{v} \in V$; the corresponding covector $\mathbf{f}^* : V \to \mathbb{C}$ may be defined as

$$\mathbf{f}^*(\mathbf{x}) \equiv \langle \mathbf{v}, \mathbf{x} \rangle \in \mathbb{C}.$$

We denote the map $\mathbf{v} \mapsto \mathbf{f}^*$ by a dagger symbol, called **Hermitian conjugation**, so that $(\mathbf{v})^\dagger = \mathbf{f}^*$. Due to the antilinearity of the scalar product, we have the property

$$(\lambda \mathbf{v})^\dagger = \lambda^* (\mathbf{v})^\dagger .$$

In the Dirac notation, one denotes covectors by the "bra" symbols such as $\langle v|$. One then may write

$$(|v\rangle)^\dagger = \langle v| ,$$

i.e. one uses the same label "v" inside the special brackets. We then have

$$(\lambda |v\rangle)^\dagger = \lambda^* \langle v| .$$

The Hermitian scalar product of vectors $|a\rangle$ and $|b\rangle$ is equal to the action of $(|a\rangle)^\dagger$ on $|b\rangle$ and denoted $\langle a|b\rangle$. Thus, the scalar product of $|a\rangle$ and $\lambda |b\rangle$ is equal to $\langle a|\,\lambda\,|b\rangle = \lambda\,\langle a|b\rangle$, while the scalar product of $\lambda\,|a\rangle$ and $|b\rangle$ is equal to $\lambda^*\,\langle a|b\rangle$. ■

Similarly to the transposed operator $\hat{A}^T$, the **Hermitian conjugate** operator $\hat{A}^\dagger$ is defined by

$$\langle \hat{A}^\dagger \mathbf{x}, \mathbf{y}\rangle \equiv \langle \mathbf{x}, \hat{A}\mathbf{y}\rangle, \quad \forall \mathbf{x}, \mathbf{y} \in V.$$

In an orthonormal basis, the matrix describing the Hermitian conjugate operator $\hat{A}^\dagger$ is obtained from the matrix of $\hat{A}$ by transposing and complex conjugating each matrix element.

Example 4: In the space of linear operators $\operatorname{End} V$, a bilinear form can be defined by

$$\langle \hat{A}, \hat{B}\rangle \equiv \operatorname{Tr}(\hat{A}^\dagger \hat{B}).$$

As we will see in the next section (Exercise 2), this bilinear form is a positive-definite scalar product in the space $\operatorname{End} V$. ■

In the following sections, we consider some applications of the Hermitian scalar product.

5.6.1 Symmetric and Hermitian operators

An operator $\hat{A}$ is **symmetric** with respect to the scalar product if

$$\langle \mathbf{u}, \hat{A}\mathbf{v}\rangle = \langle \hat{A}\mathbf{u}, \mathbf{v}\rangle, \quad \forall \mathbf{u}, \mathbf{v} \in V.$$

According to the definition of the transposed operator, the above property is the same as $\hat{A}^T = \hat{A}$.

The notion of a symmetric operator is suitable for a real vector space. In a complex vector space, one uses Hermitian conjugation instead of transposition: An operator $\hat{A}$ is called **Hermitian** if $\hat{A}^\dagger = \hat{A}$.

Symmetric as well as Hermitian operators often occur in applications and have useful properties.

Statement 1: a) All eigenvalues of a Hermitian operator are real (have zero imaginary part).

b) If $\hat{A}$ is a symmetric or Hermitian operator and $\mathbf{v}_1, \mathbf{v}_2$ are eigenvectors of $\hat{A}$ corresponding to different eigenvalues $\lambda_1 \neq \lambda_2$, then $\mathbf{v}_1$ and $\mathbf{v}_2$ are orthogonal to each other: $\langle \mathbf{v}_1, \mathbf{v}_2\rangle = 0$.

Proof: **a)** If $\mathbf{v}$ is an eigenvector of a Hermitian operator $\hat{A}$ with eigenvalue λ, we have

$$\begin{aligned}\langle \mathbf{v}, \hat{A}\mathbf{v}\rangle &= \langle \mathbf{v}, \lambda\mathbf{v}\rangle = \lambda\,\langle \mathbf{v}, \mathbf{v}\rangle \\ &= \langle \hat{A}\mathbf{v}, \mathbf{v}\rangle = \langle \lambda\mathbf{v}, \mathbf{v}\rangle = \lambda^*\,\langle \mathbf{v}, \mathbf{v}\rangle .\end{aligned}$$

Since $\langle \mathbf{v}, \mathbf{v}\rangle \neq 0$, we have $\lambda = \lambda^*$, i.e. λ is purely real.

b) We compute

$$\begin{aligned}\langle \mathbf{v}_1, \hat{A}\mathbf{v}_2\rangle &= \lambda_2\,\langle \mathbf{v}_1, \mathbf{v}_2\rangle \\ &\overset{!}{=} \langle \hat{A}\mathbf{v}_1, \mathbf{v}_2\rangle = \lambda_1\,\langle \mathbf{v}_1, \mathbf{v}_2\rangle .\end{aligned}$$

(In the case of Hermitian operators, we have used the fact that λ_1 is real.) Hence, either $\lambda_1 = \lambda_2$ or $\langle \mathbf{v}_1, \mathbf{v}_2\rangle = 0$. ■

Statement 2: If $\hat{A}$ is either symmetric or Hermitian and has an eigenvector $\mathbf{v}$, the subspace orthogonal to $\mathbf{v}$ is invariant under $\hat{A}$.

Proof: We need to show that $\langle \mathbf{x}, \mathbf{v} \rangle = 0$ entails $\langle \hat{A}\mathbf{x}, \mathbf{v} \rangle = 0$. We compute

$$\langle \hat{A}\mathbf{x}, \mathbf{v} \rangle = \langle \mathbf{x}, \hat{A}\mathbf{v} \rangle = \lambda \langle \mathbf{x}, \mathbf{v} \rangle = 0.$$

Hence, $\hat{A}\mathbf{x}$ also belongs to the subspace orthogonal to $\mathbf{v}$. ■

Statement 3: A Hermitian operator is diagonalizable.

Proof: We work in an N-dimensional space V. The characteristic polynomial of an operator $\hat{A}$ has at least one (perhaps complex-valued) root λ, which is an eigenvalue of $\hat{A}$, and thus there exists at least one eigenvector $\mathbf{v}$ corresponding to λ. By Statement 2, the subspace $\mathbf{v}^{\perp}$ (the orthogonal complement of $\mathbf{v}$) is invariant under $\hat{A}$. The space V splits into a direct sum of Span $\{\mathbf{v}\}$ and the subspace $\mathbf{v}^{\perp}$. We may consider the operator $\hat{A}$ in that subspace; again we find that there exists at least one eigenvector in $\mathbf{v}^{\perp}$. Continuing this argument, we split the entire space into a direct sum of N orthogonal eigenspaces. Hence, there exist N eigenvectors of $\hat{A}$. ■

Statement 4: A symmetric operator in a real N-dimensional vector space is diagonalizable, i.e. it has N real eigenvectors with real eigenvalues.

Proof: We cannot repeat the proof of Statement 3 literally, since we do not know *a priori* that the characteristic polynomial of a symmetric operator has all real roots; this is something we need to prove. Therefore we complexify the space V, i.e. we consider the space $V \otimes \mathbb{C}$ as a vector space over $\mathbb{C}$. In this space, we introduce a Hermitian scalar product as in Example 2 in Sec. 5.6. In the space $V \otimes \mathbb{C}$ there is a special notion of "real" vectors; these are vectors of the form $\mathbf{v} \otimes c$ with real c.

The operator $\hat{A}$ is extended to the space $V \otimes \mathbb{C}$ by

$$\hat{A}(\mathbf{v} \otimes c) \equiv (\hat{A}\mathbf{v}) \otimes c.$$

It is important to observe that the operator $\hat{A}$ transforms real vectors into real vectors, and moreover that $\hat{A}$ is Hermitian in $V \otimes \mathbb{C}$ if $\hat{A}$ is symmetric in V. Therefore, $\hat{A}$ is diagonalizable in $V \otimes \mathbb{C}$ with real eigenvalues.

It remains to show that all the eigenvectors of $\hat{A}$ can be chosen *real*; this will prove that $\hat{A}$ is also diagonalizable in the original space V. So far we only know that $\hat{A}$ has N eigenvectors in $V \otimes \mathbb{C}$. Any vector from $V \otimes \mathbb{C}$ can be transformed into the expression $\mathbf{u} \otimes 1 + \mathbf{v} \otimes \mathrm{i}$ with $\mathbf{u}, \mathbf{v} \in V$. Let us assume that $\mathbf{u} \otimes 1 + \mathbf{v} \otimes \mathrm{i}$ is an eigenvector of $\hat{A}$ with eigenvalue λ. If $\mathbf{v} = 0$, the eigenvector is real, and there is nothing left to prove; so we assume $\mathbf{v} \neq 0$. Since λ is real, we have

$$\begin{aligned} \hat{A}(\mathbf{u} \otimes 1 + \mathbf{v} \otimes \mathrm{i}) &= (\hat{A}\mathbf{u}) \otimes 1 + (\hat{A}\mathbf{v}) \otimes \mathrm{i} \\ &\overset{!}{=} \lambda \mathbf{u} \otimes 1 + \lambda \mathbf{v} \otimes \mathrm{i}. \end{aligned}$$

If both $\mathbf{u} \neq 0$ and $\mathbf{v} \neq 0$, it follows that $\mathbf{u}$ and $\mathbf{v}$ are both eigenvectors of $\hat{A}$ with eigenvalue λ. Hence, the operator $\hat{A}$ in $V \otimes \mathbb{C}$ can be diagonalized by choosing the real eigenvectors as $\mathbf{u} \otimes 1$ and $\mathbf{v} \otimes 1$ instead of the complex eigenvector $\mathbf{u} \otimes 1 + \mathbf{v} \otimes \mathrm{i}$. If $\mathbf{u} = 0$, we only need to replace the complex eigenvector $\mathbf{v} \otimes \mathrm{i}$ by the equivalent real eigenvector $\mathbf{v} \otimes 1$. We have thus shown that the eigenvectors of $\hat{A}$ in $V \otimes \mathbb{C}$ can be chosen real. ■

Exercise 1: If an operator $\hat{A}$ satisfies $\hat{A}^\dagger = -\hat{A}$, it is called **anti-Hermitian**. Show that all eigenvalues of $\hat{A}$ are pure imaginary or zero, that eigenvectors of $\hat{A}$ are orthogonal to each other, and that $\hat{A}$ is diagonalizable.

Hint: The operator $\hat{B} \equiv \mathrm{i}\hat{A}$ is Hermitian; use the properties of Hermitian operators (Statements 1,2,3).

Exercise 2: Show that $\mathrm{Tr}(\hat{A}^T\hat{A}) > 0$ for operators in a real space with a scalar product, and $\mathrm{Tr}(\hat{A}^\dagger\hat{A}) > 0$ for operators in a complex space with a Hermitian scalar product. Deduce that $\langle \hat{A}, \hat{B} \rangle \equiv \mathrm{Tr}\,(\hat{A}^T\hat{B})$ and $\langle \hat{A}, \hat{B} \rangle \equiv \mathrm{Tr}\,(\hat{A}^\dagger\hat{B})$ are positive-definite scalar products in the spaces of operators (assuming real or, respectively, complex space V with a scalar product).

Hint: Compute $\mathrm{Tr}(\hat{A}^T\hat{A})$ or $\mathrm{Tr}(\hat{A}^\dagger\hat{A})$ directly through components of $\hat{A}$ in an orthonormal basis.

Exercise 3: Show that the set of all Hermitian operators is a subspace of $\mathrm{End}\,V$, and the same for anti-Hermitian operators. Then show that these two subspaces are orthogonal to each other with respect to the scalar product of Exercise 2.

Exercise 4: Consider the space $\mathrm{End}\,V$ of linear operators and two of its subspaces: the subspace of **traceless** operators (i.e. operators $\hat{A}$ with $\mathrm{Tr}\hat{A} = 0$) and the subspace of operators proportional to the identity (i.e. operators $\lambda\hat{1}_V$ for $\lambda \in \mathbb{R}$). Show that these two subspaces are orthogonal with respect to the scalar products $\langle \hat{A}, \hat{B} \rangle \equiv \mathrm{Tr}(\hat{A}^T\hat{B})$ or $\langle \hat{A}, \hat{B} \rangle \equiv \mathrm{Tr}\,(\hat{A}^\dagger\hat{B})$.

5.6.2 Unitary transformations

In complex spaces, the notion analogous to orthogonal transformations is unitary transformations.

Definition: An operator is called **unitary** if it preserves the Hermitian scalar product:

$$\langle \hat{A}\mathbf{x}, \hat{A}\mathbf{y} \rangle = \langle \mathbf{x}, \mathbf{y} \rangle, \quad \forall \mathbf{x}, \mathbf{y} \in V.$$

It follows that a unitary operator $\hat{A}$ satisfies $\hat{A}^\dagger\hat{A} = \hat{1}$.

Exercise 2: If $\hat{A}$ is Hermitian, show that the operators $(1 + \mathrm{i}\hat{A})^{-1}(1 - \mathrm{i}\hat{A})$ and $\exp\,(\mathrm{i}\hat{A})$ are unitary.

Hint: The Hermitian conjugate of $f(\mathrm{i}\hat{A})$ is $f(-\mathrm{i}\hat{A}^\dagger)$ if $f(z)$ is an analytic function. This can be shown by considering each term in the power series for $f(z)$.

Exercise 3: Show that the determinant of a unitary operator is a complex number c such that $|c| = 1$.

Hint: First show that $\det(\hat{A}^\dagger)$ is the complex conjugate of $\det\hat{A}$.

5.7 Antisymmetric operators

In this and the following sections we work in a real vector space V in which a scalar product $\langle\cdot,\cdot\rangle$ is defined. The dimension of V is $N \equiv \dim V$.

An operator $\hat{A}$ is **antisymmetric** with respect to the scalar product if

$$\langle \mathbf{u}, \hat{A}\mathbf{v}\rangle + \langle \hat{A}\mathbf{u}, \mathbf{v}\rangle = 0, \quad \forall \mathbf{u}, \mathbf{v} \in V.$$

Exercise 1: Show that the set of all antisymmetric operators is a subspace of $V \otimes V^*$.

Exercise 2: Show that $\hat{A}^T + \hat{A} = 0$ if and only if the operator $\hat{A}$ is antisymmetric.

Remark: Exercise 2 shows that antisymmetric operators are represented by antisymmetric matrices — in an *orthonormal basis*. However, the matrix of an operator in some other basis does not have to be antisymmetric. An operator can be antisymmetric with respect to one scalar product and not antisymmetric with respect to another.

Question: Surely an antisymmetric matrix has rather special properties. Why is it that the corresponding operator is only antisymmetric *with respect to some* scalar product? Is it not true that the corresponding operator has by itself special properties, regardless of any scalar product?

Answer: Yes, it is true. It is a special property of an operator that there exists a scalar product *with respect to which* the operator is antisymmetric. If we know that this is true, we can derive some useful properties of the given operator by using that scalar product. ■

Statement 1: A 2-vector $\mathbf{a} \wedge \mathbf{b} \in \wedge^2 V$ can be mapped to an operator in V by

$$\mathbf{a} \wedge \mathbf{b} \mapsto \hat{A}; \quad \hat{A}\mathbf{x} \equiv \mathbf{a}\langle \mathbf{b}, \mathbf{x}\rangle - \mathbf{b}\langle \mathbf{a}, \mathbf{x}\rangle, \quad \forall \mathbf{x} \in V.$$

This formula defines a canonical isomorphism between the space of antisymmetric operators (with respect to the given scalar product) and $\wedge^2 V$. In other words, any antisymmetric operator $\hat{A}$ can be represented by a 2-vector $A \in \wedge^2 V$ and vice versa.

Proof: Left as exercise.

Statement 2: Any 2-vector $A \in \wedge^2 V$ can be written as a sum $\sum_{j=1}^{n} \mathbf{a}_k \wedge \mathbf{b}_k$ using n terms, where n is some number such that $n \leq \frac{1}{2}N$ (here $N \equiv \dim V$), and the set of vectors $\{\mathbf{a}_1, \mathbf{b}_1, ..., \mathbf{a}_n, \mathbf{b}_n\}$ is linearly independent.

Proof: By definition, a 2-vector A is representable as a linear combination of the form

$$A = \sum_{j=1}^{n} \mathbf{a}_j \wedge \mathbf{b}_j,$$

with *some* vectors $\mathbf{a}_j, \mathbf{b}_j \in V$ and *some* value of n. We will begin with this representation and transform it in order to minimize the number of terms.

The idea is to make sure that the set of vectors $\{\mathbf{a}_1, \mathbf{b}_1, ..., \mathbf{a}_n, \mathbf{b}_n\}$ is linearly independent. If this is not so, there exists a linear relation, say

$$\mathbf{a}_1 = \beta_1 \mathbf{b}_1 + \sum_{j=2}^{n} (\alpha_j \mathbf{a}_j + \beta_j \mathbf{b}_j),$$

with some coefficients α_j and β_j. Using this relation, the term $\mathbf{a}_1 \wedge \mathbf{b}_1$ can be rewritten as

$$\mathbf{a}_1 \wedge \mathbf{b}_1 = \sum_{j=2}^{n} (\alpha_j \mathbf{a}_j + \beta_j \mathbf{b}_j) \wedge \mathbf{b}_1.$$

These terms can be absorbed by other terms $\mathbf{a}_j \wedge \mathbf{b}_j$ ($j = 2, ..., N$). For example, by rewriting

$$\begin{aligned}
&\mathbf{a}_2 \wedge \mathbf{b}_2 + \alpha_2 \mathbf{a}_2 \wedge \mathbf{b}_1 + \beta_2 \mathbf{b}_2 \wedge \mathbf{b}_1 \\
&\quad = (\mathbf{a}_2 - \beta_2 \mathbf{b}_1) \wedge (\mathbf{b}_2 + \alpha_2 \mathbf{b}_1) \\
&\quad \equiv \tilde{\mathbf{a}}_2 \wedge \tilde{\mathbf{b}}_2
\end{aligned}$$

we can absorb the term $(\alpha_j \mathbf{a}_j + \beta_j \mathbf{b}_j) \wedge \mathbf{b}_1$ with $j = 2$ into $\mathbf{a}_2 \wedge \mathbf{b}_2$, replacing the vectors $\mathbf{a}_2$ and $\mathbf{b}_2$ by new vectors $\tilde{\mathbf{a}}_2$ and $\tilde{\mathbf{b}}_2$. In this way, we can redefine the vectors $\mathbf{a}_j, \mathbf{b}_j$ ($j = 2, ..., N$) so that the term $\mathbf{a}_1 \wedge \mathbf{b}_1$ is eliminated from the expression for A. We continue this procedure until the set of all the vectors $\mathbf{a}_j, \mathbf{b}_j$ is linearly independent. We now denote again by $\{\mathbf{a}_1, \mathbf{b}_1, ..., \mathbf{a}_n, \mathbf{b}_n\}$ the resulting linearly independent set of vectors such that the representation $A = \sum_{j=1}^{n} \mathbf{a}_j \wedge \mathbf{b}_j$ still holds. Note that the final number n may be smaller than the initial number. Since the number of vectors ($2n$) in the final, linearly independent set $\{\mathbf{a}_1, \mathbf{b}_1, ..., \mathbf{a}_n, \mathbf{b}_n\}$ cannot be greater than N, the dimension of the space V, we have $2n \leq N$ and so $n \leq \frac{1}{2}N$. ■

Exercise 3: A 2-vector $A \in \wedge^2 V$ satisfies $A \wedge A = 0$. Show that A can be expressed as a single-term exterior product, $A = \mathbf{a} \wedge \mathbf{b}$.

Hint: Express A as a sum of smallest number of single-term products, $A = \sum_{j=1}^{n} \mathbf{a}_k \wedge \mathbf{b}_k$, and show that $A \wedge A = 0$ implies $n = 1$: By Statement 2, the set $\{\mathbf{a}_i, \mathbf{b}_i\}$ is linearly independent. If $n > 1$, the expression $A \wedge A$ will contain terms such as $\mathbf{a}_1 \wedge \mathbf{b}_1 \wedge \mathbf{a}_2 \wedge \mathbf{b}_2$; a linear combination of these terms cannot vanish, since they are all linearly independent of each other. To show that rigorously, apply suitably chosen covectors $\mathbf{a}_i^*$ and $\mathbf{b}_i^*$. ■

Antisymmetric operators have the following properties.

Exercise 4: Show that the trace of an antisymmetric operator is equal to zero.

Hint: Use the property $\text{Tr}(\hat{A}^T) = \text{Tr}(\hat{A})$.

Exercise 5: Show that the determinant of the antisymmetric operator is equal to zero in an odd-dimensional space.

Remark: Note that the property of being antisymmetric is defined only with respect to a chosen scalar product. (An operator may be represented by an antisymmetric matrix in some basis, but not in another basis. An antisymmetric operator is represented by an antisymmetric matrix only in an orthonormal basis.) The properties shown in Exercises 3 and 4 will hold for any operator $\hat{A}$ such that *some scalar product exists* with respect to which $\hat{A}$ is antisymmetric. If $\hat{A}$ is represented by an antisymmetric matrix in a given basis $\{\mathbf{e}_j\}$, we may *define* the scalar product by requiring that $\{\mathbf{e}_j\}$ be an orthonormal basis; then $\hat{A}$ will be antisymmetric with respect to that scalar product.

Exercise 6: Show that the canonical scalar product $\langle A, B\rangle$ in the space $\wedge^2 V$ (see Sec. 5.5.2) coincides with the scalar product $\langle \hat{A}, \hat{B}\rangle \equiv \mathrm{Tr}(\hat{A}^T\hat{B})$ when the 2-vectors A and B are mapped into antisymmetric operators $\hat{A}$ and $\hat{B}$.

Hint: It is sufficient to consider the basis tensors $\mathbf{e}_i \wedge \mathbf{e}_j$ as operators $\hat{A}$ and $\hat{B}$.

Exercise 7:* Show that any 2-vector A can be written as $A = \sum_{i=1}^n \lambda_i \mathbf{a}_i \wedge \mathbf{b}_i$, where the set $\{\mathbf{a}_1, \mathbf{b}_1, ..., \mathbf{a}_n, \mathbf{b}_n\}$ is orthonormal.

Outline of solution: Consider the complexified vector space $V \otimes \mathbb{C}$ in which a Hermitian scalar product is defined; extend $\hat{A}$ into that space, and show that $\hat{A}$ is anti-Hermitian. Then $\hat{A}$ is diagonalizable and has all imaginary eigenvalues. However, the operator $\hat{A}$ is real; therefore, its eigenvalues come in pairs of complex conjugate imaginary values $\{\mathrm{i}\lambda_1, -\mathrm{i}\lambda_1, ..., \mathrm{i}\lambda_n, -\mathrm{i}\lambda_n\}$. The corresponding eigenvectors $\{\mathbf{v}_1, \bar{\mathbf{v}}_1, ..., \mathbf{v}_n, \bar{\mathbf{v}}_n\}$ are orthogonal and can be rescaled so that they are orthonormal. Further, we may choose these vectors such that $\bar{\mathbf{v}}_i$ is the vector complex conjugate to $\mathbf{v}_i$. The tensor representation of $\hat{A}$ is

$$\hat{A} = \sum_{i=1}^n \mathrm{i}\lambda_i \left(\mathbf{v}_i \otimes \mathbf{v}_i^* - \bar{\mathbf{v}}_i \otimes \bar{\mathbf{v}}_i^*\right),$$

where $\{\mathbf{v}_i^*, \bar{\mathbf{v}}_i^*\}$ is the basis dual to $\{\mathbf{v}_i, \bar{\mathbf{v}}_i\}$. We now define the vectors

$$\mathbf{a}_i \equiv \frac{\mathbf{v}_i + \bar{\mathbf{v}}_i}{\sqrt{2}}, \quad \mathbf{b}_i \equiv \frac{\mathbf{v}_i - \bar{\mathbf{v}}_i}{\mathrm{i}\sqrt{2}},$$

and verify that

$$\hat{A}\mathbf{a}_i = -\lambda_i \mathbf{b}_i, \quad \hat{A}\mathbf{b}_i = \lambda_i \mathbf{a}_i \quad (i = 1, ..., n).$$

Furthermore, the set of vectors $\{\mathbf{a}_1, \mathbf{b}_1, ..., \mathbf{a}_n, \mathbf{b}_n\}$ is orthonormal, and all the vectors $\mathbf{a}_i$, $\mathbf{b}_i$ are real. Therefore, we can represent $\hat{A}$ in the original space V by the 2-vector

$$A \equiv \sum_{i=1}^n \lambda_i \left(\mathbf{a}_i \wedge \mathbf{b}_i\right).$$

The set $\{\mathbf{a}_1, \mathbf{b}_1, ..., \mathbf{a}_n, \mathbf{b}_n\}$ yields the solution to the problem.

5.8 * Pfaffians

The Pfaffian is a construction analogous to the determinant, except that it applies only to antisymmetric operators in even-dimensional spaces with a scalar product.

Definition: If $\hat{A}$ is an antisymmetric operator in V and $N \equiv \dim V$ is even, the **Pfaffian** of $\hat{A}$ is the number $\mathrm{Pf}\,\hat{A}$ defined (up to a sign) as the constant factor in the tensor equality

$$(\mathrm{Pf}\,\hat{A})\mathbf{e}_1 \wedge ... \wedge \mathbf{e}_N = \frac{1}{(N/2)!}\underbrace{A \wedge ... \wedge A}_{N/2} = \frac{1}{(N/2)!}\bigwedge_{k=1}^{N/2} A,$$

where $\{\mathbf{e}_1, ..., \mathbf{e}_N\}$ is an *orthonormal* basis in V and $A \in \wedge^2 V$ is the tensor corresponding to the operator $\hat{A}$. (Note that both sides in the equation above are tensors from $\wedge^N V$.)

Remark: The sign of the Pfaffian depends on the orientation of the orthonormal basis. Other than that, the Pfaffian does not depend on the choice of the orthonormal basis $\{\mathbf{e}_j\}$. If this ambiguity is not desired, one could consider a *tensor-valued* Pfaffian, $A \wedge ... \wedge A \in \wedge^N V$; this tensor does not depend on the choice of the orientation of the orthonormal basis. This is quite similar to the ambiguity of the definition of volume and to the possibility of defining an unambiguous but tensor-valued "oriented volume." However, it is important to note that $\{\mathbf{e}_j\}$ must be a positively oriented *orthonormal* basis; if we change to an arbitrary basis, the tensor $\mathbf{e}_1 \wedge ... \wedge \mathbf{e}_N$ will be multiplied by some number not equal to ± 1, which will make the definition of $\operatorname{Pf}\hat{A}$ impossible.

Question: Can we define the Pfaffian of an operator if we do not have a scalar product in V? Can we define the Pfaffian of an antisymmetric matrix?

Answer: We need a scalar product in order to map an operator $\hat{A} \in \operatorname{End}V$ to a bivector $A \in \wedge^2 V$; this is central in the construction of the Pfaffian. If we know that an operator $\hat{A}$ is antisymmetric with respect to *some* scalar product (i.e. if we know that such a scalar product *exists*) then we can use that scalar product in order to define the Pfaffian of $\hat{A}$. In the language of matrices: If an antisymmetric matrix is given, we can postulate that this matrix represents an operator in some basis; then we can introduce a scalar product such that this basis is orthonormal, so that this operator is an antisymmetric operator with respect to this scalar product; and then the Pfaffian can be defined. ■

To make the correspondence between operators and bivectors more visual, let us represent operators by their matrices in an orthonormal basis. Antisymmetric operators are then represented by antisymmetric matrices.

Examples: First we consider a *two*-dimensional space V. Any 2×2 antisymmetric matrix $\hat{A}$ is necessarily of the form $\hat{A} = \begin{pmatrix} 0 & a \\ -a & 0 \end{pmatrix}$, where a is some number; the determinant of $\hat{A}$ is then a^2. Let us compute the Pfaffian of $\hat{A}$. We find the representation of $\hat{A}$ as an element of $\wedge^2 V$ as follows, $\hat{A} = a\mathbf{e}_1 \wedge \mathbf{e}_2$, and hence $\operatorname{Pf}\hat{A} = a$. We note that the determinant is equal to the square of the Pfaffian.

Let us now consider a four-dimensional space V and a 4×4 antisymmetric matrix; such a matrix must be of the form

$$\hat{B} = \begin{pmatrix} 0 & a & b & c \\ -a & 0 & x & y \\ -b & -x & 0 & z \\ -c & -y & -z & 0 \end{pmatrix},$$

where the numbers a, b, c, x, y, z are arbitrary. Let us compute the Pfaffian and the determinant of the operator represented by this matrix. We find the representation of $\hat{B}$ as an element of $\wedge^2 V$ as follows,

$$\hat{B} = a\mathbf{e}_1 \wedge \mathbf{e}_2 + b\mathbf{e}_1 \wedge \mathbf{e}_3 + c\mathbf{e}_1 \wedge \mathbf{e}_4 + x\mathbf{e}_2 \wedge \mathbf{e}_3 + y\mathbf{e}_2 \wedge \mathbf{e}_4 + z\mathbf{e}_3 \wedge \mathbf{e}_4.$$

Therefore,

$$\frac{1}{2!}\hat{B} \wedge \hat{B} = (az - by + cx)\, \mathbf{e}_1 \wedge \mathbf{e}_2 \wedge \mathbf{e}_3 \wedge \mathbf{e}_4.$$

(Note that the factor $\frac{1}{2!}$ cancels the combinatorial factor 2 resulting from the antisymmetry of the exterior product.) Hence, Pf $\hat{B} = az - by + cx$.

Exercise: Compute the determinant of $\hat{B}$ in the example above; show that

$$\det \hat{B} = a^2 z^2 - 2abyz + b^2 y^2 - 2bcxy + c^2 x^2 + 2acxz.$$

We see that, again, the determinant is equal to the square of the Pfaffian (which is easier to compute).

Remark: The factor $1/(N/2)!$ used in the definition of the Pfaffian is a combinatorial factor. This factor could be inconvenient if we were calculating in a finite number field where one cannot divide by $(N/2)!$. This inconvenience can be avoided if we define the Pfaffian of a tensor $A = \mathbf{v}_1 \wedge \mathbf{v}_2 + ... + \mathbf{v}_{n-1} \wedge \mathbf{v}_n$ as zero if $n < N$ and as the coefficient in the tensor equality

$$\mathbf{v}_1 \wedge ... \wedge \mathbf{v}_N \stackrel{!}{=} (\text{Pf}\, \hat{A})\mathbf{e}_1 \wedge ... \wedge \mathbf{e}_N$$

if $n = N$. For example, consider the tensor

$$A = \mathbf{a} \wedge \mathbf{b} + \mathbf{c} \wedge \mathbf{d}$$

in a four-dimensional space ($N = 4$). We compute

$$\begin{aligned} A \wedge A &= (\mathbf{a} \wedge \mathbf{b} + \mathbf{c} \wedge \mathbf{d}) \wedge (\mathbf{a} \wedge \mathbf{b} + \mathbf{c} \wedge \mathbf{d}) \\ &= 0 + \mathbf{a} \wedge \mathbf{b} \wedge \mathbf{c} \wedge \mathbf{d} + \mathbf{c} \wedge \mathbf{d} \wedge \mathbf{a} \wedge \mathbf{b} + 0 \\ &= 2\mathbf{a} \wedge \mathbf{b} \wedge \mathbf{c} \wedge \mathbf{d}. \end{aligned}$$

It is clear that the factor $2 = (N/2)!$ arises due to the presence of 2 possible permutations of the two tensors $\mathbf{a} \wedge \mathbf{b}$ and $\mathbf{c} \wedge \mathbf{d}$ and is therefore a *combinatorial factor*. We can avoid the division by 2 in the definition of the Pfaffian if we consider the tensor $\mathbf{a} \wedge \mathbf{b} \wedge \mathbf{c} \wedge \mathbf{d}$ right away, instead of dividing $A \wedge A$ by 2.■

5.8.1 Determinants are Pfaffians squared

In the examples in the previous section, we have seen that the determinant turned out to be equal to the square of the Pfaffian of the same operator. We will now prove this correspondence in the general case.

Theorem: Given a linear operator $\hat{A}$ in an even-dimensional space V where a scalar product is defined, and given that the operator $\hat{A}$ is antisymmetric with respect to that scalar product, we have

$$(\text{Pf}\, \hat{A})^2 = \det \hat{A}.$$

Proof: We know that the tensor $A \in \wedge^2 V$ corresponding to the operator $\hat{A}$ can be written in the form

$$A = \mathbf{v}_1 \wedge \mathbf{v}_2 + ... + \mathbf{v}_{n-1} \wedge \mathbf{v}_k,$$

where the set of vectors $\{\mathbf{v}_1, ..., \mathbf{v}_k\}$ is linearly independent (Statement 2 in Sec. 5.7) and $k \leq N$ is an even number.

We begin by considering the case $k < N$. In this case the exterior product $A \wedge ... \wedge A$ (where A is taken $N/2$ times) will be equal to zero because there are only k different vectors in that exterior product, while the total number of vectors is N, so at least two vectors $\mathbf{v}_i$ must be repeated. Also $\det \hat{A} = 0$ in this case; this can be shown explicitly by completing $\{\mathbf{v}_1, ..., \mathbf{v}_k\}$ to a basis $\{\mathbf{v}_1, ..., \mathbf{v}_k, \mathbf{e}_{k+1}, ..., \mathbf{e}_N\}$ such that all $\mathbf{e}_j$ are orthogonal to all $\mathbf{v}_i$. (This can be done by first completing $\{\mathbf{v}_1, ..., \mathbf{v}_k\}$ to a basis and then applying the Gram-Schmidt orthogonalization procedure to the vectors $\mathbf{e}_j$, $j = k+1, ..., N$.) Then we will have $\hat{A}\mathbf{e}_j = 0$ ($j = k + 1, ..., N$). Acting with $\wedge^N \hat{A}^N$ on the tensor $\mathbf{v}_1 \wedge ... \wedge \mathbf{v}_k \wedge \mathbf{e}_{k+1} \wedge ... \wedge \mathbf{e}_N$, we find

$$(\wedge^N \hat{A}^N)(\mathbf{v}_1 \wedge ... \wedge \mathbf{v}_k \wedge \mathbf{e}_{k+1} \wedge ... \wedge \mathbf{e}_N) = ... \wedge \hat{A}\mathbf{e}_N = 0$$

and hence $\det \hat{A} = 0$. Thus $(\mathrm{Pf}\,\hat{A})^2 = 0 = \det \hat{A}$, and there is nothing left to prove in case $k < N$.

It remains to consider the interesting case $k = N$. In this case, the set $\{\mathbf{v}_1, ..., \mathbf{v}_N\}$ is a basis in V. The Pfaffian $\mathrm{Pf}\,\hat{A}$ is the coefficient in the tensor equality

$$\frac{1}{(N/2)!} \bigwedge_{k=1}^{N/2} A = \mathbf{v}_1 \wedge ... \wedge \mathbf{v}_N \overset{!}{=} (\mathrm{Pf}\,\hat{A}) \mathbf{e}_1 \wedge ... \wedge \mathbf{e}_N,$$

where $\{\mathbf{e}_j\}$ is an orthonormal basis. In other words, $\mathrm{Pf}\,\hat{A}$ is the (oriented) volume of the parallelepiped spanned by the vectors $\{\mathbf{v}_j \,|\, j = 1, ..., N\}$, if we assume that the vectors $\{\mathbf{e}_j\}$ span a unit volume. Now it is clear that $\mathrm{Pf}\,\hat{A} \neq 0$.

Let us denote by $\{\mathbf{v}_j^*\}$ the dual basis to $\{\mathbf{v}_j\}$. Due to the one-to-one correspondence between vectors and covectors, we map $\{\mathbf{v}_j^*\}$ into the reciprocal basis $\{\mathbf{u}_j\}$. We now apply the operator $\hat{A}$ to the reciprocal basis $\{\mathbf{u}_j\}$ and find by a direct calculation (using the property $\langle \mathbf{v}_i, \mathbf{u}_j \rangle = \delta_{ij}$) that $\hat{A}\mathbf{u}_1 = -\mathbf{v}_2$, $\hat{A}\mathbf{u}_2 = \mathbf{v}_1$, and so on. Hence

$$\begin{aligned} \hat{A}\mathbf{u}_1 \wedge ... \wedge \hat{A}\mathbf{u}_N &= (-\mathbf{v}_2) \wedge \mathbf{v}_1 \wedge ... \wedge (-\mathbf{v}_N) \wedge \mathbf{v}_{N-1} \\ &= \mathbf{v}_1 \wedge \mathbf{v}_2 \wedge ... \wedge \mathbf{v}_N. \end{aligned}$$

It follows that $\det \hat{A}$ is the coefficient in the tensor equality

$$\hat{A}\mathbf{u}_1 \wedge ... \wedge \hat{A}\mathbf{u}_N = \mathbf{v}_1 \wedge ... \wedge \mathbf{v}_N \overset{!}{=} (\det \hat{A}) \mathbf{u}_1 \wedge ... \wedge \mathbf{u}_N. \tag{5.8}$$

In particular, $\det \hat{A} \neq 0$.

In order to prove the desired relationship between the determinant and the Pfaffian, it remains to compute the volume spanned by the dual basis $\{\mathbf{u}_j\}$, so that the tensor $\mathbf{u}_1 \wedge ... \wedge \mathbf{u}_N$ can be related to $\mathbf{e}_1 \wedge ... \wedge \mathbf{e}_N$. By Statement 2 in Sec. 5.4.4, the volume spanned by $\{\mathbf{u}_j\}$ is the inverse of the volume spanned by $\{\mathbf{v}_j\}$. Therefore the volume spanned by $\{\mathbf{u}_j\}$ is equal to $1/\text{Pf}\,\hat{A}$. Now we can compute the Pfaffian of $\hat{A}$ using

$$\mathbf{u}_1 \wedge ... \wedge \mathbf{u}_N = (\text{Pf}\,\hat{A})^{-1}\mathbf{e}_1 \wedge ... \wedge \mathbf{e}_N$$

together with Eq. (5.8):

$$\begin{aligned}\text{Pf}\,\hat{A} &= \frac{\mathbf{v}_1 \wedge ... \wedge \mathbf{v}_N}{\mathbf{e}_1 \wedge ... \wedge \mathbf{e}_N} = \frac{(\det \hat{A})(\text{Pf}\,\hat{A})^{-1}\mathbf{e}_1 \wedge ... \wedge \mathbf{e}_N}{\mathbf{e}_1 \wedge ... \wedge \mathbf{e}_N} \\ &= (\det \hat{A})(\text{Pf}\,\hat{A})^{-1}.\end{aligned}$$

Hence $\det \hat{A} = (\text{Pf}\,\hat{A})^2$. ■

5.8.2 Further properties

Having demonstrated the techniques of working with antisymmetric operators and Pfaffians, I propose to you the following exercises that demonstrate some other properties of Pfaffians. These exercises conclude this book.

Exercise 1: Let $\hat{A}$ be an antisymmetric operator; let $\hat{B}$ be an arbitrary operator. Prove that $\text{Pf}\,(\hat{B}\hat{A}\hat{B}^T) = \det(\hat{B})\text{Pf}\,\hat{A}$.

Hint: If $\hat{A}$ corresponds to the bivector $A = \mathbf{v}_1 \wedge \mathbf{v}_2 + ... + \mathbf{v}_{k-1} \wedge \mathbf{v}_k$, show that $\hat{B}\hat{A}\hat{B}^T$ corresponds to the bivector $\hat{B}\mathbf{v}_1 \wedge \hat{B}\mathbf{v}_2 + ... + \hat{B}\mathbf{v}_{k-1} \wedge \hat{B}\mathbf{v}_k$.

Exercise 2: Let $\hat{A}$ be an antisymmetric operator such that $\det \hat{A} \neq 0$; let $\{\mathbf{e}_i \,|\, i = 1, ..., 2n\}$ be a given orthonormal basis. Prove that there exists an operator $\hat{B}$ such that the operator $\hat{B}\hat{A}\hat{B}^T$ is represented by the bivector $\mathbf{e}_1 \wedge \mathbf{e}_2 + ... + \mathbf{e}_{2n-1} \wedge \mathbf{e}_{2n}$. Deduce that $\det \hat{A} = (\text{Pf}\,\hat{A})^2$.

Hint: This is a paraphrase of the proof of Theorem 5.8.1. Use the previous exercise and represent $\hat{A}$ by the bivector $\mathbf{v}_1 \wedge \mathbf{v}_2 + ... + \mathbf{v}_{2n-1} \wedge \mathbf{v}_{2n}$, where the set $\{\mathbf{v}_i\}$ is a basis. Define $\hat{B}$ as a map $\mathbf{e}_i \mapsto \mathbf{v}_i$; then $\hat{B}^{-1}$ exists and maps $\mathbf{v}_i \mapsto \mathbf{e}_i$. Show that $\text{Pf}\,\hat{A} = 1/(\det \hat{B})$.

Exercise 3: Use the result of Exercise 5 in Sec. 5.7 to prove that $\det \hat{A} = (\text{Pf}\,\hat{A})^2$.

Hint: For an operator $\hat{A} = \sum_{i=1}^{n} \lambda_i \mathbf{a}_i \wedge \mathbf{b}_i$, where $\{\mathbf{a}_1, \mathbf{b}_1, ..., \mathbf{a}_n, \mathbf{b}_n\}$ is a positively oriented *orthonormal* basis and $2n \equiv N$, show that $\text{Pf}\,\hat{A} = \lambda_1 ... \lambda_n$ and $\det \hat{A} = \lambda_1^2 ... \lambda_n^2$.

Exercise 4:* An operator $\hat{A}$ is antisymmetric and is represented in some orthonormal basis by a block matrix of the form

$$\hat{A} = \begin{pmatrix} 0 & \hat{M} \\ -\hat{M}^T & 0 \end{pmatrix},$$

where $\hat{M}$ is an arbitrary n-dimensional matrix. Show that

$$\text{Pf}\,\hat{A} = (-1)^{\frac{1}{2}n(n-1)} \det \hat{M}.$$

Solution: We need to represent $\hat{A}$ by a bivector from $\wedge^2 V$. The given form of the matrix $\hat{A}$ suggests that we consider the splitting of the space V into a direct sum of two orthogonal n-dimensional subspaces, $V = U_1 \oplus U_2$, where U_1 and U_2 are two copies of the same n-dimensional space U. A scalar product in U is defined naturally (by restriction), given the scalar product in V. We will denote by $\langle \cdot, \cdot \rangle$ the scalar product in U. The given matrix form of $\hat{A}$ means that we have a given operator $\hat{M} \in \operatorname{End} U$ such that $\hat{A}$ acts on vectors from V as

$$\hat{A}\,(\mathbf{v}_1 \oplus \mathbf{v}_2) = (\hat{M}\mathbf{v}_2) \oplus (-\hat{M}^T \mathbf{v}_1), \quad \mathbf{v}_1, \mathbf{v}_2 \in U. \tag{5.9}$$

We can choose an orthonormal basis $\{\mathbf{c}_i \,|\, i = 1, ..., n\}$ in U and represent the operator $\hat{M}$ through some suitable vectors $\{\mathbf{m}_i |\, i = 1, ..., n\}$ (not necessarily orthogonal) such that

$$\hat{M}\mathbf{u} = \sum_{i=1}^{n} \mathbf{m}_i \, \langle \mathbf{c}_i, \mathbf{u} \rangle\,, \quad \mathbf{u} \in U.$$

Note that the vectors $\mathbf{m}_i$ are found from $\hat{M}\mathbf{c}_i = \mathbf{m}_i$. It follows that $\hat{M}^T\mathbf{u} = \sum_{i=1}^{n} \mathbf{c}_i \, \langle \mathbf{m}_i, \mathbf{u} \rangle$. Using Eq. (5.9), we can then write the tensor representation of $\hat{A}$ as

$$\hat{A} = \sum_{i=1}^{n} \left[(\mathbf{m}_i \oplus 0) \otimes (0 \oplus \mathbf{c}_i)^* - (0 \oplus \mathbf{c}_i) \otimes (\mathbf{m}_i \oplus 0)^* \right].$$

Hence, $\hat{A}$ can be represented by the 2-vector

$$A = \sum_{i=1}^{n} (\mathbf{m}_i \oplus 0) \wedge (0 \oplus \mathbf{c}_i) \in \wedge^2 V.$$

The Pfaffian of $\hat{A}$ is then found from

$$\operatorname{Pf} \hat{A} = \frac{(\mathbf{m}_1 \oplus 0) \wedge (0 \oplus \mathbf{c}_1) \wedge ... \wedge (\mathbf{m}_n \oplus 0) \wedge (0 \oplus \mathbf{c}_n)}{\mathbf{e}_1 \wedge ... \wedge \mathbf{e}_{2n}},$$

where $\{\mathbf{e}_i \,|\, i = 1, ..., 2n\}$ is an orthonormal basis in V. We can choose this basis as $\mathbf{e}_i = \mathbf{c}_i \oplus 0$, $\mathbf{e}_{n+i} = 0 \oplus \mathbf{c}_i$ (for $i = 1, ..., n$). By introducing the sign factor $(-1)^{\frac{1}{2}n(n-1)}$, we may rearrange the exterior products so that all $\mathbf{m}_i$ are together. Hence

$$\begin{aligned} \operatorname{Pf} \hat{A} &= (-1)^{\frac{1}{2}n(n-1)} \\ &\times \frac{(\mathbf{m}_1 \oplus 0) \wedge ... \wedge (\mathbf{m}_n \oplus 0) \wedge (0 \oplus \mathbf{c}_1) \wedge ... \wedge (0 \oplus \mathbf{c}_n)}{(\mathbf{c}_1 \oplus 0) \wedge ... \wedge (\mathbf{c}_n \oplus 0) \wedge (0 \oplus \mathbf{c}_1) \wedge ... \wedge (0 \oplus \mathbf{c}_n)}. \end{aligned}$$

Vectors corresponding to different subspaces can be factorized, and then the factors containing $0 \oplus \mathbf{c}_i$ can be canceled:

$$\begin{aligned} \operatorname{Pf} \hat{A} &= (-1)^{\frac{1}{2}n(n-1)} \frac{\mathbf{m}_1 \wedge ... \wedge \mathbf{m}_n}{\mathbf{c}_1 \wedge ... \wedge \mathbf{c}_n} \frac{\mathbf{c}_1 \wedge ... \wedge \mathbf{c}_n}{\mathbf{c}_1 \wedge ... \wedge \mathbf{c}_n} \\ &= (-1)^{\frac{1}{2}n(n-1)} \frac{\mathbf{m}_1 \wedge ... \wedge \mathbf{m}_n}{\mathbf{c}_1 \wedge ... \wedge \mathbf{c}_n}. \end{aligned}$$

Finally, we have

$$\frac{\mathbf{m}_1 \wedge ... \wedge \mathbf{m}_n}{\mathbf{c}_1 \wedge ... \wedge \mathbf{c}_n} = \frac{\hat{M}\mathbf{c}_1 \wedge ... \wedge \hat{M}\mathbf{c}_n}{\mathbf{c}_1 \wedge ... \wedge \mathbf{c}_n} = \det \hat{M}.$$

This concludes the calculation. ■

A Complex numbers

This appendix is a crash course on complex numbers.

A.1 Basic definitions

A **complex number** is a formal expression $a + ib$, where a, b are real numbers. In other words, a complex number is simply a pair (a, b) of real numbers, written in a more convenient notation as $a + ib$. One writes, for example, $2 + \mathrm{i}3$ or $2 + 3\mathrm{i}$ or $3 + \mathrm{i}$ or $-5\mathrm{i} - 8$, etc. The **imaginary unit**, denoted "i", is not a real number; it is a symbol which has the property $\mathrm{i}^2 = -1$. Using this property, we can apply the usual algebraic rules to complex numbers; this is emphasized by the algebraic notation $a + ib$. For instance, we can add and multiply complex numbers,

$$\begin{aligned}(1+\mathrm{i}) + 5\mathrm{i} &= 1 + 6\mathrm{i};\\ (1-\mathrm{i})(2+\mathrm{i}) &= 2 - 2\mathrm{i} + \mathrm{i} - \mathrm{i}^2\\ &= 3 - \mathrm{i};\\ \mathrm{i}^3 = \mathrm{i}\mathrm{i}^2 &= -\mathrm{i}.\end{aligned}$$

It is straightforward to see that the result of any arithmetic operation on complex numbers turns out to be again a complex number. In other words, one can multiply, divide, add, subtract complex numbers just as directly as real numbers.

The set of all complex numbers is denoted by $\mathbb{C}$. The set of all real numbers is $\mathbb{R}$.

Exercise: Using directly the definition of the imaginary unit, compute the following complex numbers.

$$\frac{1}{\mathrm{i}} = ? \quad \mathrm{i}^4 = ? \quad \mathrm{i}^5 = ? \quad \left(\frac{1}{2} + \frac{\mathrm{i}\sqrt{3}}{2}\right)^3 = ?$$

The complex number $a - ib$ is called **complex conjugate** to $a + ib$. Conjugation is denoted either with an overbar or with a star superscript,

$$z = a + \mathrm{i}b, \quad \bar{z} = z^* = a - \mathrm{i}b,$$

according to convenience. Note that

$$zz^* = (a + \mathrm{i}b)(a - \mathrm{i}b) = a^2 + b^2 \in \mathbb{R}.$$

In order to divide by a complex number more easily, one multiplies the numerator and the denominator by the complex conjugate number, e.g.

$$\frac{1}{3+\mathrm{i}} =? = \frac{1}{3+\mathrm{i}} \cdot \frac{3-\mathrm{i}}{3-\mathrm{i}} = \frac{3-\mathrm{i}}{9-\mathrm{i}^2} = \frac{3-\mathrm{i}}{10} = \frac{3}{10} - \frac{1}{10}\mathrm{i}.$$

Exercise: Compute the following complex numbers,

$$\frac{1-\mathrm{i}}{1+\mathrm{i}} =? \quad \frac{1-\mathrm{i}}{4+\mathrm{i}} - \frac{1+\mathrm{i}}{4-\mathrm{i}} =? \quad \frac{1}{a+\mathrm{i}b} =?$$

where $a, b \in \mathbb{R}$. ■

Another view of complex numbers is that they are linear polynomials in the formal variable "i." Since we may replace i^2 by -1 and i^{-1} by $-\mathrm{i}$ wherever any power of "i" appears, we can reduce any power series in i and/or in i^{-1} to a linear combination of 1 and i.

If $z = a + \mathrm{i}b$ where $a, b \in \mathbb{R}$ then a is called the **real part**, $\mathrm{Re}\, z$, and b is the **imaginary part**, $\mathrm{Im}\, z$. In other words,

$$\mathrm{Re}\ (a+\mathrm{i}b) = a, \quad \mathrm{Im}\ (a+\mathrm{i}b) = b.$$

The **absolute value** or **modulus** of $z = a+\mathrm{i}b$ is the real number $|z| \equiv \sqrt{a^2+b^2}$.

Exercise: Compute

$$\mathrm{Re}\left[(2+\mathrm{i})^2\right] =? \quad |3+4\mathrm{i}| =?$$

Prove that

$$\mathrm{Re}\, z = \frac{z+\bar{z}}{2}; \quad \mathrm{Im}\, z = \frac{z-\bar{z}}{2\mathrm{i}}; \quad |z|^2 = z\bar{z};$$
$$|z| = |\bar{z}|; \quad |z_1 z_2| = |z_1|\,|z_2|; \quad (z_1 z_2)^* = z_1^* z_2^*$$

for any complex numbers z, z_1, $z_2 \in \mathbb{C}$.

A.2 Geometric representation

Let us draw a complex number $z = x + \mathrm{i}y$ as a point with coordinates (x, y) in the Euclidean plane, or a vector with real components (x, y). You can check that the sum $z_1 + z_2$ and the product of z with a real number λ, that is $z \mapsto z\lambda$, correspond to the familiar operations of adding two vectors and multiplying a vector by a scalar. Also, the absolute value $|z|$ is equal to the *length* of the two-dimensional vector (x, y) as computed in the usual Euclidean space.

Exercise: Show that the multiplication of $z = x + \mathrm{i}y$ by a complex number $r \equiv \cos\phi + \mathrm{i}\sin\phi$ corresponds to rotating the vector (x, y) by angle ϕ counterclockwise (assuming that the x axis is horizontal and points to the right, and the y axis points vertically upwards). Show that $|rz| = |z|$, which corresponds to the fact that the length of a vector does not change after a rotation.

A.3 Analytic functions

Analytic functions are such functions $f(x)$ that can be represented by a power series $f(x) = \sum_{n=0}^{\infty} c_n x^n$ with some coefficients c_n such that the series converges at least for some real x. In that case, the series will converge also for some complex x. In this sense, analytic functions are naturally extended from real to complex numbers. For example, $f(x) = x^2 + 1$ is an analytic function; it can be computed just as well for any complex x as for real x.

An example of a non-analytic function is the **Heaviside step function**

$$\theta(x) = \begin{cases} 0, & x < 0; \\ 1, & x \geq 0. \end{cases}$$

This function cannot be represented by a power series and thus cannot be naturally extended to complex numbers. In other words, there is no useful way to define the value of, say, $\theta(2\mathrm{i})$. On the other hand, functions such as $\cos x$, $\sqrt{x}$, $x / \ln x$, $\int_0^x e^{-t^2} dt$, and so on, are analytic and can be evaluated for complex x.

Exercise: Compute $(1 + 2\mathrm{i})\,(1 + 3\mathrm{i})$ and $(1 - 2\mathrm{i})\,(1 - 3\mathrm{i})$. What did you notice? Prove that $f(z^*) = [f(z)]^*$ for any analytic function $f(z)$.

Remark: Although $\sqrt{x}$ has no power series expansion at $x = 0$, it has a Taylor expansion at $x = 1$, which is sufficient for analyticity; one can also define $\sqrt{z}$ for complex z through the property $(\sqrt{z})^2 = z$.

Exercise: Derive an explicit formula for the square root of a complex number, $\sqrt{a + \mathrm{i}b}$, where $a, b \in \mathbb{R}$.

Hint: Write $\sqrt{a + \mathrm{i}b} = x + \mathrm{i}y$, square both sides, and solve for x and y.

Answer:

$$\sqrt{a + \mathrm{i}b} = \pm \left[\sqrt{\frac{\sqrt{a^2 + b^2} + a}{2}} + \mathrm{i}\,\mathrm{sign}(b) \sqrt{\frac{\sqrt{a^2 + b^2} - a}{2}} \right],$$

where $\mathrm{sign}(b) = 1, 0, -1$ when b is positive, zero, or negative. Note that this formula may be rewritten for quicker calculation as

$$\sqrt{a + \mathrm{i}b} = \pm \left(r + \mathrm{i}\frac{b}{2r} \right), \quad r \equiv \sqrt{\frac{\sqrt{a^2 + b^2} + a}{2}}.$$

(In this formula, the square roots in the definition of r are purely real and positive.)

A.4 Exponent and logarithm

The exponential function and the logarithmic function are analytic functions.

The **exponential** function is defined through the power series

$$e^z \equiv \exp z \equiv 1 + \frac{1}{1!}z + \frac{1}{2!}z^2 + ... = \sum_{n=0}^{\infty} \frac{z^n}{n!}.$$

This series converges for all complex z.
Exercise: Verify the **Euler formula,**

$$e^{\mathrm{i}\phi} = \cos\phi + \mathrm{i}\sin\phi, \quad \phi \in \mathbb{R},$$

by using the known Taylor series for $\sin x$ and $\cos x$. Calculate:

$$e^{2\mathrm{i}} = ? \quad e^{\pi\mathrm{i}} = ? \quad e^{\frac{1}{2}\pi\mathrm{i}} = ? \quad e^{2\pi\mathrm{i}} = ?$$

Exercise: Use the identity $e^{a+b} = e^a e^b$, which holds also for complex numbers a, b, to show that

$$e^{a+ib} = e^a(\cos b + \mathrm{i}\sin b), \quad a, b \in \mathbb{R}.$$

Calculate:

$$\exp\left[\ln 2 + \frac{\pi}{2}\mathrm{i}\right] = ? \quad \exp\left[1 + \pi\mathrm{i}\right] = ? \quad \cos\left(\frac{1}{2}\pi\mathrm{i}\right) = ?$$

The **logarithm** of a complex number z is a complex number denoted $\ln z$ such that $e^{\ln z} = z$. It is easy to see that

$$\exp\left[z + 2\pi\mathrm{i}\right] = \exp z, \quad z \in \mathbb{C},$$

in other words, the logarithm is defined only up to adding $2\pi\mathrm{i}$. So the logarithm (at least in our simple-minded approach here) is not a single-valued function. For example, we have $\ln(-1) = \pi\mathrm{i}$ or $3\pi\mathrm{i}$ or $-\pi\mathrm{i}$, so one can write

$$\ln(-1) = \{\pi\mathrm{i} + 2\pi n\mathrm{i} \mid n \in \mathbb{Z}\}.$$

Exercise: a) Calculate:

$$\ln\mathrm{i} = ? \quad \ln(-8\mathrm{i}) = ?$$

b) Show that the geometric or **polar** representation of a complex number $z = x + \mathrm{i}y = \rho e^{\mathrm{i}\phi}$ can be computed using the logarithm:

$$\rho = \exp(\mathrm{Re}\ln z) = |z|, \quad \phi = \mathrm{Im}\ln z = \arctan\frac{y}{x}.$$

Determine the polar representation of the following complex numbers: $z_1 = 2 + 2\mathrm{i}$, $z_2 = \sqrt{3} + \mathrm{i}$. Calculate also $\ln z_1$ and $\ln z_2$.

c) **Powers** of a complex number can be defined by $z^x \equiv \exp\left[x \ln z\right]$. Here x can be also a complex number! As a rule, z^x is not uniquely defined (unless x is a real integer). Calculate:

$$\sqrt{\mathrm{i}} = ? \quad \sqrt{\left(\frac{1}{2} + \frac{\sqrt{3}}{2}\mathrm{i}\right)} = ? \quad \sqrt[6]{-1} = ? \quad \mathrm{i}^{\mathrm{i}} = ? \quad 3^{2\pi\mathrm{i}} = ?$$

B Permutations

In this appendix I briefly review some basic properties of permutations.

We consider the ordered set $(1, ..., N)$ of integers. A **permutation** of the set $(1, ..., N)$ is a map $\sigma : (1, ..., N) \mapsto (k_1, ..., k_N)$ where the k_j are all different and again range from 1 to N. In other words, a permutation σ is a one-to-one map of the set $(1, ..., N)$ to itself. For example,

$$\sigma : (1, 2, 3, 4, 5) \mapsto (4, 1, 5, 3, 2)$$

is a permutation of the set of five elements.

We call a permutation **elementary** if it exchanges only two adjacent numbers, for example $(1, 2, 3, 4) \mapsto (1, 3, 2, 4)$. The **identity** permutation, denoted by id, does not permute anything. Two permutations σ_1 and σ_2 can be executed one after another; the result is also a permutation called the **product** (composition) of the elementary permutations σ_1 and σ_2 and denoted $\sigma_2\sigma_1$ (where σ_1 is executed first, and then σ_2). For example, the product of $(1, 2, 3) \mapsto (1, 3, 2)$ and $(1, 2, 3) \mapsto (2, 1, 3)$ is $(1, 2, 3) \mapsto (3, 1, 2)$. The effect of this (non-elementary) permutation is to move 3 through 1 and 2 into the first place. Note that in this way we can move any number into any other place; for that, we need to use as many elementary permutations as places we are passing through.

The set of all permutations of N elements is a group with respect to the product of permutations. This group is not commutative.

For brevity, let us write EP for "elementary permutation." Note that $\sigma\sigma =$ id when σ is an EP. Now we will prove that the permutation group is generated by EPs.

Statement 1: Any permutation can be represented as a product of some finite number of EPs.

Proof: Suppose $\sigma : (1, ..., N) \mapsto (k_1, ..., k_N)$ is a given permutation. Let us try to reduce it to EPs. If $k_1 \neq 1$ then 1 is somewhere among the k_i, say at the place i_1. We can move 1 from the i_1-th place to the first place by executing a product of $i_1 - 1$ EPs (since we pass through $i_1 - 1$ places). Then we repeat the same operation with 2, moving it to the second place, and so on. The result will be that we obtain some (perhaps a large number of) EPs σ_1, ..., σ_n, such that $\sigma_1...\sigma_n\sigma =$ id. Using the property $\sigma_i^2 =$ id, we move σ_i's to the right and obtain $\sigma = \sigma_n...\sigma_1$. ■

Any given permutation σ is thus equal to a product of EPs σ_1 to σ_n, but this representation is in any case not unique because, say, we may insert $\sigma_1\sigma_1 =$ id in any place of the product $\sigma_n...\sigma_1$ without changing the result. So the *number* of required EPs can be changed. However, it is very important (and we will prove this now) that the number of required EPs can only be changed by 2, never by 1.

In other words, we are going to prove the following statement: When a given permutation σ is represented as a product of EPs, $\sigma = \sigma_n ... \sigma_1$, the number n of these EPs is always either even or odd, depending on σ but independent of the choice of the representation $\sigma_n ... \sigma_1$. Since the parity of n (**parity** is whether n is even or odd) is a property of the permutation σ rather than of the representation of σ through EPs, it will make sense to say that the permutation σ is itself **even** or **odd**.

Statement 2: If σ is represented as a product of EPs in two different ways, namely by a product of n_1 EPs and also by a product of n_2 EPs, then the integers n_1 and n_2 are both even or both odd.

Proof: Let us denote by $|\sigma|$ the *smallest* number of EPs required to represent a given permutation σ.[1] We will now show that $|\sigma|$ is equal to the number of **order violations** in σ, i.e. the number of instances when some larger number is situated to the left of some smaller number. For example, in the permutation $(1,2,3,4) \mapsto (4,1,3,2)$ there are *four* order violations: the pairs $(4,1)$, $(4,3)$, $(4,2)$, and $(3,2)$. It is clear that the correct order can be restored only when each order violation is resolved, which requires *one* EP for each order violation.

The construction in the proof of Statement 1 shows that there exists a choice of exactly $|\sigma|$ EPs whose product equals σ. Therefore, $|\sigma|$ (the smallest number of EPs required to represent σ) is indeed equal to the number of order violations in σ.

Now consider multiplying σ by some EP σ_0; it is clear that the number of order violations changes by 1, that is, $|\sigma_0\sigma| = |\sigma| \pm 1$, depending on whether σ_0 violates the order existing in σ at the two adjacent places affected by σ_0. For example, the permutation $\sigma = (4,1,3,2)$ has four order violations, $|\sigma| = 4$; when we multiply σ by $\sigma_0 = (1,3,2,4)$, which is an EP exchanging 2 and 3, we remove the order violation in σ in the pair $(1,3)$ since $\sigma_0\sigma = (4,3,1,2)$; hence $|\sigma_0\sigma| = 3$. Since $|\sigma|$ is changed by ± 1, we have $(-1)^{|\sigma_0\sigma|} = -(-1)^{|\sigma|}$ in any case. Now we consider two representations of σ through n_1 and through n_2 EPs. If $\sigma = \sigma_{n_1} ... \sigma_1$, where σ_j are EPs, we find by induction

$$(-1)^{|\sigma|} = (-1)^{|\sigma_{n_1} ... \sigma_1|} = (-1)^{n_1} .$$

Similarly for the second representation. So it follows that

$$(-1)^{|\sigma|} = (-1)^{n_1} = (-1)^{n_2} .$$

Hence, the numbers n_1 and n_2 are either both even or both odd. ■

It follows from the proof of Statement 2 that the number $(-1)^{|\sigma|}$ is independent of the representation of σ through EPs. This number is called the **parity** of a permutation σ. For example, the permutation

$$\sigma : (1,2,3,4) \mapsto (1,4,3,2)$$

---

[1] In Definition D0 we used the notation $|\sigma|$ to mean 0 or 1 for even or odd permutations. However, the formula uses only $(-1)^{|\sigma|}$, so the present definition of $|\sigma|$ is still consistent with Definition D0.

has four order violations, $|\sigma| = 4$, and is therefore an even permutation with parity $+1$.

Definition: For a permutation σ, the **inverse permutation** σ^{-1} is defined by $\sigma^{-1}\sigma = \sigma\sigma^{-1} = \text{id}$.

Statement 3: The inverse permutation σ^{-1} exists for every permutation σ, is unique, and the parity of σ^{-1} is the same as the parity of σ.

Proof: By Statement 1, we have $\sigma = \sigma_1 ... \sigma_n$ where σ_i are EPs. Since $\sigma_i \sigma_i = \text{id}$, we can define explicitly the inverse permutation as

$$\sigma^{-1} \equiv \sigma_n \sigma_{n-1} ... \sigma_1.$$

It is obvious that $\sigma\sigma^{-1} = \sigma^{-1}\sigma = 1$, and so σ^{-1} exists. If there were two different inverse permutations, say σ^{-1} and σ', we would have

$$\sigma^{-1} = \sigma^{-1}\sigma\sigma' = \sigma'.$$

Therefore, the inverse is unique. Finally, by Statement 2, the parity of σ^{-1} is equal to the parity of the number n, and thus equal to the parity of σ. (Alternatively, we may show that $|\sigma^{-1}| = |\sigma|$.) ■

C Matrices

This appendix is a crash course on vector and matrix algebra.

C.1 Definitions

Matrices are rectangular tables of numbers; here is an example of a 4×4 matrix:

$$\begin{pmatrix} 1 & 0 & 0 & -\sqrt{2} \\ 2 & 1 & 0 & 0 \\ 3 & 2 & 1 & 0 \\ 4 & 3 & 2 & 1 \end{pmatrix}.$$

Matrices are used whenever it is convenient to arrange some numbers in a rectangular table.

To write matrices symbolically, one uses two indices, for example A_{ij} is the matrix element in the i-th row and the j-th column. In this convention, the indices are integers ranging from 1 to each dimension of the matrix. For example, a 3×2 rectangular matrix can be written as a set of coefficients $\{B_{ij} \,|\, 1 \le i \le 3,\ 1 \le j \le 2\}$ and is displayed as

$$\begin{pmatrix} B_{11} & B_{12} \\ B_{21} & B_{22} \\ B_{31} & B_{32} \end{pmatrix}.$$

A matrix with dimensions $n \times 1$ is called a **column** since it has the shape

$$\begin{bmatrix} A_{11} \\ \vdots \\ A_{n1} \end{bmatrix}.$$

A matrix with dimensions $1 \times n$ is called a row since it has the shape

$$\begin{bmatrix} A_{11} & \dots & A_{1n} \end{bmatrix}.$$

Rows and columns are sometimes distinguished from other matrices by using square brackets.

C.2 Matrix multiplication

Matrices can be multiplied by a number just like vectors: each matrix element is multiplied by the number. For example,

$$2\begin{pmatrix} u & v \\ w & x \\ y & z \end{pmatrix} = \begin{pmatrix} 2u & 2v \\ 2w & 2x \\ 2y & 2z \end{pmatrix}.$$

Now we will see how to multiply a matrix with another matrix.

The easiest is to define the multiplication of a row with a column:

$$\begin{bmatrix} a_1 & a_2 & a_3 \end{bmatrix} \begin{bmatrix} x_1 \\ x_2 \\ x_3 \end{bmatrix} = a_1x_1 + a_2x_2 + a_3x_3.$$

So the result of a multiplication of a $1 \times n$ matrix with an $n \times 1$ matrix is simply a number. The general definition is

$$\begin{bmatrix} a_1 & \dots & a_n \end{bmatrix} \begin{bmatrix} x_1 \\ \vdots \\ x_n \end{bmatrix} = \sum_{i=1}^{n} a_i x_i.$$

Let us try to guess how to define the multiplication of a column with a matrix consisting of *several* rows. Start with just two rows:

$$\begin{pmatrix} a_1 & a_2 & a_3 \\ b_1 & b_2 & b_3 \end{pmatrix} \begin{bmatrix} x_1 \\ x_2 \\ x_3 \end{bmatrix} =?$$

We can multiply each of the two rows with the column $[x_i]$ as before. Then we obtain two numbers, and it is natural to put them into a column:

$$\begin{pmatrix} a_1 & a_2 & a_3 \\ b_1 & b_2 & b_3 \end{pmatrix} \begin{bmatrix} x_1 \\ x_2 \\ x_3 \end{bmatrix} = \begin{bmatrix} a_1x_1 + a_2x_2 + a_3x_3 \\ b_1x_1 + b_2x_2 + b_3x_3 \end{bmatrix}.$$

In general, we define the product of an $m \times n$ matrix with an $n \times 1$ matrix (a column); the result is an $m \times 1$ matrix (again a column):

$$\begin{pmatrix} a_{11} & \dots & a_{1n} \\ \vdots & \vdots & \vdots \\ a_{m1} & \dots & a_{mn} \end{pmatrix} \begin{bmatrix} x_1 \\ \vdots \\ x_n \end{bmatrix} = \begin{bmatrix} \sum_{i=1}^{n} a_{1i}x_i \\ \vdots \\ \sum_{i=1}^{n} a_{mi}x_i \end{bmatrix}.$$

Exercise: Calculate the following products of matrices and columns:

$$\begin{pmatrix} -1 & 3 \\ 4 & 1 \end{pmatrix} \begin{bmatrix} -2 \\ -1 \end{bmatrix} =?$$

$$\begin{pmatrix} \sqrt{5}-1 & 2 \\ 2 & \sqrt{5}+1 \end{pmatrix} \begin{bmatrix} \sqrt{5}+1 \\ \sqrt{5}-1 \end{bmatrix} =?$$

$$\begin{pmatrix} 1 & 9 & -2 \\ 3 & 0 & 3 \\ -6 & 4 & 3 \end{pmatrix} \begin{bmatrix} -2 \\ 0 \\ 4 \end{bmatrix} =?$$

$$\begin{pmatrix} 1 & 0 & 0 & 0 \\ 2 & 1 & 0 & 0 \\ 0 & 2 & 1 & 0 \\ 0 & 0 & 2 & 1 \end{pmatrix} \begin{bmatrix} a \\ b \\ c \\ d \end{bmatrix} =?$$

$$\begin{pmatrix} 2 & 1 & 0 & 0 & \cdots & 0 \\ 1 & 2 & 1 & 0 & \cdots & 0 \\ 0 & 1 & 2 & 1 & & \vdots \\ 0 & 0 & 1 & 2 & & 0 \\ \vdots & \vdots & & & \ddots & 1 \\ 0 & 0 & & \cdots & 1 & 2 \end{pmatrix} \begin{bmatrix} 1 \\ -1 \\ 1 \\ \vdots \\ -1 \\ 1 \end{bmatrix} =?$$

Finally, we can extend this definition to products of two matrices of sizes $m \times n$ and $n \times p$. We first multiply the $m \times n$ matrix by each of the $n \times 1$ columns in the $n \times p$ matrix, yielding p columns of size $m \times 1$, and then arrange these p columns into an $m \times p$ matrix. The resulting general definition can be written as a formula for matrix multiplication: if A is an $m \times n$ matrix and B is an $n \times p$ matrix then the product of A and B is an $m \times p$ matrix C whose coefficients are given by

$$C_{ik} = \sum_{j=1}^{n} A_{ij} B_{jk}, \quad 1 \le i \le m, \quad 1 \le k \le p.$$

Exercise: Calculate the following matrix products:

$$\begin{bmatrix} 2 & 3 \end{bmatrix} \begin{pmatrix} -3 & 9 \\ 2 & -6 \end{pmatrix} =?$$

$$\begin{pmatrix} -5 & 6 \\ -6 & 5 \end{pmatrix} \begin{pmatrix} -5 & 5 \\ -6 & 6 \end{pmatrix} =?$$

$$\begin{pmatrix} \frac{\sqrt{1}+\sqrt{2}}{\sqrt{3}} & 0 \\ 0 & \frac{\sqrt{1}-\sqrt{2}}{\sqrt{3}} \end{pmatrix} \begin{pmatrix} \frac{\sqrt{1}-\sqrt{2}}{\sqrt{3}} & 0 \\ 0 & \frac{\sqrt{1}+\sqrt{2}}{\sqrt{3}} \end{pmatrix} =?$$

$$\begin{bmatrix} 0 & 1 & 2 \end{bmatrix} \begin{pmatrix} 3 & 2 & 1 \\ 2 & 1 & 0 \\ 1 & 0 & 0 \end{pmatrix} \begin{bmatrix} -2 \\ 0 \\ 0 \end{bmatrix} =?$$

$$\begin{bmatrix} w & x & y & z \end{bmatrix} \begin{pmatrix} 2 & 0 & 0 & 0 \\ 0 & 2 & 0 & 0 \\ 0 & 0 & 2 & 0 \\ 0 & 0 & 0 & 2 \end{pmatrix} \begin{pmatrix} 3 & 0 & 0 & 0 \\ 0 & 3 & 0 & 0 \\ 0 & 0 & 3 & 0 \\ 0 & 0 & 0 & 3 \end{pmatrix} \begin{bmatrix} a \\ b \\ c \\ d \end{bmatrix} = ?$$

Matrices of size $n \times n$ are called **square** matrices. They can be multiplied with each other and, according to the rules of matrix multiplication, again give square matrices of the same size.

Exercise 1: If A and B are two square matrices such that $AB = BA$ then one says that the matrices A and B **commute** with each other. Determine whether the following pairs of matrices commute:

a) $A = \begin{pmatrix} 1 & 1 \\ 0 & 2 \end{pmatrix}$ and $B = \begin{pmatrix} 3 & 0 \\ 1 & -2 \end{pmatrix}$.

b) $A = \begin{pmatrix} 2 & 0 & 0 \\ 0 & 2 & 0 \\ 0 & 0 & 2 \end{pmatrix}$ and $B = \begin{pmatrix} 3 & 1 & -1 \\ 0 & -1 & 2 \\ 2 & 8 & -7 \end{pmatrix}$.

c) $A = \begin{pmatrix} \sqrt{3} & 0 & 0 \\ 0 & \sqrt{3} & 0 \\ 0 & 0 & \sqrt{3} \end{pmatrix}$ and $B = \begin{pmatrix} 97 & 12 & -55 \\ -8 & 54 & 26 \\ 31 & 53 & -78 \end{pmatrix}$. What have you noticed?

d) Determine *all* possible matrices $B = \begin{pmatrix} w & x \\ y & z \end{pmatrix}$ that commute with the given matrix $A = \begin{pmatrix} 1 & 1 \\ 0 & 2 \end{pmatrix}$. ■

Note that a square matrix having the elements 1 at the diagonal and zeros elsewhere, for example

$$\begin{pmatrix} 1 & 0 & 0 \\ 0 & 1 & 0 \\ 0 & 0 & 1 \end{pmatrix},$$

has the property that it does not modify anything it multiplies. Therefore such matrices are called the **identity matrices** and denoted by $\hat{1}$. One has $\hat{1}A = A$ and $A\hat{1} = A$ for any matrix A (for which the product is defined).

Exercise 2: We consider real-valued 2×2 matrices.

a) The *matrix*-valued function $A(\phi)$ is defined by

$$A(\phi) = \begin{pmatrix} \cos\phi & -\sin\phi \\ \sin\phi & \cos\phi \end{pmatrix}.$$

Show that $A(\phi_1)A(\phi_2) = A(\phi_1 + \phi_2)$. Deduce that $A(\phi_1)$ commutes with $A(\phi_2)$ for arbitrary ϕ_1, ϕ_2.

b) For every complex number $z = x + \mathrm{i}y = re^{\mathrm{i}\phi}$, let us now define a matrix

$$C(z) = \begin{pmatrix} r\cos\phi & -r\sin\phi \\ r\sin\phi & r\cos\phi \end{pmatrix} = \begin{pmatrix} x & -y \\ y & x \end{pmatrix}.$$

Show that $C(z_1)$ commutes with $C(z_2)$ for arbitrary complex z_1, z_2, and that $C(z_1) + C(z_2) = C(z_1 + z_2)$ and $C(z_1)C(z_2) = C(z_1 z_2)$. In this way, complex numbers could be replaced by matrices of the form $C(z)$. The addition and the multiplication of matrices of this form corresponds exactly to the addition and the multiplication of complex numbers.

Exercise 3: The **Pauli matrices** $\sigma_1, \sigma_2, \sigma_3$ are defined as follows,

$$\sigma_1 = \begin{pmatrix} 0 & 1 \\ 1 & 0 \end{pmatrix}, \quad \sigma_2 = \begin{pmatrix} 0 & -\mathrm{i} \\ \mathrm{i} & 0 \end{pmatrix}, \quad \sigma_3 = \begin{pmatrix} 1 & 0 \\ 0 & -1 \end{pmatrix}.$$

Verify that $\sigma_1^2 = \hat{1}$ (the 2×2 identity matrix), $\sigma_1\sigma_2 = \mathrm{i}\sigma_3$, $\sigma_2\sigma_3 = \mathrm{i}\sigma_1$, and in general

$$\sigma_a\sigma_b = \delta_{ab}\hat{1} + \mathrm{i}\sum_c \varepsilon_{abc}\sigma_c.$$

b) The expression $AB - BA$ where A, B are two matrices is called the **commutator** of A and B and is denoted by

$$[A, B] = AB - BA.$$

Using the result of part a), compute $[\sigma_a, \sigma_b]$.

C.3 Linear equations

A system of linear algebraic equations, for example,

$$\begin{aligned} 2x + y &= -11 \\ 3x - y &= 6 \end{aligned}$$

can be formulated in the matrix language as follows. One introduces the column vectors $\mathbf{x} \equiv \binom{x}{y}$ and $\mathbf{b} \equiv \binom{-11}{6}$ and the matrix

$$A \equiv \begin{pmatrix} 2 & 1 \\ 3 & -1 \end{pmatrix}.$$

Then the above system of equations is equivalent to the single matrix equation,

$$A\mathbf{x} = \mathbf{b},$$

where $\mathbf{x}$ is understood as the unknown vector.

Exercise: Rewrite the following system of equations in matrix form:

$$\begin{aligned} x + y - z &= 0 \\ y - x + 2z &= 0 \\ 3y &= 2 \end{aligned}$$

Remark: In a system of equations, the number of unknowns may differ from the number of equations. In that case we need to use a rectangular (non-square) matrix to rewrite the system in a matrix form.

C.4 Inverse matrix

We consider square matrices A and B. If $AB = 1$ and $BA = 1$ then B is called the **inverse matrix** to A (and vice versa). The inverse matrix to A is denoted by A^{-1}, so that one has $AA^{-1} = A^{-1}A = 1$.
Remark: The inverse matrix does not always exist; for instance, the matrix

$$\begin{pmatrix} 1 & 1 \\ 2 & 2 \end{pmatrix}$$

does not have an inverse. For *finite-dimensional* square matrices A and B, one can derive from $AB = 1$ that also $BA = 1$. ■

The inverse matrix is useful for solving linear equations. For instance, if a matrix A has an inverse, A^{-1}, then any equation $A\mathbf{x} = \mathbf{b}$ can be solved immediately as $\mathbf{x} = A^{-1}\mathbf{b}$.

Exercise 1: a) Show that the inverse to a 2×2 matrix $A = \begin{pmatrix} w & x \\ y & z \end{pmatrix}$ exists when $wz - xy \neq 0$ and is given explicitly by the formula

$$A^{-1} = \frac{1}{wz - xy} \begin{pmatrix} z & -x \\ -y & w \end{pmatrix}.$$

b) Compute the inverse matrices A^{-1} and B^{-1} for $A = \begin{pmatrix} 1 & 1 \\ 0 & 2 \end{pmatrix}$ and $B = \begin{pmatrix} 3 & 0 \\ 1 & -2 \end{pmatrix}$. Then compute the solutions of the linear systems

$$\begin{pmatrix} 1 & 1 \\ 0 & 2 \end{pmatrix} \begin{bmatrix} x \\ y \end{bmatrix} = \begin{bmatrix} -3 \\ 5 \end{bmatrix}; \quad \begin{pmatrix} 3 & 0 \\ 1 & -2 \end{pmatrix} \begin{bmatrix} x \\ y \end{bmatrix} = \begin{bmatrix} -6 \\ 0 \end{bmatrix}.$$

Exercise 2: Show that $(AB)^{-1} = B^{-1}A^{-1}$, assuming that the inverse matrices to A and B exist.

Hint: Simplify the expression $(AB)(B^{-1}A^{-1})$.

Exercise 3: Show that

$$(\hat{1} + BA)^{-1} = A^{-1}(\hat{1} + AB)^{-1}A,$$

assuming that all the needed inverse matrices exist.

Hint: Use the property $A(\hat{1} + BA) = A + ABA = (\hat{1} + AB)A$. ■

The inverse matrix to a given $n \times n$ matrix A can be computed by solving n systems of equations,

$$A\mathbf{x}_1 = \mathbf{e}_1, \ldots, A\mathbf{x}_n = \mathbf{e}_n,$$

where the vectors $\mathbf{e}_i$ are the standard basis vectors,

$$\mathbf{e}_1 = (1, 0, \ldots, 0), \ \mathbf{e}_2 = (0, 1, 0, \ldots, 0),$$
$$\ldots, \ \mathbf{e}_n = (0, \ldots, 0, 1),$$

while the vectors $\mathbf{x}_1, \ldots, \mathbf{x}_n$ are unknown. When $\{\mathbf{x}_i\}$ are determined, their components x_{ij} form the inverse matrix.

C.5 Determinants

In the construction of the inverse matrix for a given matrix A_{ij}, one finds a formula of a peculiar type: Each element of the inverse matrix A^{-1} is equal to some polynomial in A_{ij}, divided by a certain function of A_{ij}. For example, Exercise 1a in Sec. C.4 gives such a formula for 2×2 matrices; that formula contains the expression $wz - xy$ in every denominator.

The expression in the denominator is *the same* for every element of A^{-1}. This expression needs to be nonzero in that formula, or else we cannot divide by it (and then the inverse matrix does not exist). In other words, this expression (which is a function of the matrix A_{ij}) "determines" whether the inverse matrix exists. Essentially, this function (after fixing a numerical prefactor) is called the **determinant** of the matrix A_{ij}.

The determinant for a 2×2 or 3×3 matrix is given[1] by the formulas

$$\det \begin{pmatrix} a & b \\ x & y \end{pmatrix} = ay - bx,$$

$$\det \begin{pmatrix} a & b & c \\ p & q & r \\ x & y & z \end{pmatrix} = aqz + brx + cpy - bpz - cqx - ary.$$

Determinants are also sometimes written as matrices with straight vertical lines at both sides, e.g.

$$\det \begin{pmatrix} 1 & 2 \\ 0 & 3 \end{pmatrix} \equiv \begin{vmatrix} 1 & 2 \\ 0 & 3 \end{vmatrix} = 3.$$

In this notation, a determinant resembles a matrix, so it requires that we clearly distinguish between a matrix (a table of numbers) and a determinant (which is a *single number* computed from a matrix).

To compute the determinant of an arbitrary $n \times n$ matrix A, one can use the procedure called the **Laplace expansion.**[2] First one defines the notion of a **minor** M_{ij} corresponding to some element A_{ij}: By definition, M_{ij} is the determinant of a matrix obtained from A by deleting row i and column j. For example, the minor corresponding to the element b of the matrix

$$A = \begin{pmatrix} a & b & c \\ p & q & r \\ x & y & z \end{pmatrix}$$

is the minor corresponding to A_{12}, hence we delete row 1 and column 2 from A and obtain

$$M_{12} = \begin{vmatrix} p & r \\ x & z \end{vmatrix} = pz - rx.$$

[1]I do not derive this result here; a derivation is given in the main text.

[2]Here I will only present the Laplace expansion as a computational procedure without derivation. A derivation is given as an exercise in Sec. 3.4.

Then, one sums over all the elements A_{1i} ($i = 1, ..., n$) in the first row of A, multiplied by the corresponding minors and the sign factor $(-1)^{i-1}$. In other words, the Laplace expansion is the formula

$$\det(A) = \sum_{i=1}^{n} (-1)^{i-1} A_{1i} M_{1i}.$$

A similar formula holds for any other row j instead of the first row; one needs an additional sign factor $(-1)^{j-1}$ in that case.
Example: We compute the determinant of the matrix

$$A = \begin{pmatrix} a & b & c \\ p & q & r \\ x & y & z \end{pmatrix}$$

using the Laplace expansion in the first row. The minors are

$$M_{11} = \begin{vmatrix} q & r \\ y & z \end{vmatrix} = qz - ry,$$

$$M_{12} = \begin{vmatrix} p & r \\ x & z \end{vmatrix} = pz - rx,$$

$$M_{13} = \begin{vmatrix} p & q \\ x & y \end{vmatrix} = py - qx.$$

Hence

$$\begin{aligned} \det A &= aM_{11} - bM_{12} + bM_{13} \\ &= a(qx - ry) - b(pz - rx) + c(py - qx). \end{aligned}$$

This agrees with the formula given previously.
Exercise: Compute the following determinants.
a)

$$\begin{vmatrix} 15 & -12 \\ -\frac{1}{2} & \frac{2}{5} \end{vmatrix} =? \qquad \begin{vmatrix} 1+x^2 & 1+x^2 \\ 1+x^2 & 1+x^4 \end{vmatrix} =?$$

$$\begin{vmatrix} 1 & -99 & -99 & -99 \\ 0 & 2 & -99 & -99 \\ 0 & 0 & 3 & -99 \\ 0 & 0 & 0 & 4 \end{vmatrix} =? \qquad \begin{vmatrix} 1 & 2 & 3 \\ 4 & 5 & 6 \\ 7 & 8 & 9 \end{vmatrix} =?$$

b)

$$A_2 = \begin{vmatrix} 2 & -1 \\ -1 & 2 \end{vmatrix} =? \qquad A_3 = \begin{vmatrix} 2 & -1 & 0 \\ -1 & 2 & -1 \\ 0 & -1 & 2 \end{vmatrix} =?$$

$$A_4 = \begin{vmatrix} 2 & -1 & 0 & 0 \\ -1 & 2 & -1 & 0 \\ 0 & -1 & 2 & -1 \\ 0 & 0 & -1 & 2 \end{vmatrix} =?$$

Guess and then prove (using the Laplace expansion) the general formula for determinants A_n of this form for arbitrary n,

$$A_n = \begin{vmatrix} 2 & -1 & 0 & \cdots & 0 \\ -1 & 2 & -1 & \cdots & \vdots \\ 0 & -1 & 2 & \cdots & 0 \\ \vdots & \vdots & \vdots & \ddots & -1 \\ 0 & \cdots & 0 & -1 & 2 \end{vmatrix} =?$$

Hint: Use the Laplace expansion to prove the recurrence relation $A_{n+1} = 2A_n - A_{n-1}$.

C.6 Tensor product

A matrix with rows and columns reversed is called the **transposed** matrix. For example, if

$$A = \begin{pmatrix} a & b & c \\ x & y & z \end{pmatrix}$$

is a given 2×3 matrix then the transposed matrix, denoted by A^T, is the following 3×2 matrix:

$$A^T = \begin{pmatrix} a & x \\ b & y \\ c & z \end{pmatrix}.$$

Note that a row vector becomes a column vector when transposed, and vice versa. In general, an $m \times n$ matrix becomes an $n \times m$ matrix when transposed.

The scalar product of vectors, $\mathbf{q} \cdot \mathbf{r}$, can be represented as a matrix product $\mathbf{q}^T\mathbf{r}$. For example, if $\mathbf{q} = (a, b, c)$ and $\mathbf{r} = (x, y, z)$ then

$$\mathbf{q} \cdot \mathbf{r} = ax + by + cz = \begin{bmatrix} x & y & z \end{bmatrix} \begin{bmatrix} a \\ b \\ c \end{bmatrix} = \mathbf{q}^T\mathbf{r} = \mathbf{r}^T\mathbf{q}.$$

A matrix product taken in the opposite order (i.e. a column vector times a row vector) gives a *matrix* as a result,

$$\mathbf{q}\mathbf{r}^T = \begin{bmatrix} a \\ b \\ c \end{bmatrix} \begin{bmatrix} x & y & z \end{bmatrix} = \begin{bmatrix} ax & ay & az \\ bx & by & bz \\ cx & cy & cz \end{bmatrix}.$$

This is known as the **tensor product** of two vectors. An alternative notation is $\mathbf{q} \otimes \mathbf{r}^T$. Note that the result of the tensor product is not a vector but a matrix, i.e. an object of a different kind. (The space of $n \times n$ matrices is also denoted by $\mathbb{R}^n \otimes \mathbb{R}^n$.)

Exercise: Does the tensor product commute? In a three-dimensional space, compute the matrix $\mathbf{q} \otimes \mathbf{r}^T - \mathbf{r} \otimes \mathbf{q}^T$. Compare that matrix with the vector product $\mathbf{q} \times \mathbf{r}$.

D Distribution of this text

D.1 Motivation

A scientist receives financial support from the society and the freedom to do research in any field. I believe it is a duty of scientists to make the results of their science freely available to the interested public in the form of understandable, clearly written textbooks. This task has been significantly alleviated by modern technology. Especially in theoretical sciences where no experimentally obtained photographs or other such significant third-party material need to be displayed, authors are able (if not always willing) to prepare the entire book on a personal computer, typing the text and drawing the diagrams using freely available software. Ubiquitous access to the Internet makes it possible to create texts of high typographic quality in ready-to-print form, such as a PDF file, and to distribute these texts essentially at no cost.

The distribution of texts in today's society is inextricably connected with the problem of intellectual property. One could simply upload PDF files to a Web site and declare these texts to be in public domain, so that everyone would be entitled to download them for free, print them, or distribute further. However, malicious persons might then prepare a slightly modified version and inhibit further distribution of the text by imposing a non-free license on the modified version and by threatening to sue anyone who wants to distribute *any* version of the text, including the old public-domain version. Merely a threat of a lawsuit suffices for an Internet service provider to take down any web page allegedly violating copyright, even if the actual lawsuit may be unsuccessful.

To protect the freedom of the readers, one thus needs to release the text under a *copyright* rather than into public domain, and at the same time one needs to make sure that the text, as well as any future revisions thereof, remains freely distributable. I believe that a free license, such as GNU FDL (see the next subsection), is an appropriate way of copyrighting a science textbook.

The present book is released under GNU FDL. According to the license, everyone is allowed to print this book or distribute it in any other way. In particular, any commercial publisher may offer professionally printed and bound copies of the book for sale; the permission to do so is *already granted.* Since the FDL disallows granting exclusive distribution rights, I (or anybody else) will not be able to sign a standard exclusive-rights contract with a publisher for printing this book (or any further revision of this book). I am happy that **lulu.com** offers commercial printing of the book at low cost and at the same time adheres to the conditions of a free license (the GNU FDL). The full text of the license follows.

D.2 GNU Free Documentation License

Version 1.2, November 2002

Copyright (c) 2000,2001,2002 Free Software Foundation, Inc.

59 Temple Place, Suite 330, Boston, MA 02111-1307, USA

Everyone is permitted to copy and distribute verbatim copies of this license document, but changing it is not allowed.

D.2.1 Preamble

The purpose of this License is to make a manual, textbook, or other functional and useful document free in the sense of freedom: to assure everyone the effective freedom to copy and redistribute it, with or without modifying it, either commercially or noncommercially. Secondarily, this License preserves for the author and publisher a way to get credit for their work, while not being considered responsible for modifications made by others.

This License is a kind of "copyleft", which means that derivative works of the document must themselves be free in the same sense. It complements the GNU General Public License, which is a copyleft license designed for free software.

We have designed this License in order to use it for manuals for free software, because free software needs free documentation: a free program should come with manuals providing the same freedoms that the software does. But this License is not limited to software manuals; it can be used for any textual work, regardless of subject matter or whether it is published as a printed book. We recommend this License principally for works whose purpose is instruction or reference.

D.2.2 Applicability and definitions

This License applies to any manual or other work, in any medium, that contains a notice placed by the copyright holder saying it can be distributed under the terms of this License. Such a notice grants a world-wide, royalty-free license, unlimited in duration, to use that work under the conditions stated herein. The "Document", below, refers to any such manual or work. Any member of the public is a licensee, and is addressed as "you". You accept the license if you copy, modify or distribute the work in a way requiring permission under copyright law.

A "Modified Version" of the Document means any work containing the Document or a portion of it, either copied verbatim, or with modifications and/or translated into another language.

A "Secondary Section" is a named appendix or a front-matter section of the Document that deals exclusively with the relationship of the publishers or authors of the Document to the Document's overall subject (or to related matters) and contains nothing that could fall directly within that overall subject. (Thus, if the Document is in part a textbook of mathematics, a Secondary Section may not explain any mathematics.) The relationship could be a matter of historical connection with the subject or with related matters, or of legal, commercial, philosophical, ethical or political position regarding them.

The "Invariant Sections" are certain Secondary Sections whose titles are designated, as being those of Invariant Sections, in the notice that says that the Document is released under this License. If a section does not fit the above definition of Secondary then it is not allowed to be designated as Invariant. The Document may contain zero Invariant Sections. If the Document does not identify any Invariant Sections then there are none.

The "Cover Texts" are certain short passages of text that are listed, as Front-Cover Texts or Back-Cover Texts, in the notice that says that the Document is released under this License. A Front-Cover Text may be at most 5 words, and a Back-Cover Text may be at most 25 words.

A "Transparent" copy of the Document means a machine-readable copy, represented in a format whose specification is available to the general public, that is suitable for revising the document straightforwardly with generic text editors or (for images composed of pixels) generic paint programs or (for drawings) some widely available drawing editor, and that is suitable for input to text formatters or for automatic translation to a variety of formats suitable for input to text formatters. A copy made in an otherwise Transparent file format whose markup, or absence of markup, has been arranged to thwart or discourage subsequent modification by readers is not Transparent. An image format is not Transparent if used for any substantial amount of text. A copy that is not "Transparent" is called "Opaque".

Examples of suitable formats for Transparent copies include plain ASCII without markup, Texinfo input format, LaTeX input format, SGML or XML using a publicly available DTD, and standard-conforming simple HTML, PostScript or PDF designed for human modification. Examples of transparent image formats include PNG, XCF and JPG. Opaque formats include proprietary formats that can be read and edited only by proprietary word processors, SGML or XML for which the DTD and/or processing tools are not generally available, and the machine-generated HTML, PostScript or PDF produced by some word processors for output purposes only.

The "Title Page" means, for a printed book, the title page itself, plus such following pages as are needed to hold, legibly, the material this License requires to appear in the title page. For works in formats which do not have any title page as such, "Title Page" means the text near the most prominent appearance of the work's title, preceding the beginning of the body of the text.

A section "Entitled XYZ" means a named subunit of the Document whose title either is precisely XYZ or contains XYZ in parentheses following text that translates XYZ in another language. (Here XYZ stands for a specific section name mentioned below, such as "Acknowledgements", "Dedications", "Endorsements", or "History".) To "Preserve the Title" of such a section when you modify the Document means that it remains a section "Entitled XYZ" according to this definition.

The Document may include Warranty Disclaimers next to the notice which states that this License applies to the Document. These Warranty Disclaimers are considered to be included by reference in this License, but only as regards disclaiming warranties: any other implication that these Warranty Disclaimers may have is void and has no effect on the meaning of this License.

D.2.3 Verbatim copying

You may copy and distribute the Document in any medium, either commercially or noncommercially, provided that this License, the copyright notices, and the license notice saying this License applies to the Document are reproduced in all copies, and that you add no other conditions whatsoever to those of this License. You may not use technical measures to obstruct or control the reading or further copying of the copies you make or distribute. However, you may accept compensation in exchange for copies. If you distribute a large enough number of copies you must also follow the conditions in section D.2.4.

You may also lend copies, under the same conditions stated above, and you may publicly display copies.

D.2.4 Copying in quantity

If you publish printed copies (or copies in media that commonly have printed covers) of the Document, numbering more than 100, and the Document's license notice requires Cover Texts, you must enclose the copies in covers that carry, clearly and legibly, all these Cover Texts: Front-Cover Texts on the front cover, and Back-Cover Texts on the back cover. Both covers must also clearly and legibly identify you as the publisher of these copies. The front cover must present the full title with all words of the title equally prominent and visible. You may add other material on the covers in addition. Copying with changes limited to the covers, as long as they preserve the title of the Document and satisfy these conditions, can be treated as verbatim copying in other respects.

If the required texts for either cover are too voluminous to fit legibly, you should put the first ones listed (as many as fit reasonably) on the actual cover, and continue the rest onto adjacent pages.

If you publish or distribute Opaque copies of the Document numbering more than 100, you must either include a machine-readable Transparent copy along with each Opaque copy, or state in or with each Opaque copy a computer-network location from which the general network-using public has access to download using public-standard network protocols a complete Transparent copy of the Document, free of added material. If you use the latter option, you must take reasonably prudent steps, when you begin distribution of Opaque copies in quantity, to ensure that this Transparent copy will remain thus accessible at the stated location until at least one year after the last time you distribute an Opaque copy (directly or through your agents or retailers) of that edition to the public.

It is requested, but not required, that you contact the authors of the Document well before redistributing any large number of copies, to give them a chance to provide you with an updated version of the Document.

D.2.5 Modifications

You may copy and distribute a Modified Version of the Document under the conditions of sections D.2.3 and D.2.4 above, provided that you release the Modified Version under precisely this License, with the Modified Version filling the role of the Document, thus licensing distribution and modification of the Modified Version to whoever possesses a copy of it. In addition, you must do these things in the Modified Version:

A. Use in the Title Page (and on the covers, if any) a title distinct from that of the Document, and from those of previous versions (which should, if there were any, be listed in the History section of the Document). You may use the same title as a previous version if the original publisher of that version gives permission.

B. List on the Title Page, as authors, one or more persons or entities responsible for authorship of the modifications in the Modified Version, together with at least five of the principal authors of the Document (all of its principal authors, if it has fewer than five), unless they release you from this requirement.

C. State on the Title page the name of the publisher of the Modified Version, as the publisher.

D. Preserve all the copyright notices of the Document.

E. Add an appropriate copyright notice for your modifications adjacent to the other copyright notices.

F. Include, immediately after the copyright notices, a license notice giving the public permission to use the Modified Version under the terms of this License, in the form shown in the Addendum below.

G. Preserve in that license notice the full lists of Invariant Sections and required Cover Texts given in the Document's license notice.

H. Include an unaltered copy of this License.

I. Preserve the section Entitled "History", Preserve its Title, and add to it an item stating at least the title, year, new authors, and publisher of the Modified Version as given on the Title Page. If there is no section Entitled "History" in the Document, create one stating the title, year, authors, and publisher of the Document as given on its Title Page, then add an item describing the Modified Version as stated in the previous sentence.

J. Preserve the network location, if any, given in the Document for public access to a Transparent copy of the Document, and likewise the network locations given in the Document for previous versions it was based on. These may be placed in the "History" section. You may omit a network location for a work that was published at least four years before the Document itself, or if the original publisher of the version it refers to gives permission.

K. For any section Entitled "Acknowledgements" or "Dedications", Preserve the Title of the section, and preserve in the section all the substance and tone of each of the contributor acknowledgements and/or dedications given therein.

L. Preserve all the Invariant Sections of the Document, unaltered in their text and in their titles. Section numbers or the equivalent are not considered part of the section titles.

M. Delete any section Entitled "Endorsements". Such a section may not be included in the Modified Version.

N. Do not retitle any existing section to be Entitled "Endorsements" or to conflict in title with any Invariant Section.

O. Preserve any Warranty Disclaimers.

If the Modified Version includes new front-matter sections or appendices that qualify as Secondary Sections and contain no material copied from the Document, you may at your option designate some or all of these sections as invariant. To do this, add their titles to the list of Invariant Sections in the Modified Version's license notice. These titles must be distinct from any other section titles.

You may add a section Entitled "Endorsements", provided it contains nothing but endorsements of your Modified Version by various parties—for example, statements of peer review or that the text has been approved by an organization as the authoritative definition of a standard.

You may add a passage of up to five words as a Front-Cover Text, and a passage of up to 25 words as a Back-Cover Text, to the end of the list of Cover Texts in the Modified Version. Only one passage of Front-Cover Text and one of Back-Cover Text may be added by (or through arrangements made by) any one entity. If the Document already includes a cover text for the same cover, previously added by you or by arrangement made by the same entity you are acting on behalf of, you may not add another; but you may replace the old one, on explicit permission from the previous publisher that added the old one.

The author(s) and publisher(s) of the Document do not by this License give permission to use their names for publicity for or to assert or imply endorsement of any Modified Version.

D.2.6 Combining documents

You may combine the Document with other documents released under this License, under the terms defined in section 4 above for modified versions, provided that you include in the combination all of the Invariant Sections of all of the original documents, unmodified, and list them all as Invariant Sections of your combined work in its license notice, and that you preserve all their Warranty Disclaimers.

The combined work need only contain one copy of this License, and multiple identical Invariant Sections may be replaced with a single copy. If there are multiple Invariant Sections with the same name but different contents, make the title of each such section unique by adding at the end of it, in parentheses, the name of the original author or publisher of that section if known, or else a unique number. Make the same adjustment to the section titles in the list of Invariant Sections in the license notice of the combined work.

In the combination, you must combine any sections Entitled "History" in the various original documents, forming one section Entitled "History"; likewise combine any sections Entitled "Acknowledgements", and any sections Entitled "Dedications". You must delete all sections Entitled "Endorsements."

D.2.7 Collections of documents

You may make a collection consisting of the Document and other documents released under this License, and replace the individual copies of this License in the various documents with a single copy that is included in the collection, provided that you follow the rules of this License for verbatim copying of each of the documents in all other respects.

You may extract a single document from such a collection, and distribute it individually under this License, provided you insert a copy of this License into the extracted document, and follow this License in all other respects regarding verbatim copying of that document.

D.2.8 Aggregation with independent works

A compilation of the Document or its derivatives with other separate and independent documents or works, in or on a volume of a storage or distribution medium, is called an "aggregate" if the copyright resulting from the compilation is not used to limit the legal rights of the compilation's users beyond what the individual works permit. When the Document is included an aggregate, this License does not apply to the other works in the aggregate which are not themselves derivative works of the Document.

If the Cover Text requirement of section D.2.4 is applicable to these copies of the Document, then if the Document is less than one half of the entire aggregate, the Document's Cover Texts may be placed on covers that bracket the Document within the aggregate, or the electronic equivalent of covers if the Document is in electronic form. Otherwise they must appear on printed covers that bracket the whole aggregate.

D.2.9 Translation

Translation is considered a kind of modification, so you may distribute translations of the Document under the terms of section D.2.5. Replacing Invariant Sections with translations requires special permission from their copyright holders, but you may include translations of some or all Invariant Sections in addition to the original versions of these Invariant Sections. You may include a translation of this License, and all the license notices in the Document, and any Warrany Disclaimers, provided that you also include the original English version of this License and the original versions of those notices and disclaimers. In case of a disagreement between the translation and the original version of this License or a notice or disclaimer, the original version will prevail.

If a section in the Document is Entitled "Acknowledgements", "Dedications", or "History", the requirement (section D.2.5) to Preserve its Title (section D.2.2) will typically require changing the actual title.

D.2.10 Termination

You may not copy, modify, sublicense, or distribute the Document except as expressly provided for under this License. Any other attempt to copy, modify, sublicense or distribute the Document is void, and will automatically terminate your rights under this License. However, parties who have received copies, or rights, from you under this License will not have their licenses terminated so long as such parties remain in full compliance.

D.2.11 Future revisions of this license

The Free Software Foundation may publish new, revised versions of the GNU Free Documentation License from time to time. Such new versions will be similar in spirit to the present version, but may differ in detail to address new problems or concerns. See `http://www.gnu.org/copyleft/`.

Each version of the License is given a distinguishing version number. If the Document specifies that a particular numbered version of this License "or any later version" applies to it, you have the option of following the terms and conditions either of that specified version or of any later version that has been published (not as a draft) by the Free Software Foundation. If the Document does not specify a version number of this License, you may choose any version ever published (not as a draft) by the Free Software Foundation.

D.2.12 Addendum: How to use this License for your documents

To use this License in a document you have written, include a copy of the License in the document and put the following copyright and license notices just after the title page:

Copyright (c) <year> <your name>. Permission is granted to copy, distribute and/or modify this document under the terms of the GNU Free Documentation License, Version 1.2 or any later version published by the Free Software Foundation; with no Invariant Sections, no Front-Cover Texts, and no Back-Cover Texts. A copy of the license is included in the section entitled "GNU Free Documentation License".

If you have Invariant Sections, Front-Cover Texts and Back-Cover Texts, replace the "with...Texts." line with this:

with the Invariant Sections being <list their titles>, with the Front-Cover Texts being <list>, and with the Back-Cover Texts being <list>.

If you have Invariant Sections without Cover Texts, or some other combination of the three, merge those two alternatives to suit the situation.

If your document contains nontrivial examples of program code, we recommend releasing these examples in parallel under your choice of free software license, such as the GNU General Public License, to permit their use in free software.

D.2.13 Copyright

Index

Notes

www.ingramcontent.com/pod-product-compliance
Lightning Source LLC
Chambersburg PA
CBHW031826170526
45157CB00001B/204
9781409294962